Handbook of Petroleum Science and Technology

Handbook of Petroleum Science and Technology

Edited by Natalie Mitchell

MURPHY & MOORE
www.murphy-moorepublishing.com

Published by Murphy & Moore Publishing,
1 Rockefeller Plaza,
New York City, NY 10020, USA
www.murphy-moorepublishing.com

Handbook of Petroleum Science and Technology
Edited by Natalie Mitchell

International Standard Book Number: 978-1-63987-294-7 (Hardback)

Cataloging-in-Publication Data

Handbook of petroleum science and technology / edited by Natalie Mitchell.
p. cm.
Includes bibliographical references and index.
ISBN 978-1-63987-294-7
1. Petroleum. 2. Petroleum engineering. 3. Petroleum--Geology.
I. Mitchell, Natalie.

TN870 .H36 2022
553.28--dc23

TABLE OF CONTENTS

PREFACE

This book has been an outcome of determined endeavour from a group of educationists in the field. The primary objective was to involve a broad spectrum of professionals from diverse cultural background involved in the field for developing new researches. The book not only targets students but also scholars pursuing higher research for further enhancement of the theoretical and practical applications of the subject.

Petroleum is a yellowish-black liquid which occurs naturally in the geological formations beneath the Earth's surface. It is formed when large quantities of dead organisms such as alga and zooplankton are buried under sedimentary rocks and subjected to intense pressure and heat. Petroleum is a mixture of different hydrocarbons having different molecular weights and physical properties. Petroleum science and technology refers to the various processes related to the exploration, production, transportation and refining of petroleum and its products. These activities are divided into different sectors, namely, upstream, midstream and downstream. It is an interdisciplinary field which encompasses the principles of geology, drilling technology, geostatistics, railway engineering, logistics and transport, economics, production engineering and chemistry. The various sub-fields of petroleum science and technology along with technological progress that have future implications are glanced at in this book. It includes some of the vital pieces of work being conducted across the world, on various topics related to petroleum science and technology. This book is a resource guide for experts as well as students.

It was an honour to edit such a profound book and also a challenging task to compile and examine all the relevant data for accuracy and originality. I wish to acknowledge the efforts of the contributors for submitting such brilliant and diverse chapters in the field and for endlessly working for the completion of the book. Last, but not the least; I thank my family for being a constant source of support in all my research endeavours.

Editor

1

An Experimental Investigation on Improving the Medium and Late-Age Compressive Strengths of Class G Oil Well Cement

Ali Allahverdi[1]*, Ebrahim Najafi Kani[2], and Shadi Soltani[1]

[1] Cement Research Center, School of Chemical Engineering, Iran University of Science and Technology Tehran, Iran

[2] School of Chemical, Petroleum, and Gas Engineering, Semnan University, Semnan, Iran

ABSTRACT

In this study, the effects of particle size distribution improvement and dense packing techniques on the 28- and 90-day compressive strengths of class G oil well cement have been investigated. It was observed that limited improvements in the particle size distribution of cement (regarding industrial possibilities) cannot result in a significant increase in 28- and 90-day compressive strengths. Also, in the application of the dense packing technique, silica fume and limestone powders with the grain size less than 45 micrometers cannot be utilized as substitutes for cement, because the addition of these materials simultaneously leads to the shortening of the setting times. The utilization of quartz powder as a filler exhibits partial pozzolanic properties at optimum substitution percentages, i.e. around 6 to 10% by weight; however, it can result in an almost 7% increase in the 28-day compressive strength of the hardened paste of the cement without any considerable effects on its setting times.

Keywords: Class G, Oil Well Cement, Particle Size Distribution, Dense Packing, Compressive Strength

INTRODUCTION

Cement-based materials are used in the oil and gas industry for cementing oil wells. Oil well cement is derivatives of ordinary Portland cement with special properties that are necessary due to the special conditions of oil wells, i.e. temperature, pressure, and the presence of attacking chemicals. The main role of the cement sheath is to permanently isolate all subsurface formations penetrated by a well. This permanent isolation is expected not only during the oil production but also after the plugging of wells. In any case, the engineering properties of the cement-based materials placed within the wells are of utmost importance [1-4].

The control of setting time of oil well cement during the cementing operation and at the bottom-hole conditions of wells is a very important parameter [5]. In addition, the hardened cement paste must exhibit satisfactory compressive strength and chemical durability to withstand the pressure due to the reservoir formation and the deterioration due to

***Corresponding author**

Ali Allahverdi

Email: ali.allahverdi@iust.ac.ir

the presence of attacking chemicals [4-7]. Class G Oil well cement is intended for use as basic cement from surface to a depth of 2440 m (8,000 ft) as manufactured, or can be used with accelerators and retarders to cover a wide range of well depths and temperatures [7].

Two different techniques to enhance the general engineering properties of the Portland cements include particle size distribution (PSD) improvement and dense packing using suitable supplementary cementing materials. The rate at which a cement particle hydrates when exposed to water greatly depends on its size. A small particle reacts much more quickly than a large particle and a very large particle, larger than about 50 μm, probably will never become fully hydrated, even if exposed to enough water. The PSD is therefore critical in controlling the rate at which Portland cements gain strength. To achieve a good strength development, there must be a certain amount of small particles; but, if there are too many small particles, the cement will set too quickly. The technique of dense packing, by the addition of superfine powders together with or without superplasticizer, also improves the behaviors of paste and concrete considerably. Because of the small particle size of these materials, they function as filler to improve the pore size distribution and reduce the porosity. Because of the space occupation that otherwise would have been occupied by water, more free water is released to upgrade the consistency or workability [8]. In this work, attempts were made to improve the 28- and 90-day compressive strengths of class G oil well cement. Medium and late-age compressive strengths are important, because they depend on the total amount of calcium silicate content (both alite and belite) of the cement which is a quality characteristic determining the properties of the cement including its durability performance under especial conditions. The present work is therefore concentrated on the general engineering properties (setting time and medium to late-age compressive strengths) of the studied oil well cement but not on its specific technical properties for oil well cementing. This allows the use of common laboratory techniques. The work has two individual parts. The first part deals with the effects of PSD improvement on the properties of the cement and the second part copes with the effects of dense packing technique on the properties of cement using different supplementary cementing materials including silica fume (a highly pozzolanic material), limestone (an almost inert material), and quartz powder (a partial pozzolanic material).

EXPERIMENTAL

Commercial class G oil well cement, silica fume as a by-product of ferro-silica plant, limestone, and quartz were prepared as raw materials. Limestone and quartz were ground in a laboratory ball mill and then the fractions lower than 45 micrometers were separated for being used as a fine substitute for cement. The chemical composition and specific surface area of the materials used are given in Table 1.

Figure 1 represents the PSD curve of class G oil well cement. For improving the PSD and investigating its effects on compressive strength, the coarse fraction larger than 65 micrometers, i.e. 6.7 wt% of the cement, was separated and after being ground to particles smaller than 45 micrometers was thoroughly mixed with the cement. The cement was then used to measure its 7, 28, and 90-day compressive strength and to see the effect of such an improvement in PSD on strength behavior at medium and late ages.

To evaluate the effects of dense packing on strength behavior, silica fume and powders of limestone and quartz with a grain size smaller than 45 micrometers were used; a number of binary systems were designed and prepared with the addition of 2, 4, 6,..., and 14% (relative to the mass of the cement) of them as a substitute for cement.

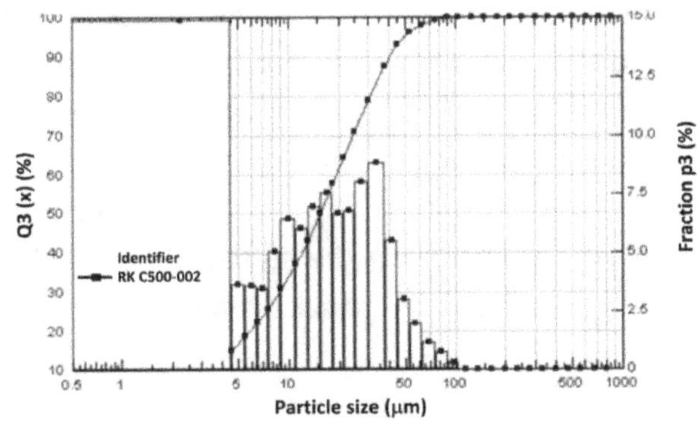

Figure 1: PSD-curve of class G oil well cement

A water-to-dry binder ratio of 0.38 was used to prepare pastes. The pastes were then cast into specimens of the size of 2×2×2 cm^3 and stored in an atmosphere of more than 95% relative humidity at 25 °C for the first 24 hrs. After demolding, the specimens were immersed in water at 25 °C for further curing until the time of testing. The blends were characterized and studied by relative visual workability and measuring their initial and final setting times, 28 and 90-day compressive strengths, and total 80 day shrinkage.

Table 1: Chemical composition (wt%) and specific surface area (Blaine, $cm^2.g^{-1}$) of materials

Oxide	SiO_2	Al_2O_3	Fe_2O_3	CaO	MgO	SO_3	K_2O	Na_2O	LOI	Blaine specific surface area
Class G oil well cement	22.43	3.86	4.48	64.26	1.75	2.07	0.59	0.75	1.95	2954
Silica fume	96.12	0.82	-	-	-	-	-	0.40	2.61	18000
Limestone	0.54	0.35	0.14	51.95	1.16	-	-	-	42.16	5210
Quartz	95.82	1.72		2.30	-	-	-	-	-	4820

RESULTS AND DISCUSSION

Particle Size Distribution Improvement

Research works have proved that PSD is an important factor in determining not only the rheological properties of the cement, but also its strength behavior. The experimental results show that an increase in the slope of PSD-curve, while keeping a suitable consistency, can considerably increase the compressive strength of cement; in fact, the higher the slope of the curve is, the superior the strength behavior of the cement becomes [9]. The highest slope achievable by industrial cement mills is around 1. However, on laboratory scales, it is possible to increase the slope to values quite higher. Here, the class G oil well cement used has a relatively suitable PSD with a mean particle size and PSD-curve slope of 21.49 micrometers and 1.15 respectively. As it can be seen, the PSD of the cement is relatively good and, by considering current capabilities on industrial scales, just small improvements in the PSD are possible. Therefore, the improvement done in the PSD of the cement is limited to just decreasing the amount of relatively coarser fraction of the cement, i.e. the fraction larger than 65 micrometers, which not only improves the total fineness of the cement, but also increases the slope of the PSD-curve. The results obtained for 7, 28, and 90-day compressive strengths after such an improvement are presented in Table 2.

Table 2: Effect of particle size distribution improvement on compressive strength

	Compressive Strength (MPa)		
	7-day	28-day	90-day
Before improvement	50.0	56.2	60.0
After improvement	55.7	59.5	60.0

As it can be seen, such an improvement in the PSD of the cement can just result in a limited increase in compressive strengths at relatively early ages. The 7-day compressive strength is increased by less than 12%, whereas 28-day compressive strength is raised by just less than 6%. The 90-day compressive strength remains unchanged. Therefore, limited improvements in PSD by regarding the industrial possibilities can result in a slight increase in 28-day compressive strength, but it does not significantly increase 90-day compressive strength.

Dense Packing Technique

Considering the relative visual workability of the pastes confirms that any substitution of cement by silica fume leads to a decrease in the workability of the cement paste due to its relatively high specific surface area and hence its high water adsorption; however, the substitution of cement by limestone or quartz powders enhances the paste workability.

Initial and Final Setting Times

Figures 2 to 4 represent the effects of filling materials on the initial and final setting times of cement. As it can be seen in Figure 2, both initial and final setting times decrease at relatively lower percentages of silica fume but slightly increase at substitution contents higher than 10%. The decrease of the setting times is due to the relatively high capacity of silica fume in water adsorption. The increase in setting times at relatively higher percentages of substitution could be attributed to the agglomeration behavior of silica fume in the absence of any dispersing agents.

Both limestone and quartz powders decrease setting times almost linearly. The addition of limestone, however, is more effective than quartz in decreasing the setting times. This is probably due to the reactivity of calcium carbonate with the tri calcium aluminate phase of the cement.

The decrease in setting times for the case of limestone and, in particular, quartz powder can also be attributed to the phenomenon of water adsorption. The particles of both limestone and quartz powder are smaller than 45 micrometers and these two materials are therefore significantly finer than cement. Such higher fineness can provide a relatively considerable capacity of water adsorption for these materials. Nevertheless, this water adsorption is quite lower than that of an ultra-fine material such as silica fume. Quartz powder is thus quite less effective in decreasing the setting times of the cement in comparison to silica fume.

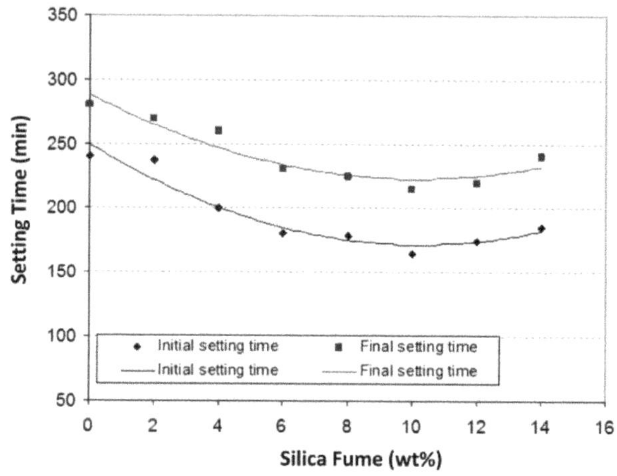

Figure 2: Effect of silica fume on the initial and final setting times of cement

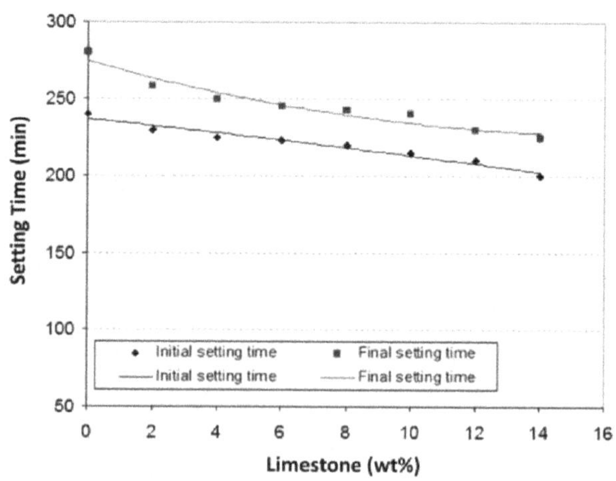

Figure 3: Effect of limestone powder on the initial and final setting times of cement

Compressive Strength

The results of 28 and 90-day compressive strengths are presented in Figures 5 and 6. All the tested filling materials are effective in increasing the compressive strengths. As it can be seen, maximum 28 and 90-day compressive strengths are achievable at substitution percentages in the range of 6 to 10% by weight for each of the filling materials used. Such an increase in compressive strength is generally due to both filling and pozzolanic effects of the additive materials used. Comparing the effects of the filling materials employed, it is seen that silica fume provides a marked improvement in both 28- and 90-day compressive strengths. In the case of silica fume, however, in addition to the filling effect, the pozzolanic reactivity of the material, i.e. reaction with calcium hydroxide and water and generating additional calcium silicate hydrates, is also a major factor in strength improvement. It is reported that the presence of ultra-fine particles of silica fume can accelerate cement hydration reactions by providing suitable nucleation centers for hydration products [10, 11]. The substitution of cement by filling materials at concentrations higher than 10 wt% decreases both 28- and 90-day compressive strengths. This is due to the fact that, firstly, a limited substitution of the cement by a filling material is to be considered for achieving dense packing and higher contents do not give rise to dense packing. Secondly, relatively higher substitution percentages leads to the effective removal of cement hydraulic phases responsible for strength behavior and the introduction of inert or less reactive additive materials.

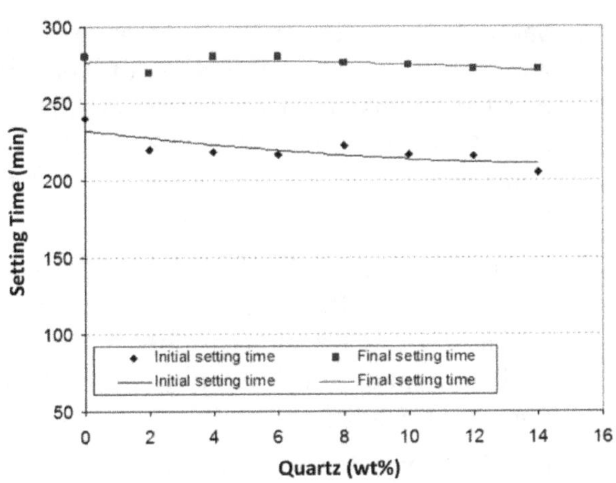

Figure 4: Effect of quartz powder on the initial and final setting times of cement

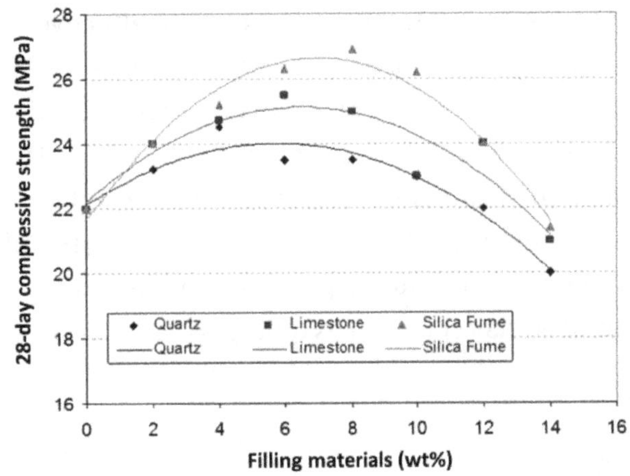

Figure 5: 28-day compressive strength of the cement blends

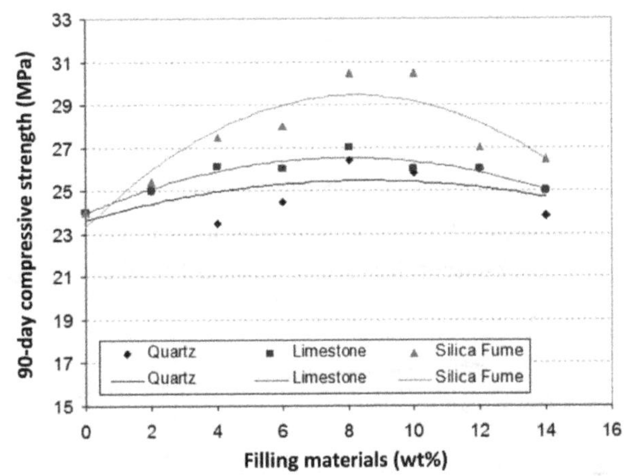

Figure 6: 90-day compressive strength of the cement blends

Shrinkage

Figures 7 to 9 represent the results of total 80-day shrinkage of the blends. In all cases, the total 80-day shrinkage decreases as the amount of filling material is increased.

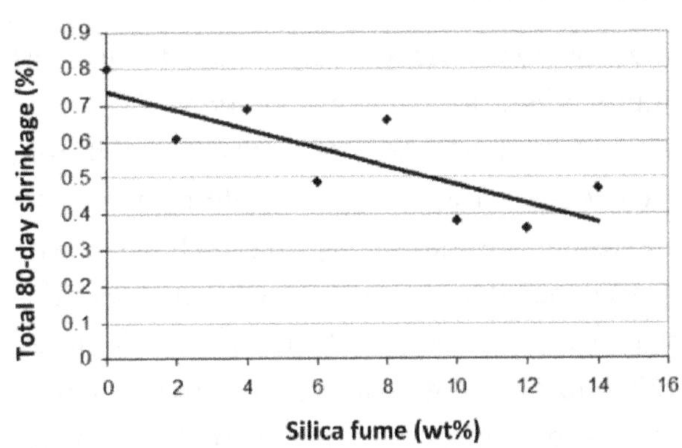

Figure 7: Total 80-day shrinkage of the cement/silica fume blend

As it can be seen, silica fume is more effective in reducing the amount of shrinkage due to its pozzolanic behavior which causes the densification of the microstructure of the hardened cement paste and therefore reducing the amount of shrinkage. In the case of limestone and quartz powders, the decrease in total 80-day shrinkage is quite less.

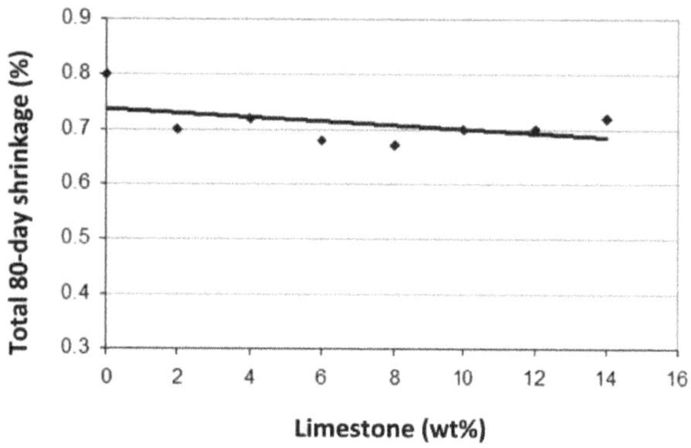

Figure 8: Total 80-day shrinkage of the cement/limestone powder blend

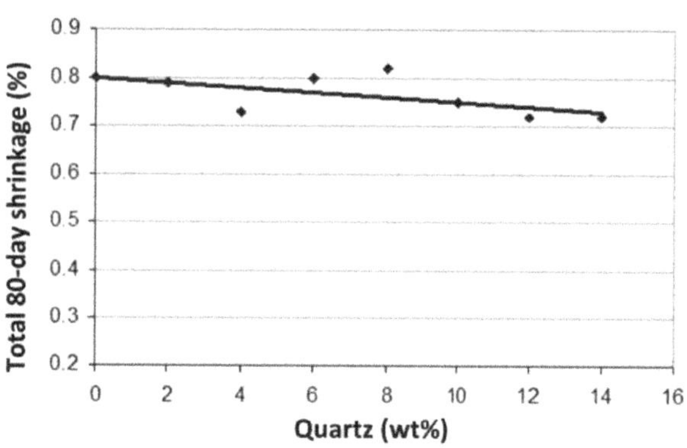

Figure 9: Total 80-day shrinkage of the cement/quartz powder blend

CONCLUSIONS

The present work has focused on the improvement in medium and late-age (28- and 90-day) compressive strengths by different techniques. The results confirm that the application of dense packing technique with quartz powder is useful without imparting significant shortening effects on setting times. The improvements can be attributed to both shrinkage control and the partial pozzolanic properties of the quartz powder. Silica fume and limestone powder with grain size smaller than 45 microns cannot be utilized as a substitute for cement to improve the compressive strength of class G oil well cement, because their addition simultaneously results in shortening setting times. The utilization of quartz powder at optimum substitution percentages, i.e. around 6 to 10% by weight, however, can increase the 28-day compressive strength of the hardened paste of cement by about 7% without any considerable effects on its setting times.

REFERENCES

[1] Andersen P. J., Johansen V., "Particle Packing and Concrete Properties, American Ceramic Society".**1995**, 12, 111-146.

[2] Heikal M., Diadamony H. E., and. Morsy M. S., "Limestone-filled Pozzolanic Cement", *Cem. & Concr. Res*. **2000**, 30, 1827-1834.

[3] Bonavetti V., Donza H., Menendez G.,. Cabera O., and Irassar E. F.," Limestone Filler Cement in Low W/C Concrete: A Rational Use of Energy", *Cem. & Concr. Res.* **2003**, 33, 865-871.

[4] Pera J., Husson S., and Guilhot B., "Influence of Finely Ground Limestone on Cement Hydration", *Cem. & Concr. Comp.* **1999**, 21, 99-105,.

[5] Svermova L., Sonebi M., and Bartos P. J. M., "Influence of Mix Proportions on Rheology of Cement Grouts Containing Limestone Powder", *Cem. & Concr. Comp*. **2003**, 25, 737-749.

[6] Saout G. L., Le´colier E., Rivereau A., and Zanni H., "Chemical Structure of Cement Aged at Normal and Elevated Temperatures and Pressures" *Part I. Class G Oilwell Cement, Cem. & Concr. Res*. **2006**, 36, 71-78.

[7] Hodne H., Saasen A., O'Hagan A. B., and Wick S. O., "Effects of Time and Shear Energy on the Rheological Behavior of oil Well Cement Slurries", *Cem. & Concr.* **2000,** *Res.*, 30, 1759-1766.

[8] Niu Q., Feng N., Yang J., and Zheng X., "Effect of Superfine Slag Powder on Cement

Properties", *Cem. & Concr. Res*. **2002**, 32, 615-621.

[9] Glasnovi A. and Hraste M., "The Grinding Equation in the Investigation of the Coarse Dispersing System of Cement", *Cem. & Concr. Res.* **1982**, 12, 415-424.

[10] Agawa K., Uchikawa H., Takemoto K. and Yasui I. V.," The Mechanism of the Hydration in the System C3S Pozzolans", *Cem. & Concr. Res*. **1980**, 8, 683-696.

[11] Davraz M. and Gunduz L., "Engineering Properties of Amorphous Silica as a New natural pozzolan for use in concrete", *Cem. & Concr. Res*. **2005**, 35, 1251-1261.

2

Size-dependent Kinetics Determination of MoS2-K2O/ CNTS Nanocatalyst in the Synthesis of Alcohols from Syngas

Saba Karimi, Rohollah M. Keaei, and Ahmad Tavasoli*

School of Chemistry, College of Science, University of Tehran, Tehran, Iran

ABSTRACT

The influence of Mo particle size on the catalytic activity and product selectivity of alkalized MoS_2 nanocatalysts has been investigated. Nanocatalysts are prepared using a microemulsion technique with water-to-surfactant ratios of 1-12. Three different techniques, including XRD, TEM, and hydrogen chemisorption were used to determine the molybdenum average particle size and their activity and selectivity in higher alcohols synthesis (HAS) carried out in a fixed bed microreactor at 330 °C and 70 bar. To fix the percentage of CO conversion, the GHSV is changed from 3.6 to 2.57 (nl/(hr.g catalyst)). The average MoS_2 particle sizes are changed from 4.5 to 11.9 nm. The experimental results showed that changing particle size from 11.9 to 4.5 nm decreased the methanol formation rate from 0.00634 to 0.00534 (mol/(hr.g catalyst)) but increased ethanol formation rate from 0.00581 to 0.00787 (mol/(hr.g catalyst)) and higher alcohols formation rate from 0.00473 to 0.00657 (mol/(hr.g catalyst)). A size-dependent kinetics model was developed to calculate the alcohol formation rates versus catalyst average particle size. The model not only matched experimental and theoretical results, but also showed that MoS_2 catalyst had size-dependent structure and for the prediction of product selectivity it was easier to use this mathematical model.

Keywords: Higher Alcohol, Rate, Selectivity, Particle Size, Size-dependent Kinetics

INTRODUCTION

Higher alcohols synthesis (HAS) from syngas, an economically attractive method for making fuels and chemicals, is interested due to the enhancement of petroleum price, environmental concerns, and additive for gasoline to increase octane number. Among different methods for higher alcohols synthesis, the molybdenum disulfide (MoS_2) catalyst mainly produces hydrocarbons but, when it is promoted with alkali metals, it can produce alcohols from syngas. The function of alkali metals is to reduce the hydrogenation ability of alkyl species to form alkanes and increase the sites active for the formation of alcohols [1-5].

Some investigations have been carried out to study the influence of catalyst physico-chemical properties on higher alcohols synthesis and gain a better understanding of the structure-sensitive effects in HAS catalysis [6-8]. A sub-category of structure-sensitive reactions regards the dependence of both catalytic activity and selectivity on catalytic metal particle size. Bezemer et al. showed that in Fischer-Tropsch

***Corresponding author**

Ahmad Tavasoli

Email: tavassolia@khayam.ut.ac.ir

synthesis, activity and selectivity towards C_{5+} are higher in the case of smaller particles. They presented sizes of 6-8 nm as the optimum range for active metal particles size [9].

The thermodynamic analysis of the effect of nanoparticle size on adsorption equilibrium and rates was initiated by Parmon [10]. Studying catalysts surface and trying to improve the properties of surface to have catalysts with better quality include a large area in recent researches. The main focus is placed on the analysis of the nanoparticle size influence on reaction rates and kinetic regularities in order to obtain kinetic expressions, which could primarily be used for engineering purposes [11, 12]. The experimental data of catalytic activity versus the average particle radius for Fischer-Tropsch synthesis by cobalt supported on carbon nanofibers showed good correspondence between the theory and experiments [13].

Tae et al. developed a mechanistic kinetic model to describe the synthesis of mixed alcohol from syngas over 17 wt.% K_2CO_3-promoted MoS_2 catalyst [14]. Also, Surisetty et al. studied the intrinsic kinetics of the higher alcohol synthesis reaction from synthesis gas over an alkalized Co-Rh-Mo catalyst supported on multi-walled carbon nanotubes [15]. However, the size-dependent intrinsic kinetic model of higher alcohol synthesis from synthesis gas based on different particle sizes of the metal particles has never been investigated.

Microemulsion, a novel technique for catalyst preparation, enables the control of metal particle size with a narrow particle size distribution, regardless of metal content [16-19]. Briefly, a microemulsion consists of nanosized water droplets surrounded by an oil phase, stabilized by a surfactant [19]. The size of the metal particles formed in water-in-oil (w/o) microemulsion can be controlled by changing the micelle size (the water-to-surfactant ratio) [19].

In this study, a theoretical analysis addressing the impact of the changes in chemical potential upon adsorption is performed. The quantitative model is illustrated by comparing the experimental data of catalytic activity versus the average particle size. Main products include methanol, ethanol, propanol, and butanol obtained by experiments using MoS_2 catalysts supported on carbon nanotubes (CNT's). Activity analysis on different particle sizes was performed for the case of consecutive mechanism suggested for overall reactions, demonstrating that significant changes could be anticipated in this case. Provided analysis suggested that thermodynamic analysis could be invoked for the explanation of nanokinetics, e.g. structure sensitivity and other kinetic regularities dependence on the size of catalytic nanoclusters.

EXPERIMENTAL

Preparation of Catalysts

Multi-walled CNT's (purity >95%) were used as the support for the preparation of the catalysts. Prior to catalyst preparation, the support was treated with 30% HNO_3 refluxed at 120 °C overnight, washed with distilled water several times, and finally dried at 120 °C for 6 hrs (the characteristics are shown in Table 1). The catalysts were prepared by a microemulsion method. Molybdenum and potassium particles were synthesized in a reverse microemulsion using a nonionic surfactant Triton X-100 and n-hexanol as the oil phase. The concentrations of molybdenum and potassium were adjusted using aqueous solutions of $(NH_4)_6Mo_7O_{24}.4H_2O$ (Merck) and K_2CO_3 at a molybdenum and potassium concentration of 15 and 8 wt.% respectively. The water-to-surfactant (W/S) molar ratios were selected as 1, 4, 8, and 12. After 15-minute vigorous stirring, microemulsions were obtained. Then, hydrazine was added to the solutions to develop nanoparticle formation in the core of the micelles by reducing the metal oxides. Next, the purified carbon nanotubes were added under stirring. Afterwards, tetrahydrofurane

(THF) was added to the solutions dropwise (1 ml/min). The mixtures were left to mature and settle slowly overnight and then decanted. The samples were filtered and washed several times with ethanol. Subsequently, the catalysts were dried at 120 °C, calcined under argon flow at 450 °C for 3 hrs, and slowly exposed to an oxygen atmosphere during the cooling step. The catalysts prepared were signified as Cat1, Cat2, Cat3, and Cat4 (see Table 1).

Analysis Methods

ICP-OES

The metal loadings of the purified support and calcined catalysts were performed using Varian VISTA-MPX inductively coupled plasma-optical emission spectrometry (ICP-OES) instrument.

BET Surface Area Measurements/BJH Pore Size Distributions

Surface area, pore volume, and pore average diameter of the support and calcined catalysts were measured using an ASAP-2010 V_2 Micrometrics system. The samples were degassed at 200 °C for 4 hrs under a 50-m Torr vacuum and their BET area, pore volume, and pore diameter was determined.

Transmission Electron Microscopy (TEM)

The morphology of the purified support and catalysts were studied by transmission electron microscopy (TEM). Sample specimens for TEM studies were prepared by ultrasonic dispersion of the catalysts in ethanol and the suspensions were dropped onto a carbon-coated copper grid. TEM investigations were carried out using a Philips CM_20 (100 kV) transmission electron microscope equipped with a NARON energy-dispersive spectrometer with a germanium detector.

X-ray Diffraction (XRD)

The phases and particle sizes of the crystals present in the catalysts were analyzed by XRD using a Philips Analytical X-ray diffractometer (XPert MPD) with monochromatized Co/Kα radiation, 2θ angles from 20° to 80°. The Debye–Scherer formula ($d = 0.9\lambda/\beta\cos\theta$, where β is full width at half maximum (FWHM), λ is the wave length of X-ray, 0.9 is a constant) was applied to MoS_2 peaks at 2θ = 43.3° (the most intense peak for MoS_2), in order to calculate the average particle sizes.

Hydrogen Chemisorption

For all the catalysts, the amount of chemisorbed hydrogen was measured using Micrometrics TPD-TPR 2900 analyzer system. 0.2 g of the calcined catalyst was first purged in a flow of argon at 140 °C to remove traces of water.

The temperature was then raised to 500 °C at a linearly programmed rate of 10 °C/min and the catalyst was reduced for 12 hrs under hydrogen flow and then cooled to 40 °C. Then, the flow of hydrogen was switched to argon for 30 min in order to remove the weakly adsorbed hydrogen.

Table 1: Textural properties of the purified CNT's and calcined catalysts

Sample	Preparation route	W/S Ratio	BET surface area (m^2/g)	Total pore volume (ml/g)	Average pore diameter (Å)
CNTs	-	-	252.6	0.59	94.12
Cat1	Microemulsion	1	165	0.468	127
Cat2	Microemulsion	4	142	0.437	128
Cat3	Microemulsion	8	136	0.426	129
Cat4	Microemulsion	12	132	0.419	129

Afterwards, the temperature programmed desorption (TPD) of the samples was performed by increasing the temperature of the samples, with a ramp rate of 40 °C/min, to a maximum temperature of 500 °C under the argon flow. The temperature programmed desorption spectrum was used to determine the molybdenum dispersion and its average crystallite size.

Reaction Setup and Experimental Procedure

Higher alcohol synthesis was performed in a tubular down flow, fixed bed microreactor system. The reactor was made up of an Inconel tube with a length and inside diameter of 450 and 22 mm respectively. The reactor temperature was controlled via a PID temperature controller. Brooks 5850 mass flow controllers were used to add H_2 and CO at the desired rate to the reactor. The reactor was packed with 0.5 g of the catalyst diluted with silicon carbide having a mesh size of 90 and housed in a molten salt bath controlled by a temperature controller. The reactor was pressurized to 30 bar with He and the reduction, as well as the sulfurization, was carried out for 6 hrs at 450 °C at a heating rate of 2 °C/min using a gas mixture containing 5 mol.% H_2S in H_2 and a flow rate of 50 ml/min. The temperature was then lowered to the reaction temperature, and the system was pressurized to the reaction condition. The feed gas mixture (H_2/CO ratio of 2) was passed through mass flow controllers and the HAS reaction was carried out in a steady state under the reaction conditions of 320 °C and 70 bar over a period of 48 hrs. In the case of Cat1 catalyst, the gas hourly space velocity is adjusted at 3.6 nl/(hr.gr catalyst). However, in order to have a constant percentage of feed conversion for all the experiments, the gas hourly space velocity is decreased to 3, 2.7, and 2.57 nl/(hr.gr catalyst) for Cat2, Cat3, and Cat4 catalysts respectively. The product gas was cooled to 0 °C and separated into gas and liquid phases at the reaction pressure. The CO, CO_2, and other gaseous products were monitored with time intervals of 1 hr. The products were collected after the completion of the run and analyzed by means of three gas chromatographs, a Shimadzu 4C gas chromatograph equipped with two subsequent connected packed columns: Porapak Q and Molecular Sieve 5Å, and a thermal conductivity detector (TCD) with argon which was used as a carrier gas for hydrogen analysis. A Varian CP 3800 with a chromosorb column and a thermal conductivity detector (TCD) were used for CO, CO_2, CH_4, and other non-condensable gases. A Varian CP 3800 with a Petrocol Tm DH100 fused silica capillary column and a flame ionization detector (FID) were used for liquid products so that a complete product distribution could be provided.

Size-dependent Kinetic Model

Alcohol synthesis from syngas over alkalized MoS_2 nanoparticles is initiated and developed by the adsorption of reactants over active sites. The following mechanism describes HAS over this catalyst [14]:

Initiation:

$$CO + s \longleftrightarrow COs \qquad K_1, k_1, k_{-1} \tag{1}$$

$$H_2 + 2s \longleftrightarrow 2Hs \qquad K_2, k_2, k_{-2} \tag{2}$$

$$COs + Hs \longleftrightarrow CHOs + s \qquad K_3, k_3, k_{-3} \tag{3}$$

$$CHOs + Hs \longleftrightarrow CH_2Os + s \qquad K_4, k_4, k_{-4} \tag{4}$$

$$CH_2Os + Hs \longleftrightarrow R_1Os + s \qquad K_5, k_5, k_{-5} \tag{5}$$

Chain growth

$$R_iOs + COs \longrightarrow R_iCOO + s \quad k_6, \quad i = 1, \ldots, 4 \tag{6}$$

$$R_iOs + 2Hs \longrightarrow R_is + H_2O + 2s \quad k_7, \quad i = 1, 2, \ldots, 8 \tag{7}$$

$$R_is + COs \longrightarrow R_iCOs + s \quad k_8, \quad i = 1, 2, \ldots 5 \tag{8}$$

$$R_iCOs + 2Hs \longrightarrow R_i + 1Os + 2s \quad (9)$$
$k_9 \quad i = 1,2,\ldots,4$

Termination

$$R_iOs + Hs \longleftrightarrow R_iOH + 2s \quad (10)$$
$k_{10}, k_{-10} \quad i = 1,2,\ldots,4$

$$R_iCOOs + Hs \longrightarrow R_1COOR_i - 1 + 2s \quad (11)$$
$k_{11} \quad i = 1,2,\ldots,4$

$$R_is + Hs \longrightarrow C_iH_{2i} + 2s \quad (12)$$
$k_{12} \quad i = 1,2,\ldots,8$

$$R_is \longrightarrow C_iH_{2i} + Hs \quad (13)$$
$k_{13} \quad i = 1,2,\ldots,5$

Based on the above mechanism, Park et al. derived a kinetic model for mixed alcohol synthesis from syngas over unsupported K/MoS_2 catalyst [14]. Based on the model derived by Park, the rate equations for the formation of alcohols (ri^{OH})), paraffins (ri^{P}), olefins (ri^{O}), esters (ri^{E}) can be written as:

$$ri^{OH} = \frac{\lambda_2 P_{H_2}^{0.5}\theta_{R_iO}}{Ads} - \frac{\lambda_5 P_i(OH)}{Ads^2} \quad (14)$$

Since the produced alcohols can be consumed in the next reactions, the rate expression for alcohol synthesis using this model has two terms. In this model P_{H_2} is hydrogen partial pressure, λ_2 and λ_5 are the kinetic parameter groups defined by equilibrium and rate constants and relate to the production and consumption of alcohols respectively; P_i(OH) is alcohol average partial pressure and *Ads* denotes the adsorption term as:

$$Ads = 1 + K_1P_{CO} + \left(K_2P_{H_2}\right)^{0.5} + K_1K_2^{0.5}K_3P_{CO}P_{H_2}^{0.5} + K_1K_2K_3K_4P_{CO}P_{H_2} + \lambda_4P_{CO}P_{H_2}^{3/2} \quad (15)$$

θ_{Rio} is surface coverage for surface species of R_iO and *i* (*i*= 1-4) is the number of carbon in alcohol.

$$\theta_{R_1O} = \frac{\lambda_4P_{CO}P_{H_2}^{\frac{3}{2}}}{Ads} \quad (16)$$

$$\theta_{R_iO} = \frac{\left[(\frac{\lambda_8}{\lambda_9})\theta_{R_{(i-1)}CO}P_{H_2} + \lambda_5P_i^{OH}\right]}{\left[(\lambda_1P_{CO} + \lambda_2P_{H2}^{\frac{1}{2}})Ads + \lambda_3P\right]} \quad (17)$$

i=2, 3, 4

$$\theta_{RiCO} = \frac{\lambda_5P_{CO}Ads\theta_{R1}}{P_{H2}} \quad i=2,3,4 \quad (18)$$

$$\theta_{(R_i)} = \frac{(\lambda_3\theta_{RiO}P_{H2})}{(\lambda_7P_{H2}^{0.5} + \lambda_8P_{CO} + k_{13}Ads)Ads} \quad (19)$$

i=1, 2, 3, 4

where,

$$[s]_0 = [s] + K_1P_{CO}[s] + (K_2P_{H_2})^{\frac{1}{2}}[s] + K_1K_2^{\frac{1}{2}}K_3P_{CO}P_{H_2}^{\frac{1}{2}}[s] + K_1K_2K_3K_4P_{CO}P_{H_2}[s] + \lambda_4P_{CO}P_{H_2}^{\frac{3}{2}}[s] \quad (20)$$

$$[s] = \frac{[s]_0}{Ads} \quad (21)$$

$$K_A = K_3K_1K_2^{\frac{1}{2}} \quad MPa^{-1.5} \quad (22)$$

$$K_B = K_4K_3K_2K_1 \quad MPa^{-2} \quad (23)$$

Using the above equations, the size-independent alcohol formation rates for methanol, ethanol, propanol, and butanol are obtained by:

$$r_{CH_3OH} = \left[\frac{\lambda_2P_{H_2O}^{\frac{1}{2}}\theta_{R_1O}}{Ads}\right] - \left[\frac{\lambda_5P_{CH_3OH}^{\frac{1}{2}}}{Ads^2}\right] \quad (24)$$

In our previous work, we used the kinetic model derived by Park, for the MoS_2-K_2O/CNT's catalyst supported on carbon nanotubes. The calculated kinetic parameters are shown in Table 2 [20]. These parameters are size-independent.

During adsorption reaction (hydrogen is adsorbed in a dissociated form with a stoichiometry of one H per surface Mo atom (stoichiometric factor equal to 0.5)) on catalysts particle, the chemical potential of surface, which expresses the incremental energy content of a system per unit particulate mass, changes.

Table 2: Calculated kinetic parameters for MoS_2-K_2O/CNT's catalyst [20]

Parameter	Value	Unit
K_1	4.261	MPa^{-1}
K_2	1.049	MPa^{-1}
K_A	0.042	$MPa^{-1.5}$
K_B	0.072	MPa^{-2}
λ_1	0.070	Mol. kg $cat^{-1}.h^{-1}.MPa^{-1}$
λ_2	18.56	Mol.kg-$cat^{-1}.h^{-1}.MPa^{-0.5}$
λ_3	3.210	Mol.kg-$cat^{-1}.h^{-1}.MPa^{-1}$
λ_4	0.0712	$MPa^{-2.5}$
λ_5	91.56	Mol.kg-$cat^{-1}.h^{-1}.MPa^{-1}$
λ_6	0.321	$MPa^{-0.5}$
λ_7	11.21	Mol.kg-$cat^{-1}.h^{-1}.MPa^{-0.5}$
λ_8	7.860	Mol.kg-$cat^{-1}.h^{-1}.MPa^{-1}$
λ_9	0.041	dimensionless
k_{13}	0.158	Mol.kg-$cat^{-1}.h^{-1}$

Thus it can be assumed that, for a cluster comprising more than 100 atoms, the properties determined by the chemical potential can be described somewhat sufficiently using continuous phase thermodynamics. The chemical potential μ_i(r) of a substance *i*, which is also dependent on the radius *r*, can be expressed by [10]:

$$\mu_i(r) = \mu_\infty + \frac{2\sigma V_M}{r} = \mu_\infty + \delta(r) \tag{25}$$

where, V_m is the partial molar volume of the substance forming the condensed phase; σ, μ_∞, and δ represent surface tension, the standard chemical potential of the bulk phase, and chemical potential increment respectively. Upon adsorption on nanoclusters, changes in the Gibbs free energy should also account for the chemical potential changes [12]:

$$\Delta G_{ads}(r) = \Delta G_{(ads)\infty} + \mu_i(r) - \mu_\infty \tag{26}$$

According to reaction mechanism, adsorption and desorption are elementary reactions in higher alcohol synthesis. Linear free energy (Brønsted-Evans-Polanyi) relationship is widely used in homogeneous and heterogeneous catalyzed reactions in the form which relates reaction constants *k* with equilibrium constants *K* in a series of analogous elementary reactions:

$$k = gK^\alpha \quad 0<\alpha<1 \tag{27}$$

where, *g* and α are Polanyi parameters and they are constant for a series of reactions and for MoS_2-K_2O/CNT's catalyst α is equal to 0.5 [12]. Equilibrium constant and Gibbs free energy are related by:

$$K(r) = \exp\left(\frac{-\Delta G(r)}{RT}\right) \tag{28}$$

Replacing Gibbs free energy from Equation 28 in Brønsted Equation 27 gives:

$$K(r) = g\exp(-\Delta G_\infty)\exp\left(\frac{-\alpha\delta_{(r)}}{RT}\right) = k\exp\left(\frac{-\alpha\frac{(2\sigma V_m)}{r}}{RT}\right) = k\exp\left(\frac{-\alpha\eta}{r}\right) \tag{29}$$

where,

$$\eta = \frac{2\sigma V_m}{RT} \tag{30}$$

In Equation 42, *k*(*r*) indicates size-dependent kinetic constant and *k* is named size-independent kinetic constant. *e*-$\alpha\eta/r$ is the size-dependent part of kinetic constant. The surface tension of metals is within a range of 1-2 J/m^2 [12]. The molar volume of catalytically active metals is within a range of 6-10 cm^3/mol. Therefore, the value of parameter η for MoS_2-K_2O/CNT's catalyst is 15.32 nm.

Based on the above discussion, in higher alcohol synthesis mechanism shown (Equations 1-13), each step has size-independent rate constant or size-independent equilibrium constant. Multiplying mentioned size-independent rate constants by *e*-

$\alpha\eta/r$ creates new rate constants which is dependent on particle size. For example, for the first step (reversible adsorption of CO over catalyst), consider $K_1(r)$ as size-dependent equilibrium constant. The forward reaction is adsorption and size-dependent rate constant is defined as:

$$k_1(r) = k_1\, e^{\frac{-\alpha\eta}{r}} \tag{31}$$

The reverse reaction is desorption and its size-dependent rate constant is defined as:

$$k_{-1}(r) = k_{-1}\, e^{\frac{(1-\alpha)\eta}{r}} \tag{32}$$

For this reaction, using Equation 31 and 32 gives the size-dependent equilibrium constant as follows:

$$K_1(r) = K\, e^{\frac{-\eta}{r}} \tag{33}$$

Equations similar to Equations 31-32 can be written for the other steps of the reaction. Substituting independent rate constants with the size-dependent rate constants used in the above equations give the size-dependent alcohol formation rates for methanol, ethanol, propanol, and butanol.

RESULTS AND DISCUSSION

Experimental Results and Model Implementation

The textural properties of the purified carbon nanotubes as well as the results of surface area measurements for the calcined catalysts are given in the Table 1. These results show that the BET surface areas of the Cat1-Cat4 catalysts are 165, 142, 136, and 132 m^2/g respectively. Moreover, the results indicate some more pore blockage by metal oxide clusters in the catalysts prepared at higher water/surfactant ratios.

The elemental compositions of the calcined catalysts measured by ICP are given in Table 3. The measured contents of the prepared catalysts are found to be slightly lower compared to the targeted values. The discrepancies may be due to the dilution of Mo, because of the hygroscopic nature of the Mo precursor.

Table 3: Chemical Composition of the Calcined Catalysts

Catalyst	Targeted Composition (wt. %)		Measured Composition (wt. %)	
	Mo	*K*	Mo K	*K*
Cat1	15	8	14.18	7.78
Cat2	15	8	14.28	7.85
Cat3	15	8	14.28	7.84
Cat4	15	8	14.30	7.90

Three techniques, including TEM, XRD, and hydrogen chemisorption were used to determine the molybdenum average particle size. Figure 1 presents the TEM image of the purified support. As shown, the treated support is comprised of an interwoven matrix of multi-walled CNT's. The TEM images of the Cat1-Cat4 catalysts are shown in Figure 2. As can be seen, very small particles are dispersed mostly inside the tubes and on the outer surface of the CNT's walls. The particle size distribution is also narrow. Indeed, the narrow inner diameter of the CNT channels (8–12 nm) restricted the insertion of particles in sizes close to the channel diameter (10 nm). Particles located outside the nanotubes have grown more. The sizes of particles are within the range of 2-7, 3-9, 5-16, and 5-23 nm for the Cat1, Cat2, Cat3, and Cat4 catalysts respectively.

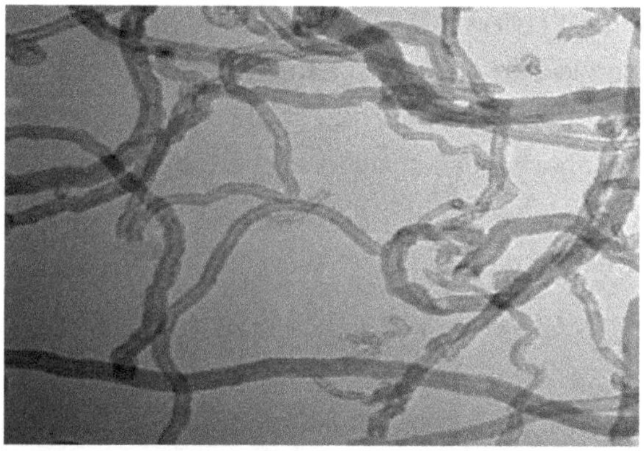

Figure 1: TEM image of the purified CNT's

Figure 3 depicts the size distribution of the particles for Cat1-Cat4 catalysts, which is determined using the population of the total particles of the catalyst based on data taken from 5 TEM pictures. According to this figure, the average particle size for the Cat1, Cat2, Cat3, and Cat4 catalysts are 4.6, 5.5, 9.1, and 11.6 nm respectively.

To determine the crystalline phases, the X-ray diffraction experiments (XRD) of the calcined catalysts were performed. Figure 4 presents the XRD spectrum of the catalysts with different preparation methods. As shown in this figure, in the case of the catalyst prepared by maximum water to surfactant ratio, i.e. Cat4, the intensity of the peaks corresponding to MoO_3 (peaks at 2θ values of 43.3°, 63.2°, and 71.9°) is higher than that of the other catalysts, indicating the greater cluster sizes of MoS_2.

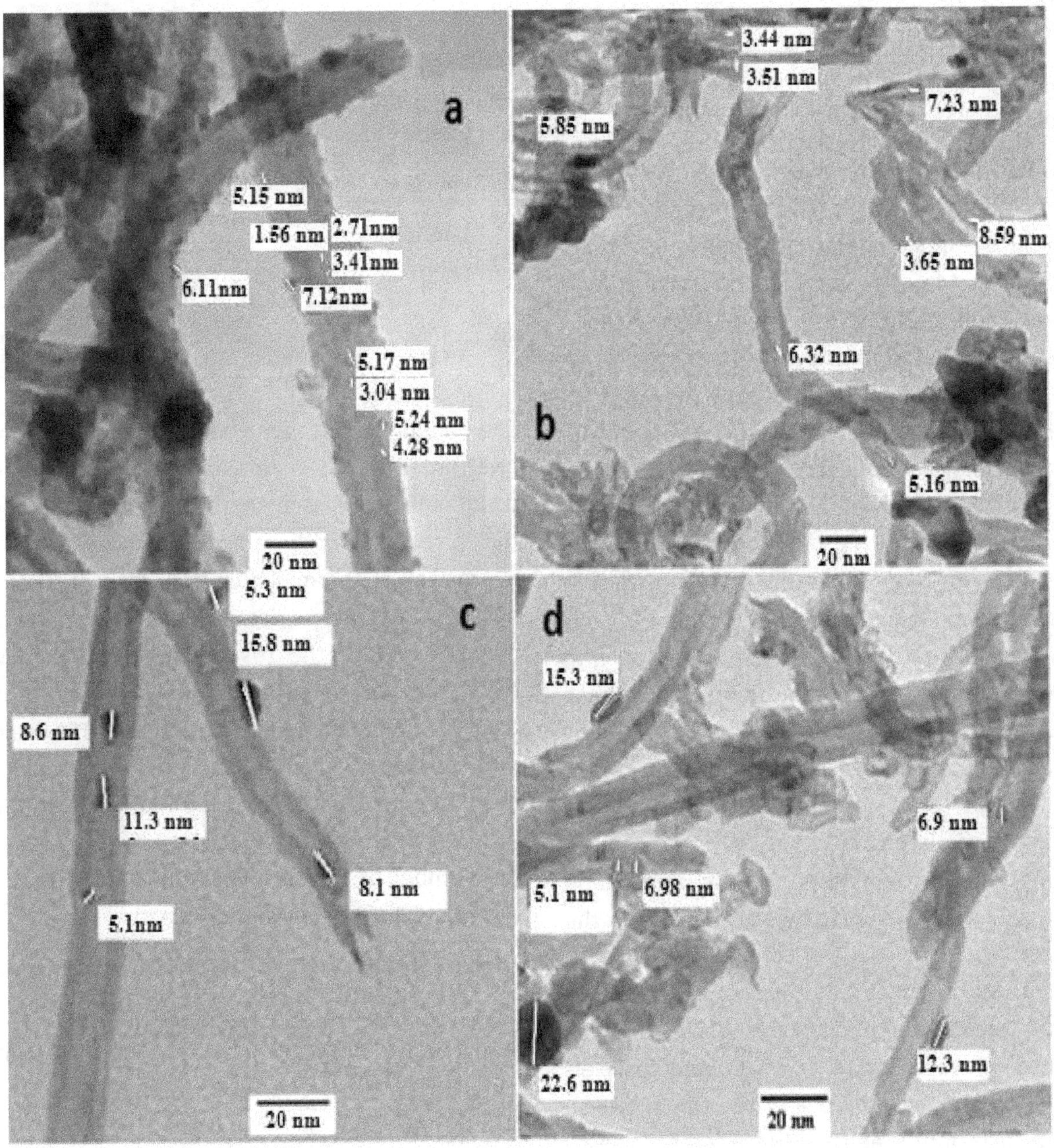

Figure 2: TEM image of the a) TEM image of Cat1 (W/S =1); b) TEM image of Cat2 (W/S =4); c) TEM image of Cat3 (W/S =8); d) TEM image of Cat4 (W/S =12).

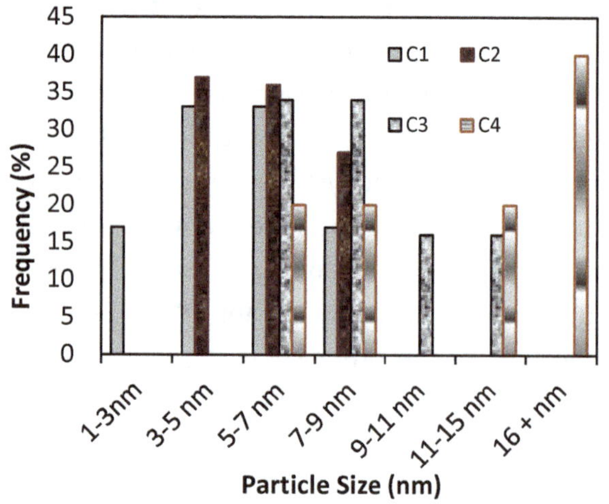

Figure 3: Particles size distribution for Cat1-Cat4 catalysts

This figure also shows that decreasing water to surfactant ratio decreases the intensity of the peaks corresponding to MoO_3, which represents a drop in MoS_2 cluster sizes at smaller W/S ratios. Table 4 shows the average sizes of MoO_3 crystals calculated from the XRD patterns, using the Debye–Scherer formula at a 2θ value of 43.3°. The data agrees reasonably well with the particle diameter obtained with the TEM results. As can be seen, the catalyst prepared at the maximum W/S ratio (a W/S ratio of 12) has the largest crystal sizes, while the catalyst prepared at the minimum W/S ratio (a W/S ratio of 1) has the smallest particle sizes. In accordance with the TEM results, these data confirm that using a microemulsion technique at low W/S ratios decreases the average crystal diameter of MoO_3 significantly. The average particle sizes are linearly correlated with the W/S ratio used during the microemulsion catalyst preparation process. In fact, nanoparticles are formed in the internal structure of the microemulsion, which is determined by the ratio of W/S. At high oil concentration, the bicontinuous phase is transformed into a structure of small water droplets within a continuous oil phase (reverse micelles), when surfactant is added. Thus the results show that the size of different droplets determines the MoS_2 particle size, depending on the amount of surfactant [19].

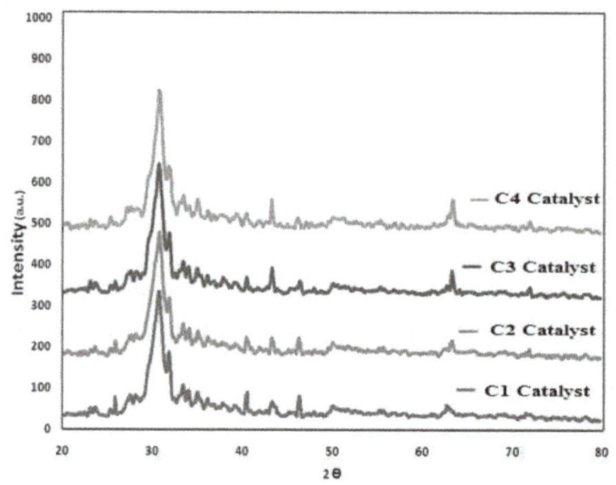

Figure 4: XRD patterns of the calcined catalysts (Cat1-Cat4)

Table 4: Particles sizes of the calcined catalysts

Catalyst	*dp* (nm) calculated by Debye–Scherer formula
Cat1	4.7
Cat2	6.0
Cat3	9.8
Cat4	12.2

The results of hydrogen chemisorption tests are given in Table 5. As shown, lowering W/S ratio considerably increased the hydrogen uptake and the percentage of dispersion. Maximum dispersion is achieved at the minimum W/S ratio of 1. The particle diameter decreases, which is in agreement with the results of XRD and TEM tests. The data agrees reasonably well with the particle diameter obtained with the TEM and XRD results. Higher dispersions and smaller molybdenum cluster sizes in the case of the catalysts prepared at low W/S ratio will increase the number of sites available for CO conversion and raises alcohols formation reaction rates.

Table 5: H_2 chemisorption results for the calcined catalysts

Catalyst	μ mole H_2 desorbed/ g. cat	Dispersion (%)	dP (nm)
Cat1	534	60.75	4.5
Cat2	453	54.86	5.9
Cat3	366	44.32	9.4
Cat4	333	40.33	11.9

The synthesis of higher alcohols was carried out after the reduction and sulfidation of the catalysts in a fixed bed reactor at 320 °C, 70 bar, and $H_2/CO=2$. The liquid products were collected at 0 °C and the exit gas was analyzed to measure the CO conversion. The analysis of the liquid products indicates that methanol, ethanol, and propanol were the major products with small amounts of butanol and other higher alcohols. The analysis of exit gas indicated that methane is the major component apart from CO, H_2, and CO_2. Table 6 presents the percentage of CO conversions during 48 hours of continuous synthesis for all the catalysts. Since the percentage of CO conversion can affect the product selectivity, to investigate the effects of particle size on the product selectivity, we changed the GHSV of feed in such a way that for all the catalytic tests the CO conversion was equal. The data in Table 6 confirms this point.

Table 6: The percentage of CO and H_2 conversions for Cat1-Cat4 catalysts

Catalyst	CO Conversion (%)	H_2 Conversion (%)
Cat1	27.70	23.9
Cat2	27.60	24.00
Cat3	27.75	24.2
Cat4	27.70	24.5

Figure 5 presents experimental methanol formation rates versus average molybdenum oxide particle sizes during 48 hours of continuous synthesis. The average molybdenum oxide particle sizes used in this figure are the average of the amounts determined by TEM, XRD, and H_2 chemisorption tests for each catalyst. As shown in this figure, the amount of methanol rises by increasing the average particle size of molybdenum oxide particles. It has been suggested that the sulfidation of the oxides present in the calcined form of the catalysts and the formation of the Mo-K-S species are responsible for the alcohol formation via CO insertion mechanism described above. Furthermore, the theoretical methanol formation rates (g CH_3OH/(hr.gr cat.)) calculated using the described model are shown in Figure 5. This figure confirms that there is very good agreement between the calculated rates and the experimental values. As depicted in Figure 8, methanol synthesis rate rises with increasing the active metal particle size, which indicates that large particles of molybdenum oxide are selective for methanol synthesis.

Figures 6 and 7 show the experimental and theoretical ethanol and higher alcohols (propanol, butanol) formation rates versus average molybdenum oxide particle sizes.

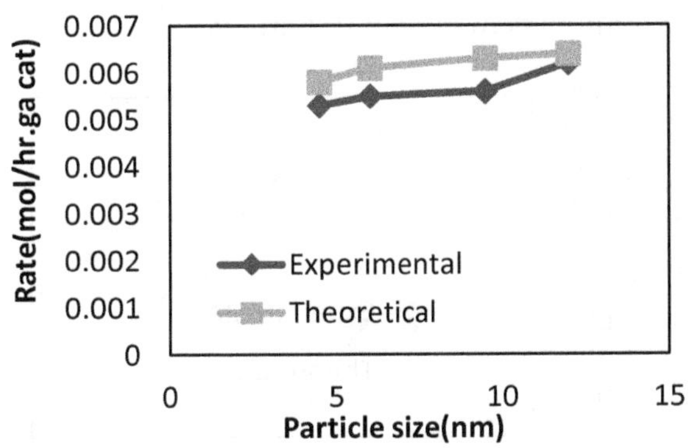

Figure 5: Methanol formation rates for Cat1-Cat4 Catalysts

In contrast to the data shown in Figure 5, ethanol and higher alcohols formation rates are higher for the catalysts with smaller average particle sizes. Smaller particles are selective for ethanol and higher alcohols formation. The decrease in the particle size of the catalytic active sites improves the standard chemical potentials of the components and appreciably affects the adsorption equilibrium. In the case of the catalysts prepared at lower W/S ratios, the drop in Mo particle sizes enhances the adsorption of CO. Ethanol and higher alcohols formation results from the adsorption of carbon monoxide over methanol as a precursor and finally hydrogenation. Increasing the partial pressure of CO on the catalyst surface can suppress the formation of methane via reactions 7 and 12 and will express the reactions 8 and 10, which in turn increase the ethanol and higher alcohols formation rates. According to the variation in chemical potential during adsorption on catalytic active phase, rate constant changes and in the presence of smaller particles this change is more noticeable. The increase in rate constant during adsorption over active sites makes difference between catalysts that have different particle sizes and thus different rates of reaction can be expected. Comparing the experimental and theoretical formation rates shows that the model successfully predicts the product formation and the distribution of the formation of alcohols within the range of the experiments covered in this study.

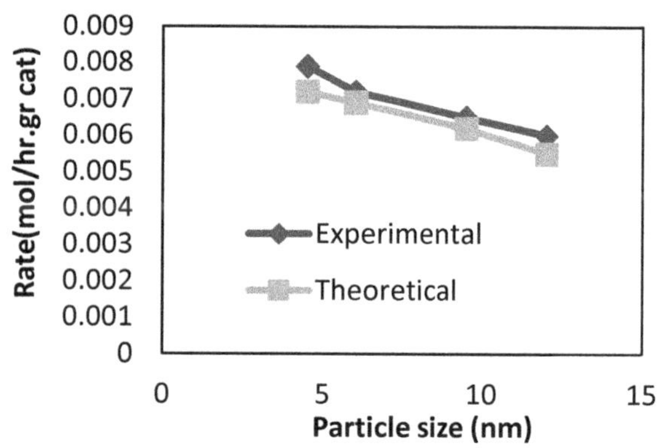

Figure 6: Ethanol formation rates for Cat1-Cat4 Catalysts

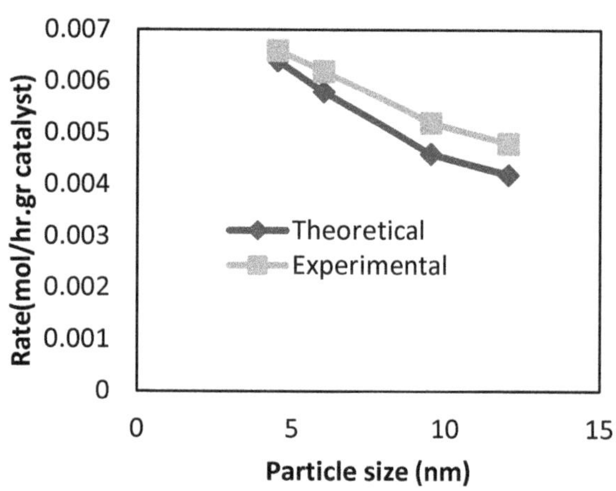

Figure 7: Higher alcohols formation rates for Cat1-Cat4 Catalysts

It should be noted that the theoretical values of rates also depend on kinetic parameters and Polanyi parameters; Polanyi parameter is usually reported to be close to 0.5. Values of kinetic parameters depend on the metal loading and metal-support interactions. Further work is needed to theoretically evaluate the values of η and thus improve the statistical analysis of the results.

CONCLUSIONS

Alkalized MoO_3 nanocatalysts supported on carbon nanotubes were prepared by a microemulsion method and used to produce higher alcohols from synthesis gas. Three different techniques, namely XRD, TEM, and hydrogen chemisorption were used to determine the molybdenum average particle sizes. The preparation of catalysts at low W/S ratios decreased the average MoS_2 particle size from 11.9 to 4.5 nm. Selectivity was found to be dependent on the particle size. The catalysts prepared at a lower water to surfactant ratio, which had lower particle sizes, severely decreased the methanol selectivity but increased ethanol and other higher alcohols selectivity. A model was derived to predict the alcohol formation rates versus catalyst average particle size. The model successfully predicted the product formation and the distribution of the formation of alcohols within the range of the experiments covered in this study. The model can be used for the design and simulation of pilot plants, semi industrial, and commercial plants as well as for the optimization and revamping of existing plants. Further work is needed to theoretically evaluate the values of kinetic parameters and Polanyi parameters and thereby improving the statistical analysis of the results.

REFERENCES

[1] Herman R., "Studies in Surface and Catalyst (Chapter 7)," Amsterdam, Elsevier, **1990**.

[2] Woo H. C. and Park K. Y., "Mixed Alcohol Synthesis from Carbon Monoxide and Hydrogen over Potassium Promoted Molybdenum Carbide Catalysts," *Applied Catalysis*, *75*, **1991**, 267-280.

[3] Jiang M., Bian G. Z., and Fu Y. L., "Effect of the K-Mo Interaction in K-MoS_2/γ-Al_2O_3 Catalysts on the Properties for Alcohol Synthesis from Syngas," *Journal of Catalysis*, *146*, **1994**, 144-154.

[4] Lee J. S., Kim S., Lee K. H., Nam I. S., et al., "Role of Alkali Promoters in K/MoS_2 Catalysts for CO-H_2 Reactions," *Applied Catalysis*, *110*, **1994**, 11-25.

[5] Tatsumi T., Muramatsu A., Yokota K., and Tominga H., "Mechanistic Study on the Alcohol Synthesis over Molybdenum Catalysts: Addition of Probe Molecules to CO-H_2," *Journal of Catalysis*, *115*, **1989**, 388-398.

[6] Surisetty V. R., Tavasoli A., and Dalai A. K., "Synthesis of Higher Alcohols from Syngas over Alkali Promoted MoS_2 Catalysts," *Applied Catalysis*, *365*, **2009**, 243-251.

[7] Iranmahboob J., Toghiani H., and Hill D. O., "Dispersion of Alkali on the Surface of Co-MoS_2/Clay Catalyst: a Comparison of K and Cs as a Promoter for Synthesis of Alcohol," *Applied Catalysis*, *247*, **2003**, 207-218.

[8] Iranmahboob J. and Hill D. O., "Alcohol Synthesis from Syngas over K_2CO_3/CoS/MoS_2 on Activated Carbon," *Catalysis Letters*, *78*, **2002**, 49-56.

[9] Leendert Bezemer G. and Bitter J. H., "Cobalt Particle Size Effects in the Fischer-Tropsch Reaction Studied with Carbon Nanofiber Supported Catalysts," *J. of American Chemical Society*, *128*, **2006**, 39-56.

[10] Parmon V. N., "Thermodynamic Analysis of the Effect of the Nanoparticle Size of the Active Component on the Adsorption Equilibrium and the Rate of Heterogeneous Catalytic Processes," *Physical Chemistry*, *413*, **2007**, 42-48.

[11] Murzin D. Y., "Kinetic Analysis of Cluster Size Dependent Activity and Selectivity," *Journal of Catalysis*, *276*, **2010**, 85-91.

[12] Dmitry Y. M., "Thermodynamic Analysis of Nanoparticle Size Effect on Catalytic Kinetics," *Chemical Engineering Science*, *64*, **2009**, 1046-1052.

[13] Dmitry Y. M., "Size-dependent Heterogeneous Catalytic Kinetics," *Journal of Molecular Catalysis A: Chemical*, *315*, **2010**, 226–230.

[14] Park T. Y., Nam I. S., and Kim Y. G., "Kinetic Analysis of Mixed Alcohol Synthesis from Syngas over K/MoS_2 Catalyst," *Ind. Eng. Chem. Res.*, *36*, **1997**, 5246-5257.

[15] Surisetty V., Dalai A., and Kozinski J., "Intrinsic Reaction Kinetics of Higher Alcohol Synthesis from Synthesis Gas over a Sulfided Alkali-Promoted Co-Rh-Mo Trimetallic Catalyst Supported on Multi-walled Carbon Nanotubes (MWCNTs)," *Energy and Fuels*, *24*, **2010**, 4130–4137.

[16] MartŁnez A. and Prieto G., "Breaking the Dispersion-reducibility Dependence in Oxide-supported Cobalt Nanoparticles," *Journal of Catalysis*, **2007**, *245*, 245-470.

[17] Kim W. Y., Haoka T., Kishida M., and Wakabayashi K., "Hydrogenation of Carbon Monoxide over Zirconia-supported Palladium Catalysts Prepared using Water-in-oil Microemulsion," *Appl. Catal. A*, **1997**, *155*, 283-289.

[18] Tago T., Hanaoka T., Dhupatemiya P., Hayashi H., et al., "Effects of Rh Content on Catalytic Behavior in CO Hydrogenation with Rh-silica Catalysts Prepared using Microemulsion," *Catal. Lett.*, **2000**, *64*(1), 27-31.

[19] Eriksson S., Nylén U., Rojas S., and Boutonnet M., "Preparation of Catalysts from Microemulsions and their Applications in Heterogeneous Catalysis," *Applied Catalysis A: General*, **2004**, *265*, 207–219.

[20] Gunturu A., KuglerE., Cropley J., and Dadyburjor D .,"A Kinetic Model for the Synthesis of High-molecular-weight Alcohols over a Sulfided Co K Mo/C Catalyst," *Ind. Eng. Chem. Res.*, **1998**, *37*, 2107-2115.

Notations

Ads: adsorption term

K_i: equilibrium constants for *i*th reaction

k_i: rate constant for *i*th reaction, kg-cat.mol^{-1}.h^{-1}

P_i: partial pressure of species i, MPa

R_i: alkyl group with carbon number i (= C_iH_{2i+1})

θ_i: surface coverage of species i

s: active site of catalyst

$\mu(r)$: chemical potential

$\mu(\infty)$: chemical potential of a metal particle of an infinite size (bulk like)

α: Polanyi parameter

Appendix

$$r_{iP} = \frac{(\lambda_7 P^{\frac{1}{2}}{}_{H_2} \theta_{Ri})}{Ads}$$

$$r_{iO} = k_{13}\, \theta_{Ri}$$

$$r_{iE} = \frac{\left[(\frac{\lambda_1}{\lambda_6}) P^{\frac{1}{2}}{}_{H_2} \theta_{R_i COO}) \right]}{Ads}$$

$$\lambda_1 = k_6 K_1 [s]_0^2$$

$$\lambda_2 = k_{10} K_2^{\frac{1}{2}} [s]_0^2$$

$$\lambda_3 = k_7 K_2 [s]_0^3$$

$$\lambda_4 = K_5 . K_4 . K_3 . K_1 . K_2^{\frac{3}{2}}$$

$$\lambda_5 = k_{10} [s]_0^2$$

$$\lambda_7 = k_{12} K_2^{\frac{1}{2}} [s]_0^2$$

$$\lambda_8 = k_8 K_1 [s]_0^2$$

$$\lambda_9 = \frac{k_8\, K_1}{k_9\, K_2\, [s]_0}$$

$$r_{C_2H_5OH} = \frac{\left[\lambda_2\, P_{H_2}^{\frac{1}{2}} \theta_{R_2O} \right]}{Ads} - \frac{\left[\lambda_5\, P^{\frac{1}{2}}{}_{C_2H_5OH} \right]}{Ads^2}$$

$$r_{C_3H_7OH} = \frac{\left[\lambda_2\, P^{\frac{1}{2}}{}_{H_2} \theta_{R_3O} \right]}{Ads} - \frac{\left[\lambda_5\, P^{\frac{1}{2}}{}_{C_3H_7OH} \right]}{Ads^2}$$

$$r_{C_4H_9OH} = \frac{\left[\lambda_2\, P^{\frac{1}{2}}{}_{H_2} \theta_{R_4O} \right]}{Ads} - \frac{\left[\lambda_5\, P^{\frac{1}{2}}{}_{C_4H_9OH} \right]}{Ads^2}$$

3

A Simulation of Managed Pressure Drilling in Iranian Darquain Oil Field

Armin Taheri Nakhost[1]* and Seyed Reza Shadizadeh[2]

[1] Department of Petroleum Engineering, Sciences & Researches Branch of Islamic Azad University, Tehran, Iran
[2] Department of Petroleum Engineering, Abadan Faculty of Petroleum Engineering, Petroleum University of Technology, Abadan, Iran

ABSTRACT

The narrow operating window between pore pressure and fracture pressure makes drilling difficult in some operations. A feasibility study of managed pressure drilling (MPD) is carried out on Iran Darquain oil field. The previous wells drilled in this field showed that mud returns were lost during drilling Gadvan formation. The present work addresses this problem by means of surface back pressure application in Darquain oil field. The methodology employed in this study is based on hydraulic analysis calculations and comparative drilling operation pressures. The DZxION MPD software performs hydraulic analysis using the API RP 13D rheological model and calculates the annular pressure drop to compare the pressures and the required back pressure, if needed. Using a mud weight of 14.31 ppg and exerting 100 psi static back pressure, the wellbore pressure profile got slightly overbalanced. When the mud pumps are in service, no back pressure is required. The problem is resolved and no kick or loss is observed using a MW of 14.31 ppg and a static back pressure of 100 psi. As a result, the managed pressure drilling technology is useful in Iran Darquian oil field through using a lower mud weight in order to overcome the circulation loss in Sarvak formation. This study is based on hydraulic analysis calculations and comparing the drilling operation pressures in Darquain oil field. For analyzing the pressure regimes throughout the well, DZxION managed pressure drilling software performs hydraulic analysis. This software calculates annular pressure drop and equivalent circulating density to compare the pressures and required back pressure, if needed. By using this method to drill the well, some advantages were gained: the mud weights used to drill the well, the number of casing strings, and the number of changing mud weights were reduced.

Keywords: Equivalent Circulating Density, Back Pressure, Annular Frictional Pressure, Constant Bottom Hole Pressure, Darquain Oil Field

INTRODUCTION

Several MPD wells have been drilled onshore and offshore worldwide so far and the range of application of MPD has enormously increased since the past few years. MPD technology has four major variations and some of them have many different methods to attain MPD, while some have just one method. These variations

***Corresponding author**
Armin Taheri Nakhost
Email: nimrarasas@gmail.com

are constant bottom hole pressure, pressurized mud cap drilling, dual gradient drilling, and HSE. MPD has been used in the USA, Canada, Mexico, South America, North-Sea, Europe, Africa, the Middle East, Australia, South East Asia, China, India, and several other parts of the world. According to some accounts and information available in the public domain, more than 350 MPD wells have been drilled offshore by the end of 2008 [7]. In 2006, drilling time and cost in Puguang province of China were reduced by half through managed pressure drilling. The use of MPD in the non-reservoir sections also provides the following advantages [2]: 1) increased ROPs and shortened drilling curves; 2) reduced bit usage; and 3) the enhancement of overall drilling performance through the reduction of vibration, drill string-related problems, and sticking potential. In 2008, the CBHP variation of MPD technique was applied in an exploratory well in Saudi Arabia. In previously drilled wells in the area, many drilling problems and wellbore stability issues were experienced. One of the possible causes of the mechanical instability could be attributed to the large fluctuations in the bottom hole pressure found in conventional drilling practices. These fluctuations are origin-nated from the stopping and starting of drilling fluid circulation during jointed pipe connections; specifically, they result from the fluctuations in the ECD, which occur when the pumps are turned on and off. The main problems during drilling in this area were associated with either increases in high nonproductive time and drilling cost caused by the partial/total loss of circulation, sloughing hole, pipe stuck, H_2S, water/oil/gas influx, and an increase in chlorides in the mud through contact with salt water formations or unsafe situations which were evident if the hydrostatic pressure exerted by the drilling fluid was not enough to control unexpected pressurized gas zones [14]. Based on these drilling problems events, CBHP was applicable and recommended as a MPD technique in this research. The MPD operation was successful and the ECD was maintained in the dynamic and static conditions between pore pressure and fracture pressure window in terms of onsite drilling, MPD software (to input data in real time such as pump rate), MW, standpipe pressure, and required choke manifold pressure. Such software can provide both early kick/loss detection, and lead time in the process to increase/decrease mud weight and circulation rate without any interruption to drilling ahead. Such services are provided by companies like Secure Drilling and AtBalance [11]. A number of studies have been involved in the feasibility study of MPD with offline hydraulic calculations. Almost all of them are done onsite [14]. Some process control methods have been utilized in many control problems [1], but a disadvantage of such methods for drilling purposes was the need to determine the control parameters in all stages (fluid rheology, cuttings loading, and temperature effects). The manipulations of cuttings loading and temperature effects make the process onerous [8]. However, for the preliminary candidate selection process, this level of accuracy is not necessary.

EXPERIMENTAL

Methodology

The Darquain oil field lies in the western banks of the Karun river in an area about 40 km of the northeast of Abadan, Iran. It is 24 km long and 10 km wide, elongated in northern-southern direction. For the scope of this study, the main source of offset data will be well DQ No. 4 and all the wells in the field, in particular, well DQ No. 2. The target reservoir was the carbonate sequence of Fahliyan formation within the so called Khami group (lower Cretaceous) found oil bearing by the exploration well DQ No. 2 and the first appraisal well DQ No. 4. The top of the reservoir is at 13290 ft TVD. The high reservoir pressures and the sour nature of some of the fluids dictate special care and attention during drilling through the cap rock and reservoir sequences to avoid potential drilling hazards [4]. It is decided to use the constant bottom hole pressure variation of managed pressure drilling technique to stay close to an agreeable pressure profile using surface backpressure. This variation is closely related to the enhanced kick and loss detection category of MPD. DZxION MPD CSM was used in order to perform offline hydraulic analysis and calculations [11]. The software can act as a preliminary screen to determine the utility of MPD for the potential MPD candidate wells. For calculating the annular and pipe pressure drop, it follows API RP 13D rheological model. Essential input parameters for this software are as follows:

1. Pp and PF data;
2. Drill string and BHA-OD's lengths;
3. Set of rheology data;
4. Mud weight, circulation rate;
5. Wellbore profile (if the well is directional);
6. Casing and open hole details (IDs and ODs).

The hydraulic calculations cannot be performed without the required input parameters mentioned above. Following the basic hydraulic analysis and calculations, it would help the user to make a better engineering decision in deciding whether to use MPD or not for a given prospect. In the method selected to perform the feasibility study of the DQ No. 5 well of Darquain field, the hydraulic calculations using API RP 13D model are performed. This can determine the ECD of each mud weight. By determining ECD and having pore pressure and fracture pressure in hand, it is possible to choose which technology is suitable for drilling the well, namely conventional drilling or managed pressure drilling.

So the first input data is the casing and drill pipe data set. The drill pipe length depends on which section of the well is to be simulated. The next step is to enter the formation pressure regime. By using overburden pressure gradient, pore pressure gradient, and Poisson's ratio in Eaton's equations, the fracture pressure gradient is obtained. The obtained fracture pressure gradient and pore pressure gradient are input to the software in pound per gallon unit. The calculated equivalent circulating density by the software is compared with the pore pressure/ fracture pressure window. Drilling fluid properties, mud rheology data, and the BHA details are next.

This is how the software determines whether this window is acceptable or not. If both the hydrostatic and dynamic pressures in the well are between the pore pressure and fracture pressures, the well does not need the MPD. If these pressures (hydrostatic of mud and dynamic pressure when pumps are on) fall below the pore pressure or exceed the fracture pressure, the software calculates the required mud weight and amount of back pressure. Afterwards, the software decides whether the MPD is applicable or not. The next section refers to the pressure gradients using the actual drilling data from well DQ No. 4 and all the important and valuable information from the offset wells. Well DQ No. 4 still remains as the reference well, though.

Fracture Pressure and Pore Pressure

The overburden gradient was calculated using the sonic log data of well DQ No. 4. The bulk density has been calculated using AGIP default formulas. The default values in the formula such as matrix bulk density, pore fluid density, and average matrix transit time were modified according to the local conditions, although there was no sufficient data. The bulk density was then integrated to calculate the overburden gradient. The fracture gradient is calculated as a function in the estimated pore and overburden gradient of the area. Depending on lithological type encountered, the K constant (a function in

Poisson's ratio) has been defined maintaining ENI AGIP strict policy operating in an unknown new area. In well DQ No. 4, all the tests were conducted as formation integrity test (FIT). Therefore, the fracture gradient curve illustrated in the gradient forecast graph was constructed using the theoretical formula using the basic rules as stated in ENI AGIP policies and manuals. The fracture pressure prediction strategy was also developed by Ben Eaton in 1975. The data required are formation overburden stress, pore pressure, and Poisson's ratio of the formation. The fracture gradient prediction equation is given by:

$$\frac{P_{frac}}{D} = \frac{\nu}{1-\nu}\left(\frac{\sigma_{ob}}{D} - \frac{P_{pore}}{D}\right) + \frac{P_{pore}}{D} \qquad (1)$$

The resulting overburden pressure gradient is integrated from the bulk density of the well DQ No. 5 and is represented in pound per gallon unit of depth. The main source of the actual pore pressure data above the reservoir section is the well DQ No. 4. In the reservoir section, the actual bottom hole formation pressure from the DST tests has been used to update the pressure gradient data from the offset wells.

Drilling Operation Window

As it can be seen in Figure 1, from surface to Pabdeh formation at about 7382 ft TVD and in the whole Fahliyan formation, there is no serious drilling problem (through the 24", 17 ½", and 8 ½" hole section) due to the wide pressure margin. Figure 1 illustrates the drilling operation window of Darquain field. A significant loss has been observed during drilling lower 12 ¼" hole that has marginal pressure. Thus, the focus of this study is only on this section.

Steps of the Study

The steps involved either candidate selection or a feasibility study that can be divided into the following main categories: defining the purpose, procuring information, performing hydraulic analysis, and selecting the method [10]. First of all, establishing the purpose of the study has a higher precedence compared to the remaining steps. Heavy losses occur during drilling the lower 12 ¼" hole section; thus suggesting a way solving this drilling problem seems satisfactory. Therefore, curing the loss of circulation through that way provides advantages, including cost effectiveness (due to less mud loss) and eliminating excess casing string and saving time (because of fewer drilling problems and less rig cost).

All available data from the well being drilled (such as pressure regimes, drill string and BHA details, mud weight and rheology, and well bore geometry) are used in this study. Tables 1 to 7 are the input data of the software.

By using API RP 13D rheological model, the annular frictional pressure, ECD changes, and the required mud weight are calculated. The feasibility of the option, hydraulic analysis, constraints of the rig, and availability of the equipment assist choosing the best method along the different MPD variation.

RESULTS AND DISCUSSION

All the data mentioned above were input to the software. After running casing 13 3/8" to 7382 ft TVD and cementing, drilling continued with a 12 ¼" bit without major kick or loss problems according to the planned mud weight (ENI AGIP, 2005) [4]. When the bit reaches 13181 ft TVD (lower 12 ¼" hole section), the operator changes the MW from 13.34 ppg to 14.68 ppg because of high pore pressure expected in Gadvan. This is continued to the planned 9 5/8 " casing setting depth (13304 ft TVD). Figure 2 illustrates this procedure. According to the Figure 2, the pore pressure at 13304 ft TVD is about 9983 psi, and the column pressure of 14.68 ppg of mud is 10156 psi. When circulating the mud at 767 gpm, the BHP increases to 10259 psi. Although the well does not flow, as it generates positive differential pressures of 173 psi and 276 psi in static and dynamic conditions respectively, mud returns are lost throughout the Sarvak formation during drilling and making connection. It is critical to reduce the mud weight so as to possibly cure the problem. The minimum mud weigh that can be used to drill this interval (about last 123 ft of Gadvan formation) is an EMW of 14.43 ppg. For safety reasons, a mud weight of 14.45 ppg was used to drill this section. Hence a little overbalanced was expected at the bottom of the hole; however, when the pump turned on at 767 gpm, the mud returns could be found in Sarvak formation.

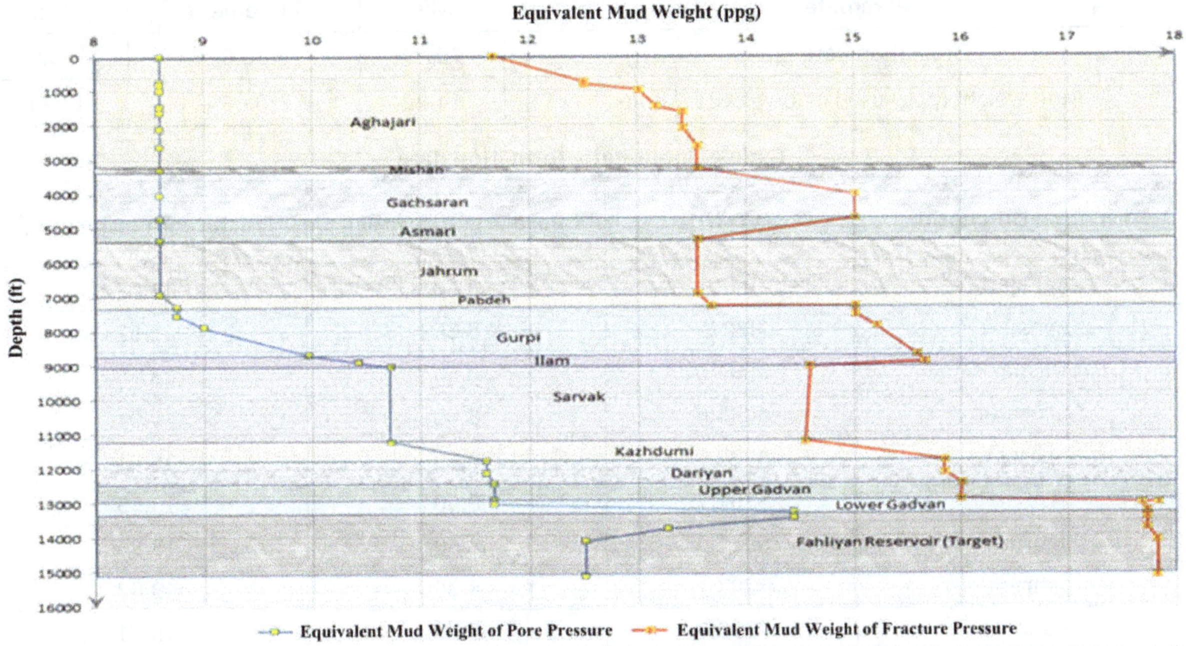

Figure 1: Pore pressure and fracture pressure profile of Well DQ No. 5

Table 1: Input data, drill string and BHA details, used in Fahliyan formation

Drill String Description (From Bit to Top)	ID (in)	OD (in)	Length (ft)	Distance from Bit (ft)
DC	2.81	8.00	30.80	30.80
St. Stab	2.81	12.25	4.82	35.62
DC	2.81	8.00	62.73	98.35
St. Stab	2.81	12.25	5.41	103.76
DC	2.81	8.00	180.90	284.66
Jar	2.81	8.00	16.53	301.19
DC	2.81	8.00	30.80	331.99
HWDP	3.00	5.00	460.78	792.77
DP	4.27	5.00	12511.00	13304.50

Table 2: Input data, casing design data

Description (Go From Bottom to Top)	Hole Dia. (in)	Casing OD (in)	Casing ID (in)	Depth From (ft)	Depth To (ft)
Open Hole	12.25	9.625	8.535	7382.25	13304.45
Surface Casing	17.50	13.375	12.415	0.00	7382.25
Conductor	24.00	18.625	17.755	0.00	820.25

Table 3: Input data, drilling fluid and circulation data

Rotational Speeds	Fann Viscometer Dial Readings		
θ_3	2		
θ_6	3		
θ_{100}	20		
θ_{200}	30		
θ_{300}	40		
θ_{600}	60		
Parameter	**Min**	**Increment**	**Max**
Circulation Rate (gpm)	707.0	10.0	767
Mud weight (ppg) used in lower 12 ¼" hole	14.22	0.03	14.31

Table 4: Input data, formation data

Formation Description	TVD (ft)	Pore Pressure (ppg)	Fracture Pressure (ppg)
Aghajari	0.00	8.60	11.67
Aghajari	820.2	8.60	12.51
Mishan	3281.0	8.60	13.55
Gachsaran	4035.6	8.60	15.01
Gachsaran	4727.9	8.60	15.01
Asmari	5367.7	8.60	13.55
Jahrum	6922.9	8.60	13.55
Jahrum	7287.1	8.75	13.67
Pabdeh	7290.0	8.75	15.01
Pabdeh	7533.1	8.75	15.01
Gurpi	7874.0	9.00	15.22
Gurpi	8697.9	9.96	15.59
Ilam	8924.3	10.42	15.67
Ilam	9058.8	10.72	14.59
Sarvak	11224.3	10.72	14.54
Kazhdumi	11782.0	11.60	15.84
Dariyan	12139.7	11.60	15.84
Upper Gadvan Lower Gadvan	12943.5 13051.0	11.67 11.67	16.00 17.84
Lower Gadvan	13288.0	14.43	17.73
Fahliyan Reservoir	13304.4	14.43	17.73
Fahliyan Reservoir	13780.2	13.26	17.73
Fahliyan Reservoir	15151.6	12.51	17.83

Table 5: Input data, used mud systems (ENI AGIP, 2005)

	Mud system	Density range (ppg)	Mud volume (ft^3)
24" hole section at 820 ft	Fresh water, bentonite (FW-GE)	8.75 -9.17	17700
17½" hole at 7382 ft	Salt water, polymer-lignosulfonate system (SW-PO-LS)	9.17-11.5	77700
12¼" hole at 13304 ft	Salt water, polymer-lignosulfonate system (SW-PO-LS)	12.5-14.7	47700

Table 6: Input data, mud characteristics (ENI AGIP, 2005)

		Hole Phases		
		24"	**17 ½"**	**12 ¼"**
Mud Type	Units	FW-GE	SW-PO-LS	SW-PO-LS
From	ft	0	820	7382
To	ft	820	7382	13304
Mud Density	ppg	8.75 -9.17	9.17-11.5	12.5-14.7
Viscosity	sec^{-1}	70	50-60	50-60
PV	cps	15-20	15-20	15-20
YP	lb/100 ft^2	61	18-22	19-25
Gel 10"	lb/100 ft^2	NA	2-4	2-4
Gel 10'	lb/100 ft^2	NA	4-6	4-6
PH	-	9.5-10	9-10	9-10
Filtrate API	cc/30'	NA	< 8	4-6
Pm	cm^3 0.02N H_2SO_4	NA	1	1
Pf	cm^3 0.02N H_2SO_4	NA	0.7	0.7

Table 7: Hydraulic program, 17 ½" section from 820 to 7382 ft RKB

		Pump data						Bit Data				
Depth (ft)	Mud Weight (ppg)	Flow Rate (gpm)	Pressure (psi)	Force (HHP)	Annular Velocity (ft/sec)	Nozzles (1/32 in.)	TFA (in^2)	Pressure at bit (psi)	% pressure at bit	Jet velocity (ft/sec)	Pressure (HHP/in^2)	Impact Force (kg)
1640	9.2	898	1251	662	1.306	3*18+1*16	0.941	793	61.0	305	1.7	594
3821	10.1	898	1507	792	1.306	3*18+1*16	0.941	867	55.0	305	1.8	648
4921	10.2	898	1678	886	1.306	3*18+1*16	0.941	867	49.6	305	1.8	648
7218	10.2	898	2361	1241	1.306	3*18	0.745	1381	56.0	387	2.9	820
7382	10.2	898	2446	1288	1.306	3*18	0.745	1381	54.0	387	2.9	820

Conventional Drilling

When a MW of 14.68 ppg is used, there is no other way to reduce loss amount and stop it, even if the pump is turned off. Thus this mud weight is rejected. By using a MW of 14.45 ppg, the circulation is lost under dynamic conditions using a pump rate of 767 gpm (see Figure 3). The only way to eliminate the problem in this situation is to reduce the pump rate to about 545 gpm; nevertheless, this rate might increase the risk of undesirable hole cleaning. It generates excessive frictional pressure of the cuttings, compensates for the pump rate reducing action, and might increase the ECD greater than that of the 767 gpm condition. Therefore, this mud weight could not treat loss of returns. These results make the use of a lower mud weight and the application of surface back pressure by choke manifold or back pressure pump inevitable in order to compensate for the bottom hole pressure due to a lower mud weight.

Constant Bottom Hole Pressure Solution

In the simulator, a MW of 14.31 ppg is used to drill this section. It is obvious that this mud is about 0.12 ppg lower than the pore pressure of the Gadvan formation at 13304 ft TVD. The static BHP of this mud is 9900 psi, while the pore pressure is about 9983 psi (see Figure 4). The well will certainly flow during drilling and connection, and thus the minimum 83 psi back pressure should be exerted at surface through choke manifold to compensate for the BHP. When the bit is reaming the formation at a pump rate of 767 gpm, the pressure throughout the wellbore is in a margin of safety and kick or loss scenario does not happen. The main issue is when a connection is to be made and the mud pumps are to be turned off. This is called the transition from dynamic to static condition.

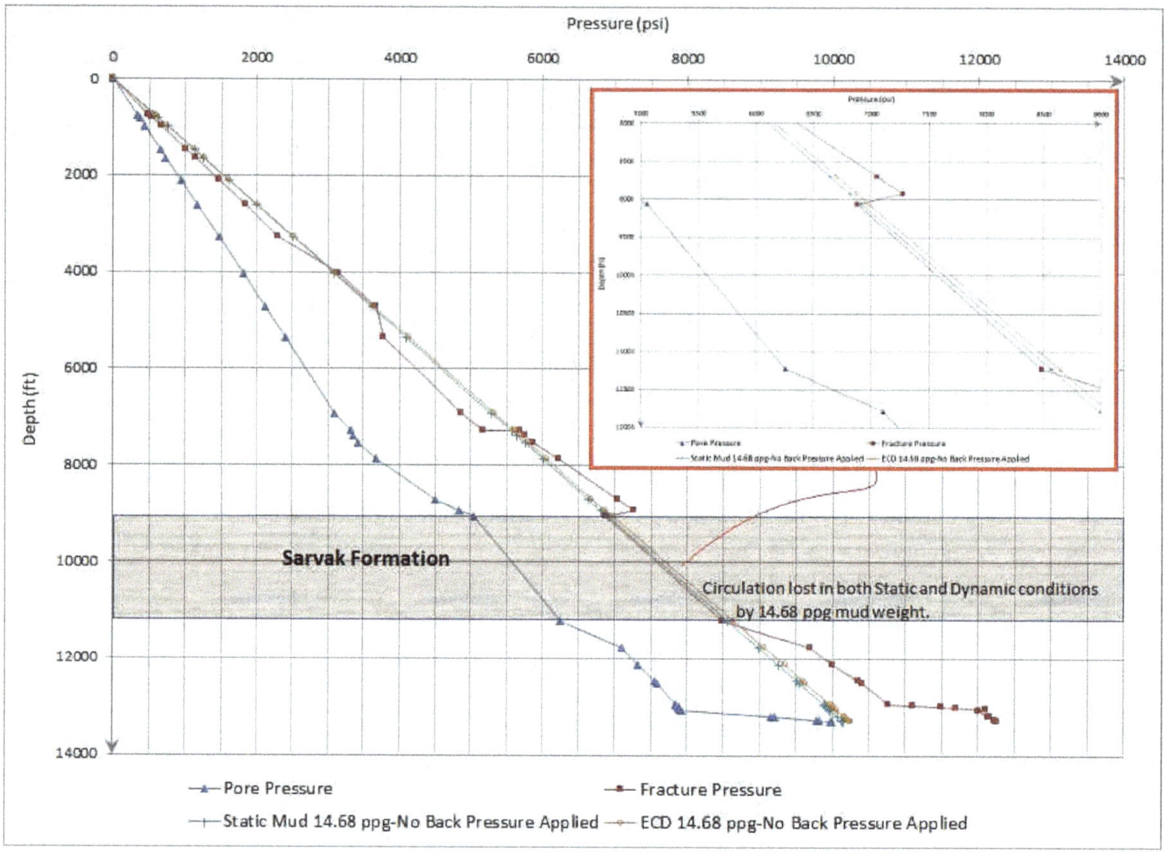

Figure 2: Operation window, static and dynamic BHP using a MW of 14.68 ppg at a pump rate of 767 gpm

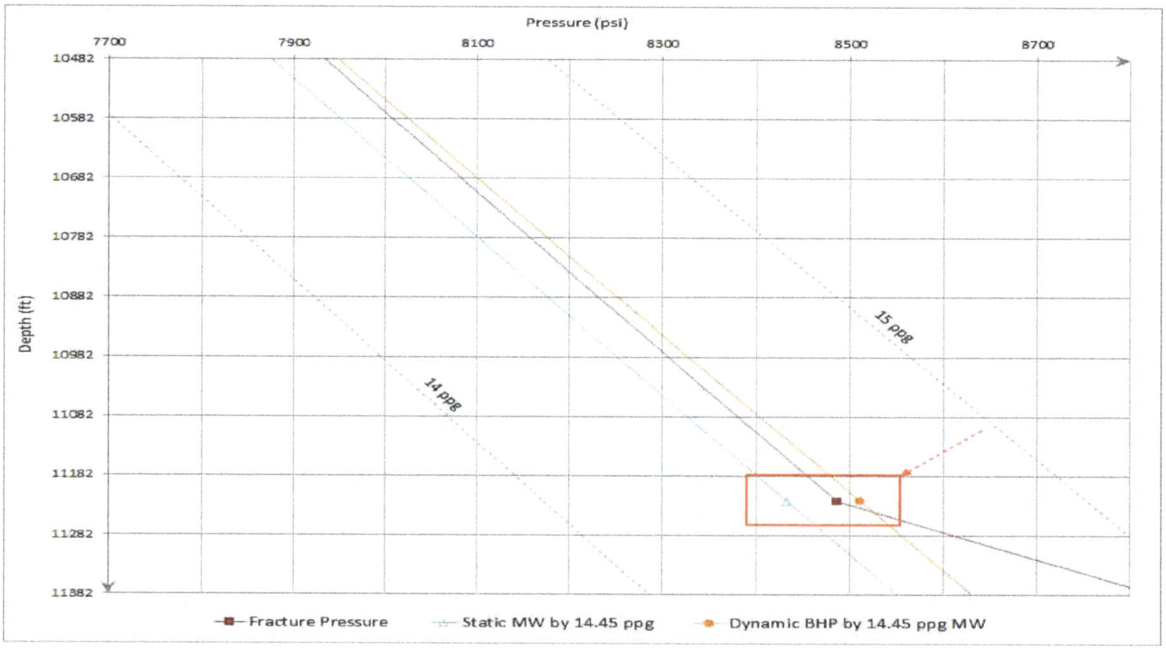

Figure 3: Static and dynamic BHP using a MW of 14.45 ppg at a pump rate of 767 gpm

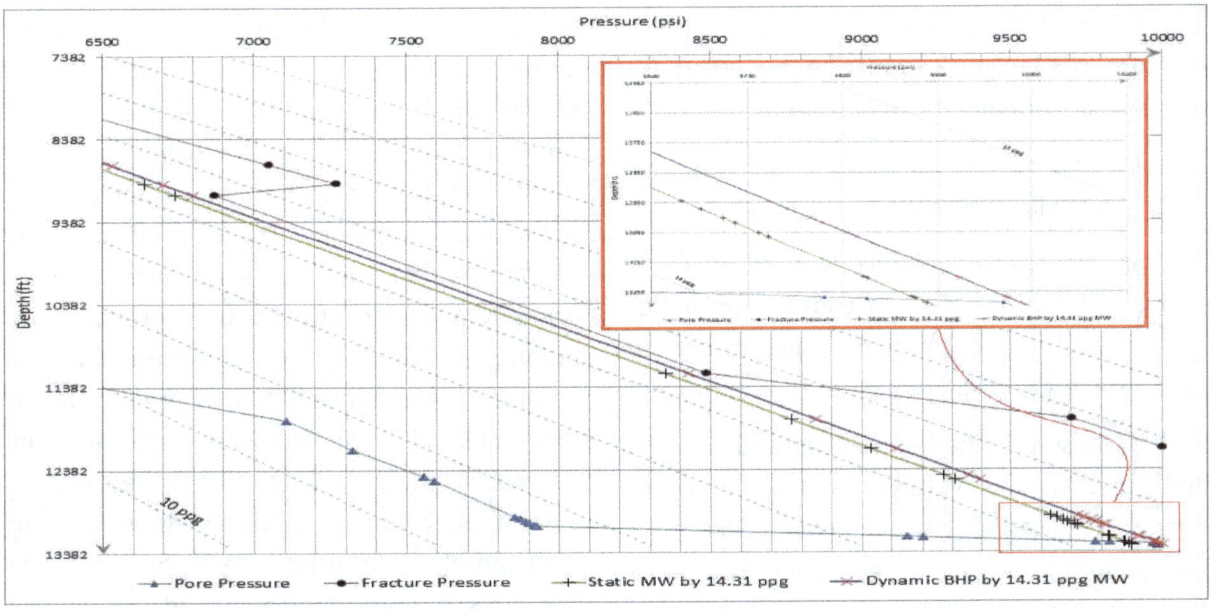

Figure 4: Static and dynamic BHP using a MW of 14.31 ppg at a pump rate of 767 gpm

Choke Opening Design

To shut down the pump for any reasons or if the pump is shut down suddenly, the BHP falls down the formation pressure and the well flows. Then, transition from dynamic to static conditions should follow a scheduled process to overcome the problem. Imagine that the pump rate is reduced; thus the BHP decreases stepwise until the pump is shut down. To avoid flow while the pump rate is reduced, the back pressure should slowly be exerted to the surface simultaneously to compensate for BHP. The choke opening is reduced until the annular pressure reaches the desired pressure at the next pump rate on the schedule; then, the pump rate is reduced to the one matching that annular pressure. Table 8 and Figure 5 show the design of the reducing pump rate while increasing choke back pressure to maintain the bottom hole pressure constant. It can be deduced from the bottom part of Figure 6 that the well flows if the pump is suddenly shut down from 767 gpm to a static condition. However, if the back pressure schedule is followed, the colored lines between "static-no BP" (green dotted line) and "static-100 psi BP" (red dotted line) are obtained. The result is that all of the stages from dynamic to static conditions are accompanied with a reduction in choke opening (means BP applied) and simultaneously pump rate is reduced; as a consequence, the bottom hole pressure is always constant.

Tripping in and Tripping out

Due to the marginal pressure window while drilling the last 123 ft of Gadvan formation, the surge and swab pressures should be taken into consideration. The lowest and highest criteria of allowable pressures while tripping are pore pressure in Gadvan and fracture pressure in Sarvak respectively. Since the BHP under both static and dynamic conditions are nearly equal to Gadvan pore pressure (about 17 psi over-balanced), the maximum pipe velocity while tripping out the hole should be simulated. The fracture pressure at bottom Sarvak is 8487 psi and should not exceed a maximum of 35 psi during tripping in. The mud clinging constant for annulus between 5" drill pipe and 12 ¼" open hole is equal to 0.36. The maximum allowable downward pipe velocity to have no returns loss to Sarvak is simulated to be about 1.75 ft/s. Also, the maximum allowable upward pipe velocity to avoid well flow is simulated to be about 0.58 ft/s. It is recommended that a good condition bit be selected for drilling this interval to avoid any bit problem in pulling out of hole.

Thus when a MW of 14.68 ppg is used and there is return loss in Sarvak formation, two ways are available to continue the operation:

1-Casing the bottom Sarvak and then using a higher MW will be possible;

2-Using the lowest possible MW to drill the Gadvan formation safely by choking the outlet of the wellbore to generate a pressure drop over a surface choke.

Each of these ways has a series of disadvantages and benefits. The first way requires a higher budget for casing and cementing excessive casing string, and as a consequence reduces borehole size. Therefore, it places some constraints on completion and production operations. The risk of stuck pipe due to high differential pressure is also high. The NPT increases and the total project cost per foot rises. In the second way, with an initial fixed cost of implementing MPD process, the need for excessive casing string is eliminated. Casing setting depth is extended and the wellbore reaches the reservoir with a higher hole size; this is suitable for performing completion and production operations. The risks of kick, loss, and differential stuck pipe are reduced due to a lower MW and controlled back pressure. The number of mud weight changes throughout the process is also reduced. All of these advantages can reduce NPT with no drilling problems, and eventually lower cost per foot is needed. The biggest disadvantage of this method is surge/ swab problem. By using a velocity of 1.75 ft/s and 0.58 ft/s in tripping in and tripping out respectively, it take a longer time to complete the operation. About 2 hours to trip in and 6 and half hours to trip out, which is too much high and can increase the total cost per foot of project.

Table 8: Schedule of reducing choke opening in shutting down the mud pump process with a mud weight of 14.31 ppg

Pump rate (gpm)	767	677	587	497	407	317	137	0
Applied Back pressure (psi)	0	17	37	52	67	78	94	100

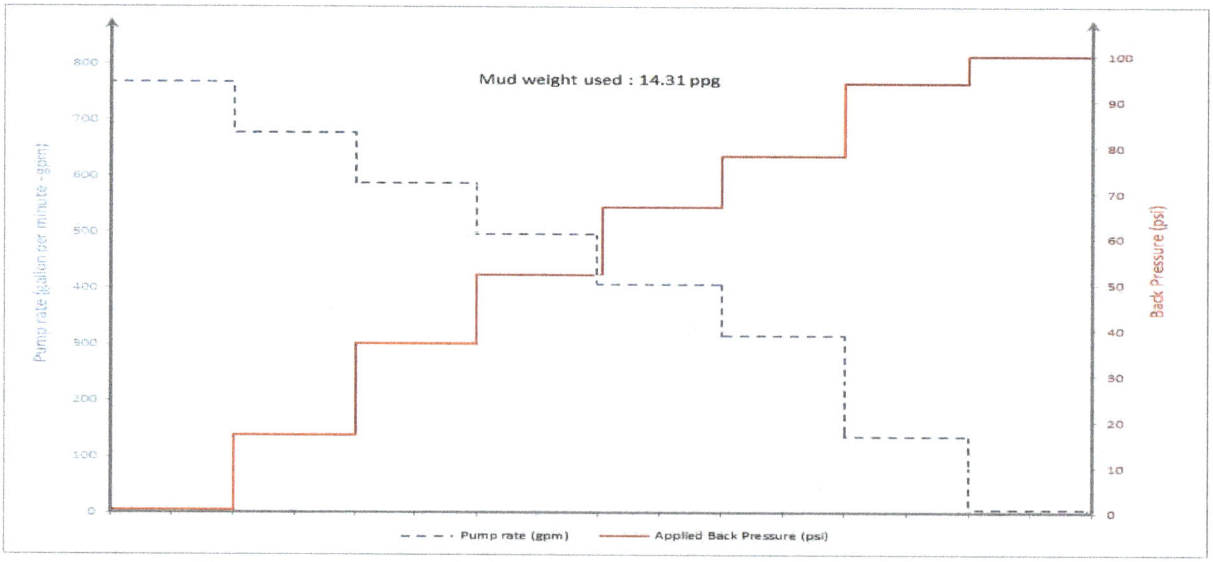

Figure 5: Pump rate and back pressure schedule to maintain constant BHP

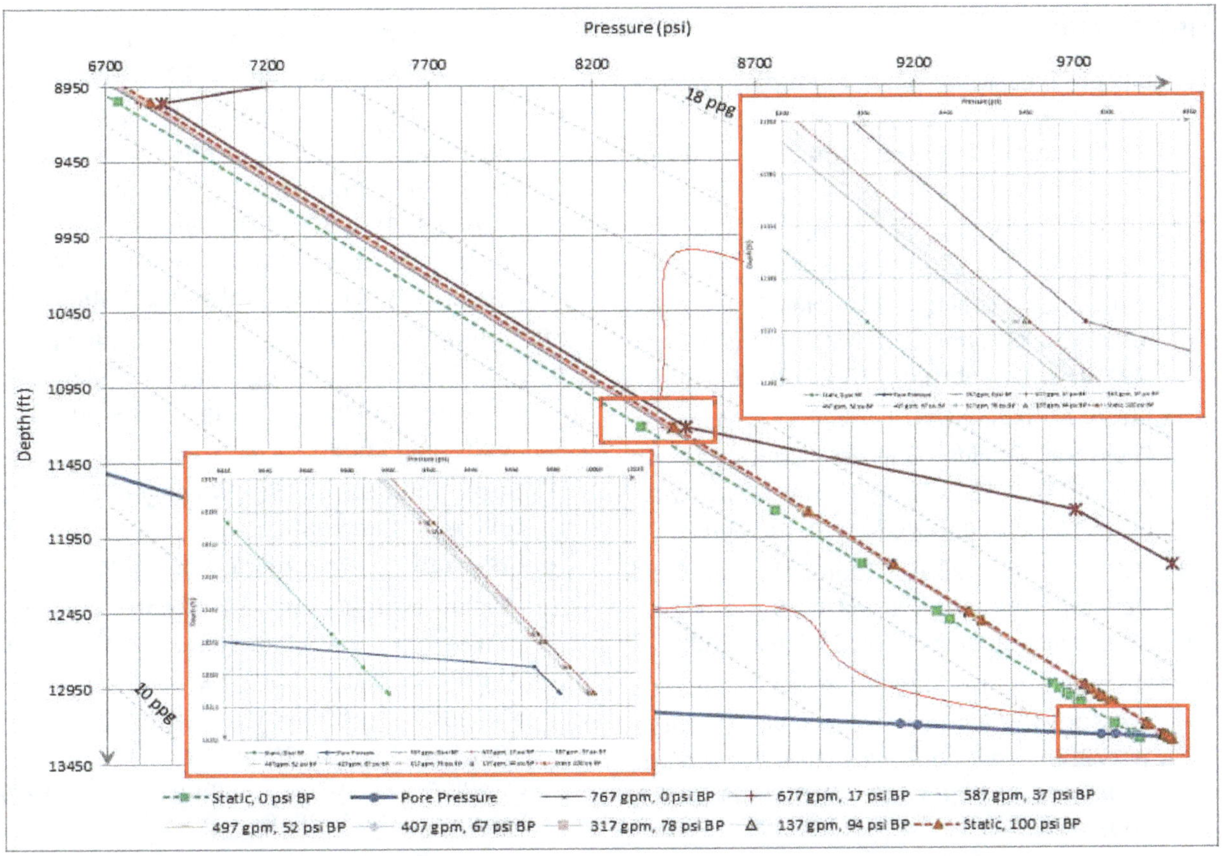

Figure 6: Pressure profile by following scheduled choke opening to maintain constant BHP

CONCLUSIONS

The feasibility study of implementing CBHP variation in Darquain oil field was done;

It was more challenging when the hole was simultaneously exposed to Gadvan formation with a pore pressure very close to the fracture pressure of the other exposed formation (Sarvak). A MW of 14.68 ppg induced circulation loss in Sarvak;

A MW of 14.45 ppg lowered the differential pressure and possibly brought formation fluid into the wellbore when the mud pumps were on;

By using the MPD software, a MW of 14.31 ppg was selected to drill the bottom 12 ¼" hole section;

Under static conditions, applying a surface back pressure of 100 psi by following the scheduled pump rate-choke opening eliminated the problem.

ACKNOWLEDGMENTS

The authors wish to express their gratitude to ENI AGIP Co. for geological and drilling reports and creating the opportunity to conduct this research.

NOMENCLATURE

AFP : Annular frictional pressure

BHP : Bottom hole pressure

BP : Back pressure

CBHP : Constant bottom hole pressure

ECD : Equivalent circulating density

EMW : Equivalent mud weight

Fp : Fracture pressure

gpm : Gallon per minute

MW : Mud weight

NPT : Nonproductive time

Pp : Pore pressure

ppg : Pound per gallon

TVD : True vertical depth

REFERENCES

[1] Balchen, J. G., Mumme, K. I., *Process Control: Structures and applications*, New York, VanNostrand Reinhold Company Inc, **1988,** 539.

[2] Chen, S., Xinming, N., Steve N., Holt, C., "Managed Pressure Drilling Reduces China Hard-Rock Drilling by Half," 105490, *SPE/IADC Drilling Conference*, Amsterdam, Netherlands, February, 20-22, **2007**.

[3] Chustz, M. J., Smith, L. D., Dell, D., "Managed Pressure Drilling Success Continues on Auger TLP," 112662, *IADC/SPE Conference*, Orlando, Florida, USA, March 4-6, **2008**.

[4] ENI AGIP Co. 2005, "Darquain Oil Field Development: Well DQ No. 5 Geological and Drilling Reports," NIOC PEDEK, **2009**.

[5] Hannegan, D. M. and Fisher, K., "Managed Pressure Drilling in Marine Environments," 10173*, International Petroleum Technology Conference*, Doha, Qatar, November 21-23, **2005**.

[6] Hannegan, D. M. and Wanzer, G., "Well Control Considerations-Offshore Applications of Underbalanced Drilling Technology," 79854, *SPE/IADC Drilling Conference*, Amsterdam, February 19-21, **2005**.

[7] Hannegan, D. M., "Case Studies-Offshore Managed Pressure Drilling," 101855, *SPE Annual Technical Conference and Exhibition*, San Antonio, Texas, September 24-27, **2006**.

[8] Iversen, F., Gravdal, J. E., Dvergsnes, E. W., Nygaard, G., Gjeraldstveit, H., Carlsen, L. A., Intl. Research Inst. Of Stavanger, And Low, E., Munro, C., Torvund, S., "Statoil Feasibility Study of Managed-pressure Drilling with Automatic Choke Control in Depleted HP/HT Field," 102842, *SPE Annual Technical Conference and Exhibition*, San Antonio, Texas, September 24-27, **2006**.

[9] Miller, A., Boyce, G., Moheno, L., Arellano, J., and Murillo, J., "Innovative MPD Techniques Improve Drilling Success in Mexico," 104030, *First International Oil Conference and Exhibition in Mexico*, Cancun, Mexico, September 1-2, **2006**.

[10] Rehm, B., Schubert, J., Haghshenas, A., Paknejad, A.S., Hughes, J. "Managed Pressure Drilling," Texas, Houston Gulf Publishing, **2009**.

[11] Sagar Nauduri, A. S., "Managed Pressure Drilling Candidate Selection", Ph.D. Dissertation, Texas A&M University, **2009**.

[12] Sagar Nauduri, A. S., Medley, G. H., Schubert, J., "MPD Candidate Identification: To MPD or Not To MPD," 130330, *SPE/AIDC Managed Pressure Drilling and Underbalanced Operations Conference and Exhibition*, Kuala Lumpur, Malaysia, February ,24-25, **2010**.

[13] Sagar Nauduri, A. S., Medley, G. H., Schubert, J., "MPD: Beyond Narrow Pressure Windows," 122276, *AIDC/SPE Managed Pressure Drilling and Underbalanced Operations Conference and Exhibition*, San Antonio, Texas, February, 12-13, **2009**.

[14] Vieira, P., Arnone, M., Russel, B., Cook, I., Moyse, K., Torres, F., Qutob, H., Weatherford International, and Yuesheng, C., Qing, C., "Sino Saudi Gas Constant Bottom hole Pressure: Managed-Pressure Drilling Technique Applied in an Exploratory Well in Saudi Arabia," 113679, *SPE Managed Pressure Drilling and Underbalanced Operations Conference and Exhibition*, Abu Dhabi, UAE, January 28-29, **2008**.

4

Crude Glycerol as a Substrate for Sulfate-reducing Bacteria from a Mature Oil Field and its Potential Impact on Souring

Sidnei Cerqueira dos Santos[1]*, Martin Gerald Liebensteiner[2], Leila Cristiane Silva das Virgens de Souza[1], Catia Larissa Santos Ramos[1], Luiz Lázaro Franco Batista[1], Josilene Borges Torres Lima Matos[1], Cristina Maria Quintella[3], and Paulo Fernando de Almeida[1]

[1] Laboratory of Biotechnology and Ecology of Microorganisms, Department of Bio-Interaction Sciences, Federal University of Bahia, Brazil

[2] Laboratory of Microbiology, Wageningen University, Wageningen, The Netherlands

[3] Laboratory of Kinetic and Molecular Dynamic, Institute of Chemistry, Federal University of Bahia, Salvador, Brazil

ABSTRACT

Crude glycerol (CG) is an abundantly available and cheap by-product from biodiesel production. Value-added applications for CG are highly wanted by industry and several processes such as the use of CG for enhanced oil recovery have been proposed. The aim of this study was to evaluate the sulfide production of sulfate-reducing bacteria (SRB) indigenous to oil reservoirs using CG as substrate. The samples of CG were obtained from a biodiesel production plant, processing castor beans, soybeans, cotton, and waste oils and fats. Growth tests were performed in Postgate medium, with different types and concentrations of CG, and a mixed inoculum of SRB isolated from the produced water of a mature oil well of Bahia (Brazil). The experiment was monitored by measuring the concentration of sulfide using a colorimetric method. The results showed that SRB grew and produced more than 250 ppm sulfide at CG concentrations of 2%. However, at CG concentrations of 3% or higher, the biogenic production of sulfide was reduced. The study demonstrates that CG will likely stimulate SRB in oil fields, whenever CG is present at lower concentrations. Maintaining CG concentrations inhibitive to SRB will not certainly be achievable throughout oil reservoirs. Dosing CG to oil fields may lead to problems associated with souring in longer terms. The utilization of CG by SRB could in turn be interesting for other biotechnological processes, e.g. metal recovery processes based on precipitation with biologically formed sulfide.

Keywords: Souring, Sulfate-reducing Bacteria, Crude Glycerol, Inhibition

INTRODUCTION

Crude glycerol (CG) is the main by-product of biodiesel production. It accounts for up to 10% of the product and contains between 40 and 90% pure glycerol [1,2]. The growth of biodiesel production has resulted in an excess of glycerol and consequently in a decline of the market price for glycerol [3,4]. This glycerol is considered as waste due to the presence of impurities such as alcohol, salts, and heavy metals,

***Corresponding author**

Sidnei Cerqueira dos Santos

Email: sidnei.cerqueiradossantos@gmail.com

which make it unsuitable for direct industrial use and its purification has a higher cost than obtaining it from other sources [5]. This scenario indicates that the commercial viability of biodiesel calls for new and sustainable solutions for the consumption of crude glycerol [6].

New strategies are being developed with the aim to increase the economic value of this by-product [4,6] such as the use of CG as fluid for oil recovery in the micro-reservoir cell. The latter could achieve an oil recovery factor of 80% and can thus be regarded as a promising method for the petroleum industry [7]. However, there is still limited knowledge about the microorganisms utilizing the injected CG and what products are formed during the process. The injection of crude glycerol in oil wells could possibly stimulate the growth and sulfide production of SRB, which are widely distributed in anoxic environments like oil reservoirs [8,9]. Microbial sulfide production in oil reservoirs has been a major problem in the petroleum industry (called souring), been lowering the productivity and quality of produced hydrocarbons, been increasing corrosion, and causing safety, health, and environmental problems [10,11]. A potentially stimulating effect of CG on SRB which are indigenous to oil reservoirs would therefore intensify the problem of souring and thus have a negative impact on its usability for enhanced oil recovery purposes.

The aim of this study was to evaluate the use of crude glycerol by oil reservoir inhabiting sulfate-reducing bacteria. A mixed culture of sulfate-reducing bacteria isolated from a mature oil field (Bahia, Brazil) was incubated with different sorts of crude glycerol from biodiesel production as the only carbon and energy source and the biological formation of sulfide was followed over time.

MATERIALS AND METHODS

Produced Water (PW)

The samples of the produced water were collected from a mature oil field of Bahia (Brazil) in 2009. The reservoir temperature was 44 °C, the initial pressure was 55 bar [12] and the salt concentration of the collected samples was 3.5%. The pH and sulfide concentration of the produced water were analyzed after centrifugation.

Quantification and Selection of SRB

The quantification of sulfate-reducing bacteria from the produced water samples was done by the most probable number (MPN) method, according to the American Society for Testing and Material [13]. MPN estimates were calculated from statistical tables [14]. Serial dilutions were performed in modified Postgate E medium, containing the following (g/l): NaCl, 35.0; $C_6H_5Na_3O_7$, 6.38; $MgCl_2.6H_2O$, 1.83; KH_2PO_4, 0.5; NH_4Cl, 1.0; Na_2SO_4, 1.0; $CaCl_2$, 1.0; yeast extract, 1.0; ascorbic acid, 0.1; sodium thioglycolate, 0.01; $FeSO_4.7H_2O$, 0.5; agar, 1.5; sodium lactate, 1.75 ml (60%, w/w); resazurin, 2.0 ml (0.025%, w/v). The pH was adjusted to 7.5-8.0 using HCl or NaOH and the medium was sterilized at 121 °C for 20 min. The samples were incubated in an anaerobic chamber (Bactron VI, Shellab, Sheldon Manufacturing Inc.) at 38 °C for 21 days. The growth of sulfate-reducing bacteria was indirectly followed by the formation of iron sulfides (appearing black), only occurring if sulfate reduction takes place.

Aliquots of 0.1 ml of positive MPN cultures were transferred to solid Postgate medium (agar, 15 g/l), and streaked using the spread plate method [15]. The plates were incubated in the anaerobic chamber at 38 °C. The selection of SRB was made based on macroscopic features and the formation of black precipitates in the colonies. A mixed culture of sulfate-reducing bacteria, called APB55, was composed by the selection of five different pure colonies. The morphological characteristics of the particular strains in the mixed culture were confirmed by microscopy.

Identification of Sulfate-reducing Bacteria (SRB)

Fluorescence in situ hybridization (FISH) technique was used to identify the presence of sulfate-reducing bacteria in APB55, according to standard procedures [16]. We used the indocarbocyanine (Cy3)-labeled 16S rRNA oligonucleotide probes SRB-385 (5'-CGGCGTCGCTGCGTCAGG-3') for the SRB group of the δ-Proteobacteria [17], DSV-698 (5'-GTTCCTCCAGATATCTACGG-3') for the Desulfovibrionaceae family [18], and SRB-129 (5'-CAGGCTTGAAGGCAGATT-3') for the Desulfobacter genus [19]. The total amount of cells was visualized by using DAPI (4',6-diamino-2-phenyl indole), a DNA-binding stain. Fluorescence microscopy equipped with two sets of filters (U-MWU2-330/420nm and U-MSWG2-480/590nm) was used to detect DAPI and Cy-3 stained cells respectively [20]. The slides were analyzed using a microscope (Olympus®, BX51) coupled to a camera (Olympus®, Q-Color). Image analyses were carried out using the software Image Pro-Plus 5.1.

Crude Glycerol (CG)

The samples of crude glycerol were collected in the Petrobras Biofuel plant, situated in Candeias (Bahia, Brazil). The CG samples differed regarding the raw materials used for biodiesel production, namely oil from castor beans, soybeans, cotton, and waste oils and fats (WOF) as raw materials. The CG samples were sterilized in an oven at 170 °C for 1 hour [21]. The chemical composition of the crude glycerol samples was determined according to the analytical methods mentioned below.

Evaluation of Sulfide Production

The assay of biogenic sulfide production was carried out in test tubes containing 20 ml of modified Postgate E medium containing crude glycerol as the sole carbon and energy source (leaving out sodium citrate and sodium lactate). Four different samples of CG from biodiesel production were used at concentrations of 1-5% (w/v). Sterilized commercial glycerol (99.5%, Merck) was used as a control. Inocula of 10^3-10^5 cells/ml of APB55 were added to the test tubes. Epifluorescent microscopy counting acridine orange (3,6-bis dimethylamino acridinium chloride) stained cells was carried out to standardize the inocula of APB55 [22]. Before staining, the culture was washed with Tween 80 (0.1%) and centrifuged at 7500 g for 10 minutes. The slides were analyzed with an Olympus® BX51 microscope. The test tubes were incubated in an anaerobic chamber at 38 °C for 96 hours. Aliquots of the incubated tubes were collected to analyze the sulfide production. The tests were performed with duplicates. The sulfide production was expressed by using the arithmetic mean and the range of duplicates.

Analytical Methods

For the pH measurements a pH-Fix 0-14 tape (Macherey-Nagel) was used. Sulfide was determined with the N,N-dimetil-p-fenilenediamina method [23]. The sulfide content of the cultures was measured at 660 nm on a spectrophotometer (PerkinElmer, Vitor 1420). The standard curve was prepared from the dilution of a standard solution of sodium sulfide and results were expressed in parts per million (ppm). The glycerol concentration of the CG samples was determined by an enzymatic assay measuring triglycerides (Doles), which is also sensitive to glycerol, and the absorbance was measured at 510 nm. Aldehydes were quantified by extracting them with acetonitrile (Merck) followed by a derivatization step with an acid solution of 2,4-dinitrophenylhydrazine. The carbonyl hydrazones were separated on a X-Terra MS C18 column (2.1 x 250 mm) and analyzed in an Agilent 1100 LC-DAD-MS system (Agilent, Waldbronn) Bruker Esquire 3000 plus (Bruker, Billerica). The mobile phase was water (25%) and acetonitrile (75%). The quantification of the aldehydes, as the respective hydrazones, was done by external calibration curves and the absorbance of compounds was monitored at 365 nm [24].

RESULTS AND DISCUSSION

Quantification and Identification of SRB

The sulfide concentration of the produced water sample was around 20 ppm and the pH was 7.0. The quantification of sulfate-reducing bacteria from the produced water samples by the MPN method showed a low microbial cell density of 2.4×10^3 cells/ml. This result combined with the low sulfide concentrations measured confirmed the low activity of SRB in the produced water sample. The composed SRB culture APB55 showed the presence of SRB belonging to the family of Desulfovibrionaceae and the genus *Desulfobacter* by using FISH and SRB-385, DSV-698, and SRB-129 probes (Figure 1).

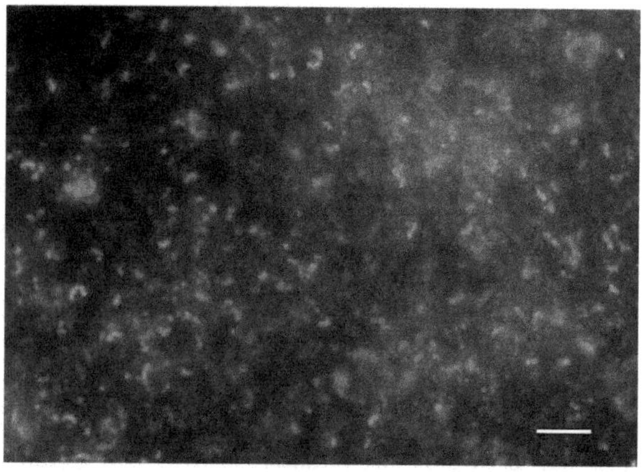

Figure 1: Mixed culture of SRB (APB55) visualized by fluorescence in situ hybridization (FISH) using SRB-385 probe and Cy3 as a fluorochrome; Bar represents 5 μm.

Biogenic Production of Sulfide

The results obtained from APB55 showed that the culture produced sulfide during its growth, when CG was used at concentrations of 1 (data not shown) and 2% (Figure 2A). The sulfide level reached more than 250 ppm for 2% CG, which equated to the complete utilization of the entire sulfate available in the medium. The sulfide production by APB55 with CG originating from soybeans was similar to the control (commercial glycerol) (Figure 2A). However, the initial sulfide production (after 25 hours) was almost two times lower compared to the commercial glycerol. All the other types of CG resulted in a delayed sulfidogenic activity (WOF- and cotton-type CG) of the culture or lowered final concentrations (WOF) of sulfide.

At concentrations of 3% (Figure 2B), 4% (data not shown), and 5% CG (Figure 2C), the biogenic production of sulfide was lowered (castor beans- and soybeans-type CG) or even completely inhibited (WOF- and cotton-type CG). At 3% CG, the sulfide level was reduced from more than 250 ppm (control) to values close to 30 ppm.

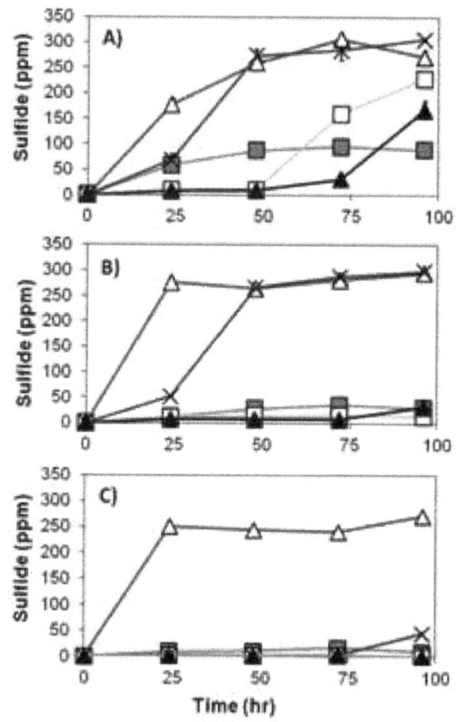

Figure 2: Sulfide production by a mixed culture of sulfate-reducing bacteria (APB55, 10^5 cells/ml inoculum) using crude glycerol from biodiesel production plants using different source materials: ■, castor beans; ▲, cotton; □, waste oils and fats (WOF); X, soybeans; Δ,control (pure glycerol). The concentration of the substrates varied as follows: A) 2%, B) 3%, and C) 5% crude glycerol (w/v). Means ± range (bars), *n* = 2.

Only crude glycerol from soybeans reached the same end concentration of sulfide compared with the control, though having a slower start-up

(Figure 2B). Overall, the inhibition of biogenic sulfide formation was most pronounced at a CG concentration of 5%. Then, sulfide levels were reduced to approximately 10 ppm. Only in cultures with CG from soybeans, the amount of sulfide reached around 50 ppm (Figure 2C).

Moreover, the inoculum density (10^3 and 10^4 cells/ml of APB55 compared to 10^5 cells/ml) had an influence on biogenic sulfate formation (data not shown). The lower the initial cell density was, the higher the inhibitive effect of CG on sulfate reduction became compared to control with commercial glycerol.

Chemical Composition of Crude Glycerol

The concentration of pure glycerol in different crude glycerol samples ranged from 2 to 40 mg/ml (Table 1). Especially cotton- and soybeans-type CG showed very low concentrations of glycerol. These values were below the expected ones reported in literature [2,7]. The lower glycerol content in cotton- and soybeans-type CG samples may be related to the presence of polymers of glycerol formed during the production of biodiesel [25]. These glycerol complexes are probably not, or to a lower degree, reactive in the assay, underestimating the total glycerol concentration. The quantification of the aldehydes using liquid chromatography identified some carbonyl compounds (Table 1). Some of these compounds have earlier been reported to inhibit the growth of SRB such as acrolein and formaldehyde [26-28]. The concentrations of acrolein were highest in CG from soybeans, whereas formaldehyde was most enriched in CG from castor beans (Table 1).

Crude glycerol is a cheap carbon source and could be seen as potential substrate for many biotechnological processes. Next to other applications, crude glycerol has successfully been used for microbial enhanced oil recovery purposes. However, several questions, including which microorganisms are involved and, foremost, what impact such a treatment has on the sulfate-reducing community and consequently on souring, have still remained open.

Table 1: Chemical composition of crude glycerol samples, deriving from biodiesel plants processing different source materials.

Composition (mg/ml)	Crude glycerol			
	Castor Beans	**Cotton**	**Waste oils and fats**	**Soybeans**
Glycerol	20.98	2.57	41.87	2.66
Formaldehyde	0.12	-	0.05	-
Acetaldehyde	1.26	0.11	0.70	0.03
Acrolein	0.10	0.23	0.18	12.60
Propion aldehyde	-	0.63	0.19	0.10
Butiraldehyde	-	0.31	0.16	0.06
Benzaldehyde	-	0.51	-	-
Isovaler aldehyde	-	0.22	0.11	-
Valeraldehyde	-	1.70	0.52	0.21
o-tolualdehyde	-	0.17	0.15	-
m-tolualdehyde	0.20	0.58	-	0.08
p-tolualdehyde	0.04	3.38	0.83	0.29

The main problems of souring are the reduction of the productivity and quality of produced hydrocarbons and increased corrosive processes due to the activity of sulfate reducers [10,11]. In our study, SRB indigenous to oil fields were used in order to mimic the native in situ microbiology and to have a stronger predictive power in this question. Two groups of sulfate reducers were isolated from the produced water and used for further experiments; they belonged to the Desulfovibrionaceae family and the genus *Desulfobacter*. Voordouw et al. [29] observed that all SRB detected from different oil fields using the 16S rRNA gene belong to the Desulfovibrionaceae or Desulfobacteriaceae families. *Desulfovibrio* is a regularly found genus of SRB in produced waters from oil fields and several species have been isolated from this environment [30,31].

The biological conversion of 1 mM glycerol to acetate theoretically reduces 0.75 mM sulfate to sulfide. However, the complete oxidation of glycerol coupled to sulfate reduction reduces 1.75 mM sulfate. Sulfide formation higher than stoichiometrically expected values may also be related to the broad variety of compounds present in CG samples besides glycerol (such as organic acids). These compounds may also contribute as potential electron donors to the reduction of sulfate, and consequently a higher amount of sulfide will be formed compared to the amount stoichiometrically expected from glycerol oxidation only.

In this work, crude glycerol was shown to promote the biological reduction of sulfate to sulfide by a mixed culture of SRB, deriving from an oil field. Several sulfate-reducing bacteria are able to use glycerol as an electron donor. Several *Desulfovibrio* species were reported to grow with glycerol [32-34]. The ability to use glycerol was also demonstrated for *Desulfospira joergensenii*, isolated from marine sediments [35] and members of the genus *Desulfosporosinus* originating from acid sediments [36,37].

Our study shows that increased concentrations of crude glycerol inhibit the biogenic production of sulfide. This could possibly be related to the presence of toxic compounds in CG. The chromatographical analyses of the CG samples identified aldehydes that may be associated with the inhibition of microorganisms, as has been shown for acrolein and formaldehyde [26,27] (Table 1). Acrolein is the main product of dehydration of glycerol during transesterification processes. It is also found in biofuel plants when fats and oils are directly used in catalytic thermolysis reactions [38]. Formaldehyde is the main product formed by the catalytic oxidation of methanol [39], which is usually used in the transesterification process. In this way, both acrolein and formaldehyde can be formed during the biodiesel production. These compounds are also used in biocides to control the growth of gram-positive and gram-negative bacteria, fungi, and yeast [26-28]. Nevertheless, in this study, a direct relationship between the measured concentrations of acrolein and formaldehyde and the inhibition of sulfidogenesis was not observed (comparing Table 1 and Figures 2B/C).

Alkali metals may also have inhibitory effects on microorganisms. The inhibitory concentration of sodium ion (Na^+) is between 3 to 16 g/l [40] and that of potassium ion (K^+) is 2 to 12 g/l [41]. Chaves et al. [42] detected high values of sodium (0.36 to 19 g/l) and potassium (<0.08 to 92 g/l) in the CG samples from vegetable waste oil and commercial soybeans oil. The authors also observed that there was a large variation in the concentrations of Na^+ and K^+ in the CG samples. This indicates that much of the catalyst used is concentrated in the crude glycerol phase. The high concentration of Na^+ and K^+ may be related to the choice of catalyst used in the biodiesel production, which are usually sodium hydroxide (NaOH) or potassium hydroxide (KOH).

We suppose that organic waste compounds from biodiesel production (such as acrolein and formaldehyde) and the high concentration of catalysts (NaOH or KOH) in CG act directly or via an associated way on the metabolism of SRB. This is in line with other studies where crude glycerol (in contrast to pure glycerol) showed to be toxic to *Clostridium pasteurianum* [43-44].

Further studies have to be conducted for a better understanding of the exact inhibitory effect of higher concentrations of CG on the biogenic formation of sulfide.

ACKNOWLEDGEMENTS

The authors thank the Coordenação de Aperfeiçoamento de Pessoal de Nível Superior (CAPES), Petróleo Brasileiro S.A., and the Fundação de Amparo à Pesquisa do Estado da Bahia (FAPESB). The research was part of the Ph.D. program of the Northeast

Brazil Biotechnology Network (RENORBIO).

CONCLUSIONS

A definite prediction about the potential of sulfide formation after CG injection in oil fields is not possible from this study. However the results show that CG is utilized by indigenous SRB and may therefore contribute to an increase in souring of oil reservoirs. A continuous dosing of sufficiently high CG concentrations to inhibit sulfidogenesis seems impracticable and could not be maintained throughout an oil reservoir.

On the other hand, the study suggests CG (at lower concentrations) as a good electron donor for SRB (Figure 2A). CG might, for instance, serve as a cheap and abundantly available source for metal precipitation processes based on biologically formed sulfide such as the processes used for the bioremediation of acid mine drainages and the biorecovery of metals.

REFERENCES

[1] Hansen C. F., Hernandez A., Mullan B. P., Moore K., et al., "A Chemical Analysis of Samples of Crude Glycerol from the Production of Biodiesel in Australia, and the Effects of Feeding Crude Glycerol to Growing-finishing Pigs on Performance, Plasma Metabolites and Meat Quality at Slaughter," *Anim. Prod. Sci.*, **2009**, *49*, 154-161.

[2] Hazimah A. H., Ooi T. L., and Salmiah A., "Recovery of Glycerol and Diglycerol from Glycerol Pitch," *J. Oil Palm Res.*, **2003**, *15*, 1-5.

[3] Sandun F., Sushil A., Kiran K., and Ranjitha B., "Glycerol Based Automotive Fuels from Future Biorefineries," *Fuel*, **2007**, *86*, 2806-2809.

[4] Yazdani S. S. and Gonzalez R., "Anaerobic Fermentation of Glycerol: a Path to Economic Viability for the Biofuels Industry," *Curr. Opin. Biotechnol.*, **2007**, *18*, 213-219.

[5] Johnson D. T. and Taconi K. A., "The Glycerin Glut: Options for the Value-added Conversion of Crude Glycerol Resulting from Biodiesel Production," *Environ. Progress*, **2007**, *2*, 338-348.

[6] Silva G. P., Mack M., and Contiero J., "Glycerol: A Promising and Abundant Carbon Source for Industrial Microbiology," *Biotechnol. Adv.*, **2009**, *27*, 30-39.

[7] Quintella C. M., Borges S. M. S., Almeida P. M. M., Musse A. P. S., et al., "Secondary Recovery of Crude Oil with Glycerin, Co-product of the Biodiesel Production," *Anais Rio Oil & Gas Expo. and Conference*, Rio de Janeiro, **2006**.

[8] Muyzer G. and Stams A. J. M., "Ecology and Biotechnology of Sulphate-reducing Bacteria," *Nature Rev. Microbiol.*, **2008**, *6*, 441-454.

[9] Postgate J. R., *The Sulphate-reducing Bacteria*, Cambridge: Cambridge University Press, **1984**.

[10] Hubert C., Nemati M., Jenneman G., and Voordouw G., "Corrosion Risk Associated with Microbial Souring Control Using Nitrate or Nitrite," *Appl. Microbiol. Biotechnol.*, **2005**, *68*, 272-282.

[11] Coetser S. E. and Cloete T. E., "Biofouling and Biocorrosion in Industrial Water Systems," *Crit. Rev. Microbiol.*, **2005**, *31*, 213-232.

[12] Estublier A., Dino R., Schinelli M. C., Barroux C., et al., "CO_2 Injection in Buracica-long-term Performance Assessment," *Energy Procedia.*, **2011**, *4*, 4028-4035.

[13] American Society of Testing and Material, *Standard Test Method for Sulphate-reducing Bacteria in Water and Water-formed Deposits*, West Conshohocken: American Society of Testing and Material Annual Book of Standards (ASTM, D4412), **2002**.

[14] *American Public Health Association, Standard Methods for the Examination of Water and Wastewater* (13th ed.), Washington: American Public Health Association, **1973**.

[15] Clesceri S., Greenberg A. E., and Eaton A. D., *Standard Methods for the Examination of Water and Wastewater* (20th ed.),

Washington: American Public Health Association, American Water Works Association, Water Environmental Federation, **1998**.

[16] Daims H., Stöcker K., and Wagner M., Fluorescence In Situ Hybridization for the Detection of Prokaryotes, in *Advanced Methods in Molecular Microbial Ecology*, Osborn A. M., Smith C. J., ed. Abingdon: BIOS Scientific Publishers, **2005**, 213-239.

[17] Amann R. I., Devereux R., Key R., and Stahl D. A., "Molecular and Microscopic Identification of Sulfate-reducing Bacteria in Multispecies Biofilms," *Appl. Environ. Microbiol.*, **1992**, *58*, 614-623.

[18] Manz W., Eisenbrecher M., Neu T. R., and Szewyk U., "Abundance and Spatial Organization of Gram-negative Sulfate Reducing Bacteria in Activated Sludge Investigated by In Situ Probing with Specific 16S rRNA Targeted Oligonucleotides," *FEMS Microbiol. Ecol.*, **1998**, *25*, 43-61.

[19] Devereux R., Kane M. D., Winfrey J., and Stahl D. A., "Genus and Group-specific Hybridization Probes for Determinative and Environmental Studies of Sulfate Reducing Bacteria," *Syst. Appl. Microbiol.*, **1992**, *15*, 601-609.

[20] Santos A. F. J., Batista L. L. F., Lima J. B. T., Almeida R. C. C., et al., "Evaluation of the Fluorescence in Situ Hybridization Technique for Detection of Eubacteria and Sulfate-reducing Bacteria from Samples of Water in Oil Fields," *Chem. Eng. Trans.*, **2010**, *20*, 139-144.

[21] Gherna R. L., "Preservation," in *Manual of Methods for General Microbiology, Gerhard P.*, ed. Wshington: American Society for Microbiology, **1981**, 208-217.

[22] Francisco D. E., Mah R. A., and Rabin A. C., "Acridine Orange Epifluorescence Technique for Counting Bacteria in Natural Waters," *Trans. Am. Microsc. Soc.*, **1973**, *92*, 416-421.

[23] Trüper H. G. and Schlegel H. G., "Sulfur Metabolism in Thiorhodeaceae. I. Quantitative Measurements on Growing Cells of Chromatium Okenii," *Antonie van Leeuwenhoek*, **1964**, *30*, 225-238.

[24] Bastos L. C. S. and Pereira P. A. P., "Influence of Heating Time and Metal Ions on the Amount of Free Fatty Acids and Formation Rates of Selected Carbonyl Compounds During the Thermal Oxidation of Canola Oil," *J. Agric. Food Chem.*, **2010**, *58*, 12777-12783.

[25] "Journal of Plastic," Internet: www.jorplast.com.br, Nov. 9, 2010 [Nov. 9, 2010].

[26] Feron V. J., Til H. P., de Vrijes F., Wouterson R. A., et al., "Aldehydes: Occurrence, Carcinogenicity Potential, Mechanism of Action and Risk Assessment," *Mutat. Res.*, **1991**, *259*, 363-385.

[27] Ma T. H. and Harris M. M., "Review of the Genotoxicity of Formaldehyde," *Mutat. Res.*, **1988**, *196*, 37-59.

[28] Fink T.R. and Worner P., "Pollution Prevention Opportunities in Oil and Gas Production, Drilling, and Exploration," Internet: http://infohouse.p2ric.org/ref/03/02975.pdf, **1993** [Dec. 23, 2014].

[29] Voordouw G., Armstrong S. M., Reimer M. F., Fouts B., et al., "Characterization of 16S rRNA Genes from Oil Field Microbial Communities Indicates the Presence of a Variety of Sulfate-reducing, Fermentative, and Sulfide-oxidizing Bacteria," *Appl. Environ. Microbiol.*, **1996**, *62*, 1623-1629.

[30] Magot M., Basso O., Tardy-Jacquenod C., and Caumette P., "Desulfovibrio Bastinni sp. nov. and Desulfovibrio Gracilis sp. nov., Moderatety Halophilic, Sulfate-reducing Bacteria Isolated from Deep Subsurface Oilfield Water," *Int. J. Syst. Evol. Microbiol.*, **2004**, *54*, 1693-1697.

[31] Miranda-Tello E., Fardeau M. L., Fernandéz L., Ramiréz F., et al., "Desulfovibrio Capillatus sp. nov., a Novel Sulfate-reducing Bacterium Isolated from an Oil Field Separator located in the Gulf of Mexico," *Anaerobe*, **2003**, *8*, 97-103.

[32] Stams A. J. M., Hansen T. A., and Skyring G. W., "Utilization of Amino Acids as Energy Substrates by Two Marine Desulfovibrio Strains," *FEMS Microbiol. Ecol.*, **1985**, *31*, 11-15.

[33] Kremer D. R. and Hansen T. A., "Glycerol and Dihydroxyacetone Dissimilation in Desulfovibrio Strains," *Arch. Microbiol.*, **1987**, *147*, 249-256.

[34] Esnault G., Caumette P., and Garcia J. L., "Characterization of Desulfovibrio Giganteus sp. nov., a Sulfate-reducing Bacterium Isolated from a Brackish Coastal Lagoon," *Syst. Appl. Microbiol.*, **1988**, *10*, 147-151.

[35] Finster K., Coates J. D., Liesack W., and Pfennig N., "Desulfuromonas Thiophila sp. nov., a New Obligately Sulfur-reducing Bacterium from Anoxic Freshwater Sediment," *Int. J. Syst. Bacteriol.*, **1997**, *47*, 754-758.

[36] Alazard D., Joseph M., Battaglia-Brunet F., Cayol J. L., et al., "Desulfosporosinus Acidiphilus sp. nov.: a Moderately Acidophilic Sulfate-reducing Bacterium Isolated from Acid Mining Drainage Sediments," *Extremophiles*, **2010**, *14*, 305-312.

[37] Sánchez-Andrea I., Stams A. J., Amils R., and Sanz J. L., "Enrichment and Isolation of Acidophilic Sulfate-reducing Bacteria from Tinto River Sediments," *Environ. Microbiol. Reports.*, **2013**, *5*, 672-678.

[38] Parshall G. W. and Ittel S. D., "Homogeneous Catalysis: the Application and Chemistry of Catalysis by Soluble Transition Metal Catalysts," New York: John Willey and Sons, **1992**.

[39] Allison J. N., Goddard III W. A., "Oxidative Dehydrogenation of Methanol to Formaldehyde," *J. Cat.*, **1985**, *92*, 127-135.

[40] Feijoo G., Soto M., Mendez R. E., and Lema J. R., "Sodium Inhibition in the Anaerobic Digestion Process: Antagonism and Adaptation Phenomena," *Enz. Microbial. Technol.*, **1995**, *17*, 180-188.

[41] Bashir B. H. and Matin A., "Combined Effect of Calcium and Sodium on Potassium Toxicity in Anaerobic Treatment Processes," *Electron. J. Environ. Agric. Food Chem.*, **2004**, *4*, 670-676.

[42] Chaves E. S., Ramos J. C., Fontana K. B., Modolon S., et al., "Simple and Fast Method for the Determination of Na and K in Raw Glycerin from Biodiesel Production by Flame Atomic Emission Spectrometry," *Br. J. Anal. Chem.*, **2010**, *1*, 24-59.

[43] Jensen T., Kvist T., Mikkelsen M. J., Christensen P. V., et al., "Fermentation of Crude Glycerol from Biodiesel Production by Clostridium Pasteurianum," *J. Ind. Microbiol. Biotechnol.*, **2012**, *39*, 709-717.

[44] Dabrock B., Bahl H., and Gottschalk G., "Parameters Affecting Solvent Production by Clostridium Pasteurianum," *Appl. Environ. Microbiol.*, **1992**, *58*, 1233-1239.

Effect of Calcium Promoters on Nanostructured Iron Catalyst for Fischer-Tropsch Synthesis

Yahya Zamani*, Akbar Zamaniyan, Farzad Bahadoran, and Mehrdad Shojaei

Research Institute of Petroleum Industry (RIPI), National Iranian Oil Company, West Blvd. of Azadi Sports Complex, P.O.BOX 14665-137, Tehran, Iran

ABSTRACT

The Fischer-Tropsch synthesis (FTS) has been recognized as a heterogeneous surface-catalyzed polymerization process. During this process, CH_x monomers formed via the hydrogenation of adsorbed CO on transition metals produce hydrocarbons and oxygenates with a broad range of chain lengths and functional groups. A series of Fe/Cu Fischer-Tropsch synthesis catalysts incorporated with a calcium promoter were prepared by a microemulsion method. The composition of the final nanosized iron catalysts in terms of the atomic ratio is as follows: 100Fe/4Cu, 100Fe/4Cu/2Ca, 100Fe/4Cu/4Ca. XRD, BET, TEM, and TPR techniques were used to study the catalysts phase, structure, and morphology. Fischer-Tropsch synthesis (FTS) reaction test was performed in a fixed bed reactor. All the promoted catalysts, compared to the unpromoted catalysts, have higher rates of FT and the secondary reaction for CO_2 production. The formation of methane and light hydrocarbons is restrained with increasing the amount of calcium. The 100Fe/4Cu/2Ca shows the best performance between the prepared catalysts.

Keywords: Nanoparticle Iron Catalyst, Fischer-Tropsch Synthesis, Calcium Promoter

INTRODUCTION

High energy cost is the main driving force behind currently increasing interest in the Fischer-Tropsch synthesis (FTS) for the conversion of natural gas to liquids (GTL). The catalytic synthesis of hydrocarbons from CO and H_2 syngas mixtures leads to a large variety of products such as paraffins, olefins, alcohols, and aldehydes. Several metal catalysts can be used for the FTS; however, only iron and cobalt catalysts appear to be economically feasible on an industrial scale [1]. Cobalt catalysts yield mainly straight-chain hydrocarbons [2,3], while iron catalysts are more useful either when the H_2/CO ratio is low because of the water-gas shift (WGS) activity of Fe or for the production of alkenes, oxygenates, and branched hydrocarbons, which depends on the promoters and process conditions employed [2,4]. Most of the studies reported have attempted to improve catalyst performance by promoting with additives such as K [5,6], Mn [7], Cr [5], Ru [8], and Pt [9]. Among promoters, potassium has been used as a promoter for iron catalysts. Potassium can also increase the catalytic activity in FTS and WGS reactions [10,11]. Copper is normally added to Fe-based Fischer-Tropsch synthesis (FTS) catalysts as a chemical promoter; it is added to enhance hematite reducibility [10]. Very stable

***Corresponding author**

Yahya Zamani

Email: yahyazamani@yahoo.com

activity and high selectivity to light olefin formation have been observed for Mn-promoted Fe catalysts [12,13]. A positive effect of other transition metals such as La, Mo, Ta, V, and Zr on the catalyst activity for both CO hydrogenation and WGS activity has also been reported [14-17]. Although the studies on the Fe - based FT catalysts are extensive, the investigations on the effect of calcium on the catalyst are scarce and the effect of calcium on the nanosized iron-based catalysts has not been investigated. In this study, a microemulsion method has been developed to prepare three nanosized iron catalysts by preparing nanosized iron with copper and calcium oxide separately from their solutions. The objective of this work is to investigate the effect of calcium promoter on catalyst morphology, activity, and product selectivity in Fischer-Tropsch synthesis. The catalysts were tested in a fixed bed stainless steel reactor at FTS conditions.

EXPERIMENTALS

Catalyst Preparation

Nanostructured iron catalysts were prepared by a microemulsion method. A water solution of metal precursors, $FeCl_3 \bullet 6H_2O$ and $Cu(NO_3)_2 \bullet 4H_2O$ were added to a mixture of 2-propanol and chloroform with a ratio of 1:1 and sodium dodecyl sulfate (SDS) as a surfactant. Hydrazine (25-30%) in the aqueous phase was added as a precipitating agent and stirred for 4 hours. The solid was recovered by centrifugation and washed thoroughly with distilled water, ethanol, and acetone. Finally, the samples were dried overnight at 110 °C, and subsequently calcined in air at 380 °C for 4 hrs. Nanostructured calcium oxide was prepared like nanostructured Fe-Cu. At the next step, they were mixed together. The promoted catalysts were dried at 110 °C for 15 hrs and calcined at 380 °C for 4 hrs in air [16,17]. The catalyst compositions were designated in terms of the atomic ratios as: 100Fe/4Cu, 100Fe/4Cu/ 2Ca, and 100Fe/4Cu/4Ca. All the samples were pressed into pellets, crushed, and sieved to obtain particles with 30-40 mesh.

Catalyst Characterization

BET surface area and the pore volume of the catalysts were determined by N_2 physisorption using a Micromeritics ASAP 2010 automated system. A 0.3 g catalyst sample was degassed in the system at 100 °C for 1 hr and then at 300 °C for 2 hrs prior to analysis. The analysis was done using N_2 adsorption at 196 °C. The average particle size of the calcined powders was measured by LEO 912AB TEM. The XRD spectra of the fresh catalyst were conducted with a Philips PW1840 X-ray diffractometer with monochromatized Cu (Kα) radiation for determining iron phases. Temperature programmed reduction (TPR) profiles of the calcined catalysts were recorded using a Micromeritics TPD-TPR 290 system. The TPR of 50 mg of each sample was performed in 5% hydrogen/argon gas mixture. The samples were heated from 50 to 900 °C at a heating rate of 10 °C/min. The H_2 reduction process illustrated three stages in the temperature range between 200-900 °C.

Reactor System and Operation Procedure

The catalytic reaction experiments were conducted in a fixed bed stainless steel reactor. The flow rate of inlet gases and reactor pressure were controlled by electronic mass flow and pressure controllers respectively. A four heating zone furnace with a temperature controller and indicator supplied the required reaction heat. The reactor was loaded by 1 g of catalyst. The catalyst was reduced in a 10% H_2/N_2 flow for 3 hrs. The catalyst activation was followed in a stream of synthesis gas with H_2/CO =1 and SV= 1.5 $nl.h^{-1}.gcat^{-1}$ for 24 hrs at atmospheric pressure and a temperature of 270 °C. Following the activation process, the reactor pressure and temperature raised to 17 bar and 290°C respectively and the reaction was initiated in synthesis gas stream with H_2/CO =1 and GHSV=3 nl.h^{-1}.gCat^{-1}. The products were analyzed by two gas chromatographs (Varian CP 3800), one of which was equipped with two subsequent connected columns. The two packed columns connected to two thermal conductivity detectors (TCD) were used for analyzing H_2, CO, CO_2, CH_4, and other non-condensable gases and the other one with a Petrocol Tm DH100 fused silica capillary column attached to a flame ionization detector (FID) was used for analyzing organic liquid products [18,19]. The activities and product selectivity were assessed after 72 hrs from the initial time.

RESULTS AND DISCUSSION

Alkaline elements are used as promoters because they can modify the adsorption pattern of the reactants (H_2 and CO) on the active sites. The overall effects of these promoters, such as potassium, on the behavior of the iron-based FTS catalysts, namely CO chemisorptions enhancement, have been justified as a consequence of the iron tendency to withdraw electronic density from potassium. Therefore, the strength of the Fe-CO bond was enhanced [18]. While alkali loading is high, CO dissociation proceeds faster than carbon hydrogenation, which causes an excessive carbon deposition and consequently deactivates the catalyst surface ultimately [20]. Table 1 shows the results of catalysts surface area. By adding calcium, the BET surface area and pore volume of the catalysts decrease as it promotes the aggregation of the catalyst crystallites and blocks up the pore volume of the catalyst.

Nanostructured iron catalysts were characterized by X-ray diffraction (XRD) after calcinations. Figure 1 shows the XRD patterns of the catalysts prepared by the microemulsion method. The addition of Ca did not cause any obvious changes and no phase containing metals were detected. All the catalysts showed the Fe_2O_3 crystalline phase; however, their structure seemed to be like cubic hematite structured Fe_2O_3 crystal in JCPDS database.

Table 1: Catalysts surface area

Catalysts	BET Surface area(m^2/g)	Pore Volume (cm^3/g)
100Fe/4Cu	49.3	0.26
100Fe/4Cu/2Ca	45.5	0.23
100Fe/4Cu/4Ca	41.2	0.19

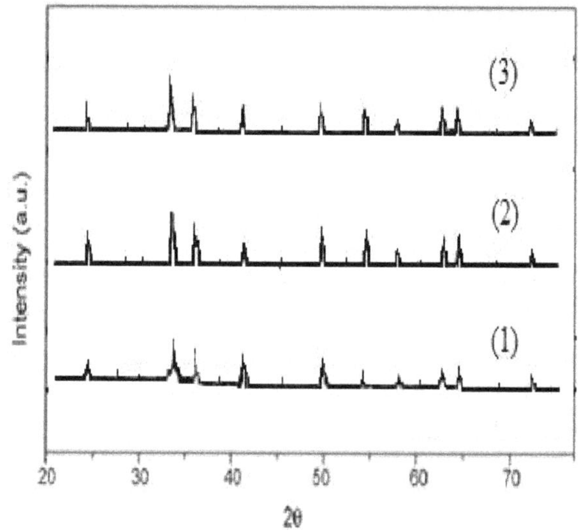

Figure 1: XRD spectra of the fresh catalysts: (1) 100Fe/4Cu; (2)100Fe/4Cu/2Ca; (3) 00Fe/4Cu/4Ca.

The morphology of the catalysts was illustrated by TEM images as shown in Figure 2. Although TEM revealed that the nanoparticle diameter was in the range of 10-40 nm, difference between the catalysts with diverse ratios of metal oxides was obscured.

Figure 3 shows the H_2-TPR profiles of the nanosized iron catalysts. H_2-TPR determines the reduction behavior of the catalysts. The first stage is ascribed to the transformations of CuO to Cu, and the second stage is attributed to the transformation of Fe_2O_3 to Fe_3O_4, whereas the third stage represents the transformation of Fe_3O_4 to Fe.

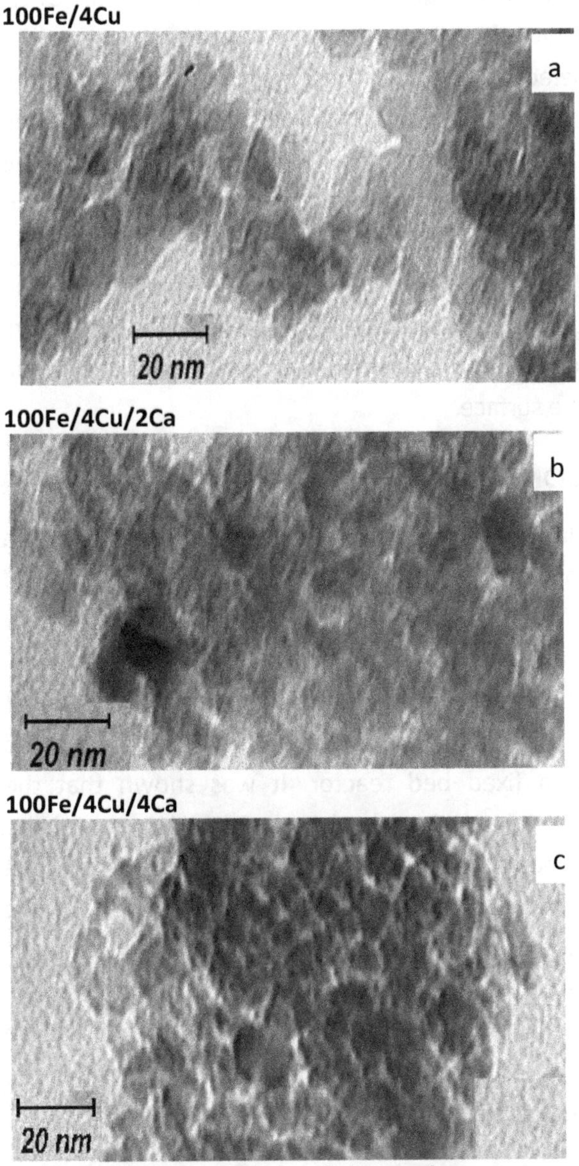

Figure 2: TEM micrographs of the catalysts

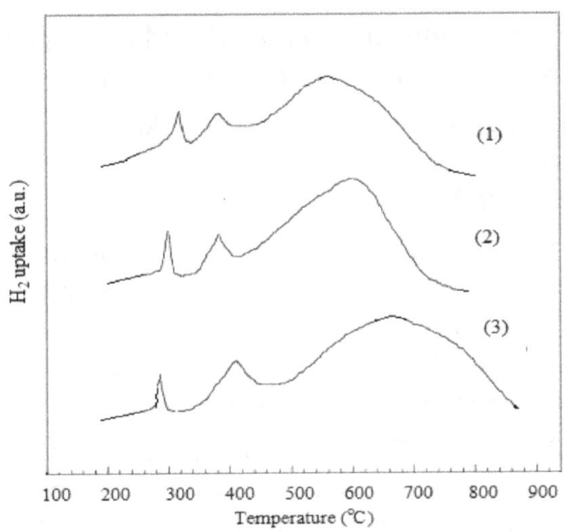

Figure 3: H_2-TPR profiles of the catalysts: (1) 100Fe/4Cu; (2) 100Fe/4Cu/2Ca; (3) 100Fe/4Cu/4Ca.

An increase in the concentration of the promoter led to an increase in the reduction temperature and reaction time. Figure 4 shows the rates of FTS and WGS.

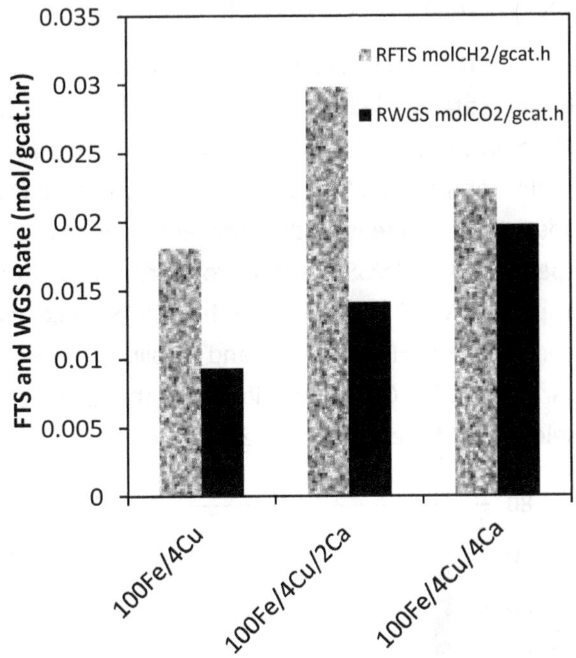

Figure 4: R_{FTS} and R_{WGS} of the catalysts

The addition of promoter increases the catalyst activity, whereas the addition of copper accelerates the deactivation of the catalyst; the addition of calcium as a promoter apparently improves the FTS activity of iron-based catalysts. However, the promoted iron catalysts have higher rates of FTS and WGS than the unpromoted catalyst.

H_2O plays an important role in changing the iron phase composition during the FTS process. A reversible WGS reaction accompanying the FTS reaction over iron-based catalyst is recognized well (Equations 1 and 2) [21]:

$$r_{WGS} = r_{CO_2} \tag{1}$$

and

$$r_{FTS} = r_{CO} - r_{WGS} \tag{2}$$

As the WGS reaction needs the water produced by the FTS, an inequality is always observed:

$$r_{WGS} \leq r_{FTS} \quad (3)$$

During the FTS process, one part of H_2O produced by FTS reaction is consumed by WGS reaction, whereas the remainder is left in the reactor and keeps certain pressure of water vapor.

Fe_3O_4 is the active site for WGS reaction on iron-based catalyst [22], while calcium is an only electronic promoter. The incorporation of calcium promoter into iron-based catalyst can promote CO adsorption, increases the concentration of CO species, shifts WGS reaction forwards and thus improves WGS activity. The high WGS activity decreases the H_2O pressure and stabilizes the iron carbides [23]. CO conversion and the products selectivity of the catalysts are displayed in Figure 5.

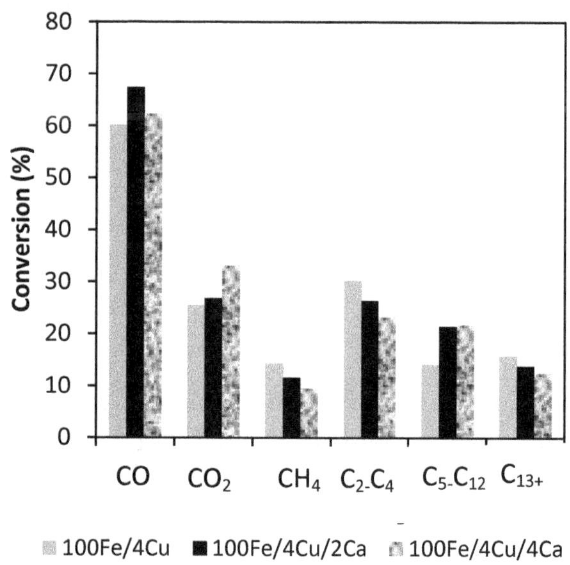

Figure 5: Product selectivity of the catalysts; Reaction condition: Time on Stream 72 hrs, 290 °C, 1.7 MPa, $H_2/CO = 1$, and SV= 3nl.gCat^{-1}.hr^{-1}.

Figure 5 shows the selectivity in gaseous and light hydrocarbons (methane and C_2-C_4) and heavy hydrocarbons (C_{5+}). It should be noted that the selectivity of oxygenates was negligible (<3%) in all the cases. The addition of the Ca promoter with a content of 2% led to an increase in CO conversion, but higher concentration up to 4% decreased the CO conversion. The higher hydrocarbon selectivity increased with increasing calcium oxide content. However, the selectivities of methane and light weight hydrocarbons (C_2-C_4) showed an opposite trend with respect to C_{5+} selectivity. All of these results imply that the hydrogenation reaction was decreased while the promoter was added to the catalyst. Both the amount of the promoter and the reaction conditions influenced the product selectivity. The results showed that the addition of the promoter facilitated the CO dissociative adsorption, leading to a higher deposition of carbon species on the surface.

CONCLUSIONS

Nanostructured iron catalysts were prepared by a microemulsion method. A series of unpromoted and CaO-promoted iron nanocatalysts were studied using different characterization techniques. The effect of calcium oxide on the Fischer-Tropsch synthesis activity and selectivity was investigated in a fixed bed reactor. It was shown that the activity and C_{5+} selectivity increased, but methane selectivity decreased with increasing the amount of calcium oxide, which was attributed to the increased reducibility. The changes in the catalytic performances could be ascribed to the effect of the promoter on H_2 and CO adsorption, which further affected the FTS performances of the catalysts significantly.

ACKNOWLEDGMENT

We are grateful to the Research and Development of National Iranian Oil Company (NIOC) for the financial support of the current work.

REFERENCES

[1] Rao V. U. S., Stiegel G. J., Cinquegrane G. J., and Srivastava R. D., "Ron-based Catalysts for Slurry-phase Fischer-Tropsch Process: Technology Review," *Fuel Process. Techno.*, **1992**, *30*, 83-107.

[2] O'Brien R. J., Xu L., Spicer R. L., Bao S., et al., "Activity and Selectivity of Precipitated Iron Fischer-Tropsch Catalysts," *Catal. Today*, **1997**, *36*, 325-334.

[3] Panpranot J., Kaewkun S., Praserthdam P.,

and Goodwin J. G., "Effect of Cobalt Precursors on the Dispersion of Cobalt on MCM-41," *Catal. Lett.*, **2003**, *91*, 95-102.

[4] Mills G. A., "Catalysts for Fuels from Syngas," *IEA Coal Research*, London, **1988**, 34-44.

[5] Duvenhage D. J. and Coville N. J., "Effect of K, Mn and Cr on the Fischer-Tropsch Activity of Fe:Co/TiO_2 Catalysts," *Catal. Lett.*, **2005**, *104*, 129-133.

[6] Yang Y., Xiang H. W., Xu Y. Y., Bai L., et al., "Effect of Potassium Promoter on Precipitated Iron-manganese Catalyst for Fischer-Tropsch Synthesis," *Appl. Catal. A*, **2004**, *266*, 181-194.

[7] Li T., Yang Y., Zhang C., An X., et al., "Effect of Manganese on an Iron-based Fischer-Tropsch Synthesis Catalyst Prepared from Ferrous Sulfate," *Fuel.*, **2007**, *86*, 921-928.

[8] Li S., Krishnamoorthy S., Li A., Meitzner G. D., et al., "Promoted Iron-based Catalysts for the Fischer-Tropsch Synthesis: Design, Synthesis, Site Densities, and Catalytic Properties," *J. Catal.*, **2002**, *206*, 202-217.

[9] Yu W., Wu B., Xu J., Tao Z., et al., "Effect of Pt Impregnation on a Precipitated Iron-based Fischer-Tropsch Synthesis Catalyst," *Catal. Lett.*, **2008**, *125*, 116-122.

[10] Li S., Li A., Krishnamoorthy S., and Iglesia E., "Effects of Zn, Cu, and K Promoters on the Structure and on the Reduction, Carburization, and Catalytic Behavior of Iron-based Fischer-Tropsch Synthesis Catalysts," *Catal. Lett.*, **2001**, *77*, 197-205.

[11] Fazlollahi F., Sarkari M., Gharebaghi H., Atashi H., et al., "Preparation of Fe-Mn/K/Al_2O_3 Fischer-Tropsch Catalyst and Its Catalytic Kinetics for the Hydrogenation of Carbon Monoxide," *Chin. J. Chem. Eng.*, **2013**, *21*, 507-519.

[12] Bai L., Xiang H. W., Li Y. W., Han Y. Z., et al., "Slurry Phase Fischer-Tropsch Synthesis over Manganese-promoted Iron Ultrafine Particle Catalyst," *Fuel.*, **2002**, *81*, 1577-1581.

[13] Tao Z., Yang Y., Wan H., Li T., et al., "Effect of Manganese on a Potassium-promoted Iron-based Fischer-Tropsch Synthesis Catalyst," *Catalysis Letters.*, **2007**, *114*, 161-168.

[14] Nattaporn L., James G., and Edgar L., "Fe-based Fischer-Tropsch Synthesis Catalysts Containing Carbide-forming Transition Metal Promoters," *Journal of Catalysis*, **2008**, *255*, 104-113.

[15] Feyzi M., Irandoust M., and Mirzaei A. A., "Effects of Promoters and Calcinations Conditions on the Catalytic Performance of Iron-manganese Catalysts for Fischer-Tropsch Synthesis," *Fuel. Pro. Tech.*, **2011**, 92, 1136-1143.

[16] Nakhaei Pour A., Taghipoor S., Shekarriz M., Shahri S. M. K., et al., "Fisher-Tropsch Synthesis with Fe/Cu/La/SiO_2 Nanostructured Catalyst," *J. Nanosci. Nanotech.*, **2008**, *8*, 1-5.

[17] Zamani Y., Bakavoli M., Rahimizadeh M., Mohajeri A., et al., "Synergetic Effect of La and Ba Promoters on Nanostructured Iron Catalyst in Fischer-Tropsch Synthesis," *Chin. J. Catal.*, **2012**, *33*, 1119-1124.

[18] Zamani Y., Yousefian S. H., Pour A. N., Moshtari B., et al., "A Method for the Regeneration of Used Fe-ZSM_5 Catalyst in Fischer-Tropsch Synthesis," *Chem. Eng. Trans.*, **2010**, *21*, 1045-1051.

[19] Nakhaei Pour A., Kamali Shahri S. M., Bozorgzadeh H. R., Zamani Y, et al., "Effect of Mg, La and Ca Promoters on the Structure and Catalytic Behavior of Iron-based Catalysts in Fischer-Tropsch Synthesis," *Appl. Catal. A: Gen.*, **2008**, *348*, 201-208.

[20] Herranz T., Rojas S., Pe´ rez-Alonso F. J., Ojeda M., et al., "Hydrogenation of Carbon Oxides over Promoted Fe-Mn Catalysts Prepared by the Microemulsion Methodology," *Appl. Catal. A*, **2006**, *311*, 66-75.

[21] Bukur D. B., Nowicki L., Manne R. K., and Lang X. S., "Activation Studies with a Precipitated Iron Catalyst for Fischer-Tropsch Synthesis: II Reaction Studies," *J. Catal.*, **1995**, *155*, 366-375.

[22] Li S., O'Brien R. J., Meitzner G. D., Hamdeh

H., et al., "Structural Analysis of Unpromoted Fe-based Fischer-Tropsch Catalysts Using X-ray Absorption Spectroscopy," *Appl. Catal. A: Gen.*, **2001**, *219*, 215-222.

[23] Yang Y., Xiang H. W., Xu Y. Y., Bai L., et al., "Effect of Potassium Promoter on Precipitated Iron-manganese Catalyst for Fischer-Tropsch Synthesis," *Appl. Catal. A: Gen.*, **2004**, *266*, 181-194.

6

A New Cement Additive to Improve the Physical Properties of Oil Well Cement and to Enhance Zonal Isolation

Hamid Soltanian[1]*, Abdolrahim Khojastefar[2], and Ghorban Ali Sobhi[1]

[1] Exploration Department, Research Institute of Petroleum Industry (RIPI), West Blvd. of Azadi Sports Complex, Tehran, Iran.

[2] Department of Chemical and Petroleum Engineering, Sharif University of Technology, Tehran, Iran.

ABSTRACT

The main purpose of cementing of the well is to provide effective zonal isolation for the entire life of wells. The standard cements cannot prevent loss of zonal isolation when the stress level in the well cement is severely fluctuated. Since elastic cements improve elastic properties, recently, a number of studies have been done in this area. For this purpose, a nanoparticle, named EX-RIPI, was designed as an additive to improve cement elasticity. A series of experiments were carried out to investigate the effects of this additive on the elastic properties of the cement. The experimental results showed that the addition of EX-RIPI to the cement slurries decreased Young's modulus from 2.86 GPa to 1.94 GPa but increased Poisson's ratio from 0.04 to 0.25. Consequently, EX-RIPI can be used as a suitable additive for improving the mechanical properties of cement used for a long time zonal isolation. To illustrate the performance of this additive, the properties of the cement stone was compared with one of available additives in the market (i.e., Flexstone).

Keywords: Elastic Cement, Young's Modulus, Poisson's Ratio, Compressive Strength, Cement Additive

INTRODUCTION

Casing is cemented in wellbore to provide casing support, to protect fresh water invasion from contamination, to prevent casing collapse which may be caused by formation creep to provide zonal isolation of producing zones, and to isolate casing from corrosive brines [1].

The main purpose of primary cementing is to provide an effective zonal isolation during the entire life of the well. This zonal isolation must be continued during the life of the well even after well abandonment. No fluid invasion must be occurred in the cement matrix, casing/ cement interface, and cement/formation interface [2]. The loss of zonal isolation in the absence of chemical attack can be due to either mechanical failure of the cement or debonding of casing from cement or debonding of cement from the formation [3]. In general, when wellbore temperature pressure changes in 2D cylindrical coordinate system, two principal stresses, which are called radial and tangential stresses, are imposed on the cement. Cement

*Corresponding author
Hamid Soltanian
Email: Soltanianh@ripi.ir

failure type is determined by evaluating these stresses and divided into either tensile or shear failure. Tensile failure causes cracks while shear failure leads to cement plastic deformation [4]. The loss of zonal isolation can result in the following problems:

- Loss of gas reservoir;
- Unsafe operation;
- Production of premature gas or water;
- Well shutdown [5].

Neat and standard cements usually have a high compressive strength and show brittle behavior. This brittle behavior is determined by low tensile strength and cement toughness. On the other hand, the high Young's modulus and low Poisson's ratio of these cements reduce their deformation capacity. Such cements can be used where there is a small change in the cement stress level [6]. The formation of micro-annulus can be prevented by using expanding agents in the cement. New cement systems containing expanding agents and flexible additives lead to the reduction of cement Young's modulus in comparison with conventional cements [7]. As a result, these systems increase the cement elasticity. Poisson's ratio is another parameter which plays a key role in cement sheath integrity. A higher cement Poisson's ratio results in a more qualified sheath. Elastic cements have a higher Poisson's ratio usually greater than 0.2 [8]. In fact, a synergy exists between elastic cements and expanding agents. The expanding cements will move toward casing when these cements are more flexible than the formation [9]. Le-Roy Delage et al. investigated the effects of a flexible additive and its concentration on cement mechanical properties [3]. The results of their work can be summarized as follows; flexible cements have a lower Young's modulus and higher Poisson's ratio values in comparison with conventional cements. Moreover, using higher flexible agent concentrations improves the elasticity of cement. Walter Morris et al. used a polymeric fiber in cement to improve its mechanical properties [6]. Their study indicated that fiber reinforced cements had a high tensile strength, good elastic behavior, and improved toughness. Boukhelifa et al. developed a large scale laboratory test for sealants in annular geometry [7]. They used different cement systems in their work. This study showed that cement mechanical properties (elasticity, strength, and expansion) and rock properties were key parameters in maintaining the cement sheath integrity during the life of the well. They found that the expanding and flexible systems showed the best properties in the loading stage. Barry Wray et al. compared the integrity of two different cement sheaths during high pressure high temperature (HPHT) well operations [10]. These systems were standard cement, and flexible cement. In order to prepare elastic cement they used elastomer and fiber in the cement for improving elasticity and increasing tensile strength of the cement [11]. The results showed that the standard cement debonded from casing during well completion and stimulation. However, when elastic cement was used in the well, casing debonding from cement was not observed. Consequently this cement created a better zonal isolation in comparison with the standard cement [12,13].

The elastic properties of the cement stone can be improved by adding nano-sized particles in low volumetric fractions to the cement formulation. This study includes the laboratory examination of the effect of a nano-particle, named EX-RIPI, on the physical properties of heavy-weight oil well cement.

EXPERIMENTAL SET-UP

In order to design proper cement for using in oil/gas wells, a series of experiments were done. These tests were classified in two categories, namely preliminary and compressive strength tests. Preliminary tests were done to measure the routine cement slurry properties such as slurry weight, free water, rheology, fluid loss, and thickening time according to API RP 10B-2 [14]. Slurry weight was measured by mud balance. HPHT filter press apparatus was used to determine fluid loss. The rheological properties of slurry were measured by viscometer. Finally, pressure-temperature consistometer was used to determine the slurry thickening time [15]. An ultrasonic cement analyzer (UCA) was used to measure the compressive strength at well condition. The transit time of ultra-sonic wave through the cement sample is the basis of compressive strength measurement with UCA. American petroleum institute (API) compressive strength tester model 4207 was used to measure the uniaxial compressive strength of the set cement [16]. In this test, axial and lateral strains can be determined at a specific stress by four dis-placement sensors. This apparatus provides stress-strain curve of the set cement. Cement Young's modulus and Poisson's ratio were determined from this curve. This apparatus yields the stress-strain curve through the following steps:

1. The cylindrical set cement sample is placed on the lower part of hydraulic press;
2. The loading rate is set by a control unit at 0.5 $MPa.sec^{-1}$;
3. The sample is engaged with the upper part of hydraulic press;
4. Three displacement sensors are set around the sample to measure the average diamet-ric strain;
5. A displacement sensor is set in axial direction to measure the axial strain;
6. Loading continues until the sample fails. Data during loading are analyzed by data processor unit and are then sent to a PC for plotting.

EX-RIPI Additive (Elastic Additive)

EX-RIPI is a nano-particle specifically designed to improve the tensile strength and elastic properties of cement. Also, it allows cement sheath to withstand stresses imposed by varying conditions throughout the life of a well. EX-RIPI slurries provide enhanced mechanical properties This additive spreads point or non-uniform loads through the cement sheath and causes the cement sheath to experience uniform loads. The spreading effect of EX-RIPI prevents casing collapse in moving salt formation. The physical properties of this nanoparticle are shown in Table 1.

Table 1: Physical properties of EX-RIPI

Specific gravity	1.37
Additive form	Powder
Color	Dark brown
Applicable temperature	16 to 160 °C
Proper concentration	3 to 7% by weight of cement
Solubility in water	0.3 g/100ml H_2O, at 20 °C

EX-RIPI also can affect slurry properties. These effects can be itemized as slurry gelation, rapid initial set, expansion increase, viscosity increase due to high surface area of the particles constituting EX-RIPI, and density decrease due to its low specific gravity.

RESULTS AND DISCUSSION

In this study, a cement formulation for salt formation cementing was used and was modified. During the test Finally, EX-RIPI was added to the modified cement. The properties of the cement slurry and stone were compared with commercially available elastic cement. The formulations for these cement slurries are shown in Table 2. Table 3 shows the cement slurry properties measured during the preliminary tests. Compressive strength and its progression were determined with UCA.

Figures 1, 2, 3, and 4 show that the compressive strength is gradually increased in the cement samples. The ultimate compressive strengths of these cements are shown in Table 4.

Table 2: Cement formulation

	Cement class E (g)	Water (cc)	Retarder D13 (g)	Highdense (g)	Salt (g)	Dispersant (g)	EX-RIPI (g)	Flexstone (g)
Base cement	1000	545	3.1	964	203	-	-	-
Modified cement	1000	380	3.2	577	142	-	-	-
Elastic cement	1000	380	3.1	550	142	1	90	-
Flexstone cement	1000	385	3.1	552	141	-	-	90

Table 3: Cement slurry properties

Cement type	Plastic viscosity (Pa.s)	Yield point (kPa)	Density ($kg.m^{-3}$)	Free water (cc)	Thickening time (min)
Base cement	0.0615	4.55	2387.8	15	151
Modified cement	0.249	11.49	2387.8	1.5	136
Elastic cement	0.254	10.77	2339.7	0.5	125
Flexstone cement	0.239	12.02	2338.9	10	145

Table 4: Measured cement compressive strengths with UCA

Cement type	Compressive strength (MPa)
Base cement	14.87
Modified cement	25.50
Elastic cement	16.47
Flexstone cement	21.05

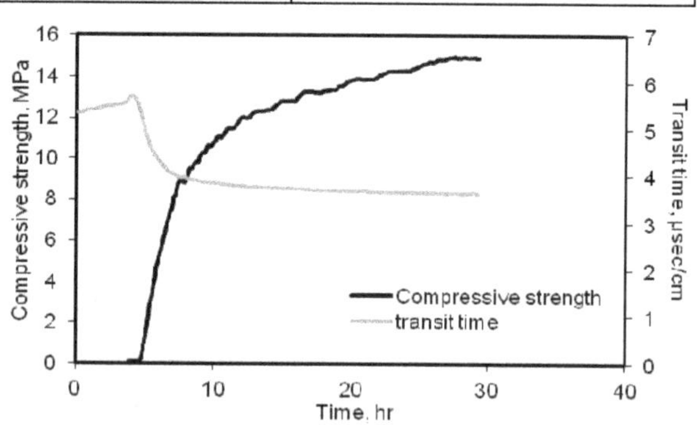

Figure 1: Compressive strength progression of the base cement

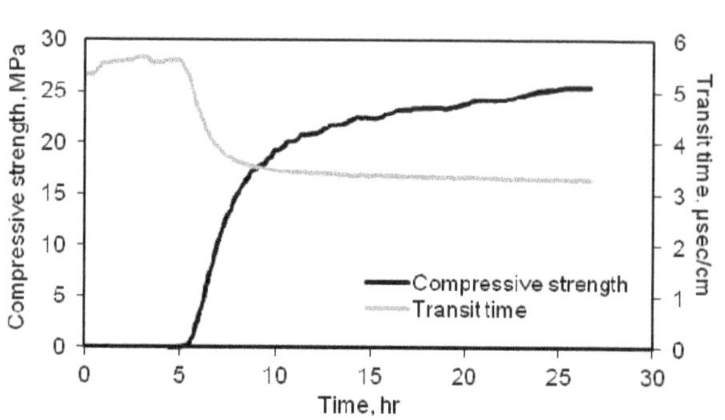

Figure 2: Compressive strength progression of the modified cement

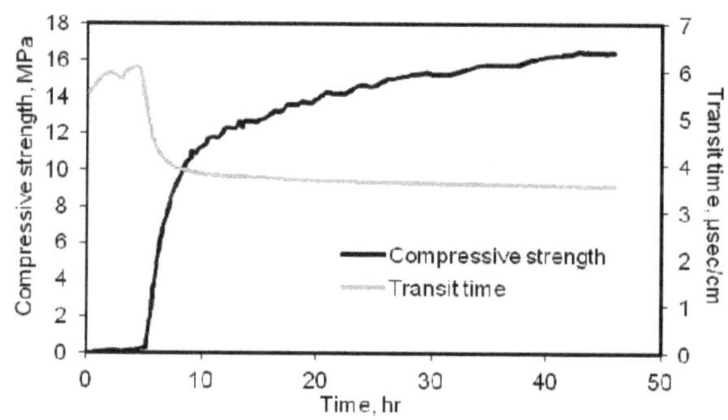

Figure 3: Compressive strength progression of the elastic cement

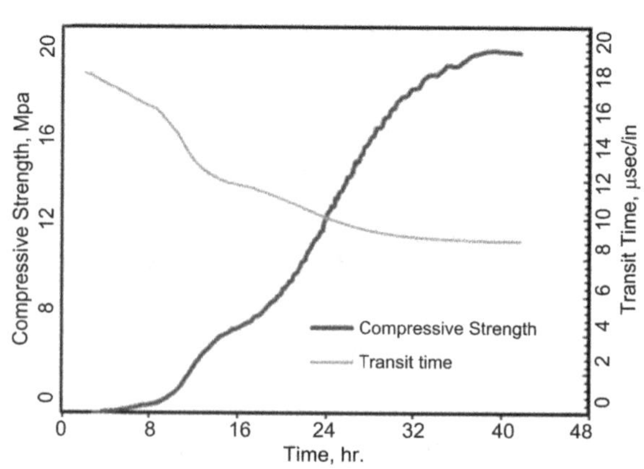

Figure 4: Compressive strength progression of the Flexstone cement

For measuring uniaxial compressive, the cement samples were cured at 210 °F and 10680 psi for 3 days; sample curing was done with a curing chamber. Uniaxial compressive strength, elastic (Young's) modulus, and Poisson's ratio were measured with API compressive strength tester, model 4207; the corresponding results are shown in Tables 5, 6, 7, and 8. Stress-strain curves are also indicated in Figures 5, 6, 7, and 8. The base cement formulation was modified due to its high free water and low yield point. The low yield point leads to the settlement of high density additives. In order to modify the cement formulation, the amounts of water, salt, and high density additive were decreased. When the water content of the slurry is decreased, the water cement ratio (WCR) is also reduced. WCR reduction causes the compressive strength, and therefore the Young's modulus, of the modified cement to become greater than that of the base cements. The stress-diametric strain curves of these two cements are approximately similar and lie near the stress axis (Figures 5 and 6). Since Poisson's ratio is the ratio of Young's modulus to the slope of the linear part of stress-diametric strain curve, it can be obvious that the Poisson's ratios of these two cements are very low and nearly equal to each other.

Table 5: Results of uniaxial compressive strength test for the base cement

Cement diameter (mm)	66.8
Cement height (mm)	166.4
Peak stress (MPa)	12.75
Failure type	Axial
Loading rate (MPa.sec^{-1})	0.5
Young's modulus (GPa)	2.43
Poisson's ratio	0.03

Table 6: Results of uniaxial compressive strength test for the modified cement

Cement diameter (mm)	65.8
Cement height (mm)	168.4
Peak stress (MPa)	23.58
Failure type	Axial
Loading rate (MPa.sec^{-1})	0.5
Young's modulus (GPa)	2.86
Poisson's ratio	0.04

The modified cement and cement containing EX-RIPI have the same WCR; therefore, it is beneficial to compare their mechanical properties. The cement containing EX-RIPI has a lower compressive strength and Young's modulus in comparison with the modified cement. Moreover, the stress-strain curve of this cement shows that the Poisson's ratio of the cement containing EX-RIPI is much higher than the modified cement (Figure 7). Such mechanical property changes indicate that EX-RIPI improves the cement elasticity. In addition, the EX-RIPI performs much better to improve the elastic properties of the cement stone than the commercially available additives.

Table 7: Results of uniaxial compressive strength test for the elastic cement

Cement diameter (mm)	65.5
Cement height (mm)	168
Peak stress (MPa)	13.07
Failure type	Axial
Loading rate (MPa.sec^{-1})	0.5
Young's modulus (GPa)	1.94
Poisson's ratio	0.25

Table 8: Results of uniaxial compressive strength test for the Flexstone cement

Cement diameter (mm)	65.8
Cement height (mm)	167.4
Peak stress (MPa)	16.00
Failure type	Axial
Loading rate (MPa.sec^{-1})	0.5
Young's modulus (GPa)	2.33
Poisson's ratio	0.18

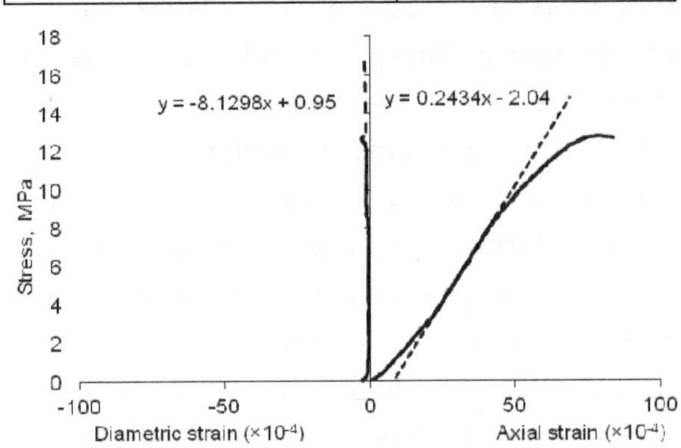

Figure 5: Stress-strain curve of the base cement

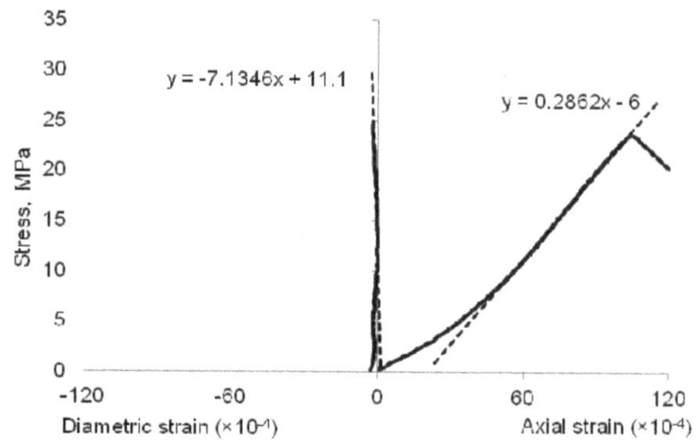

Figure 6: Stress-strain curve of the modified cement

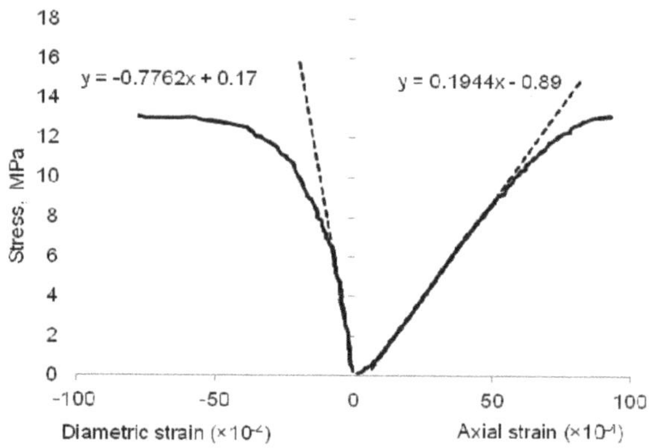

Figure 7: Stress-strain curve of the elastic cement

Using EX-RIPI nano-particles as an additive has significant effects on the compressive strength and elasticity properties of the cement stone because the compressive strength is primarily controlled by the total porosity. The nano-particles will fill the pores, which leads to increasing strength and improving the microstructure of the cement if they are uniformly dispersed. EX-RIPI has elastic properties and adding elastic additives to the cement composition can improve the elastic properties of the cement stone, which leads to an increase in Young's modulus and Poisson's ratio. The cement stone matrix consists of two elastic and rigid phases; under the exposed tensions, the rigid phase resists and the elastic phase deforms. By increasing the elastic phase the whole system tends to increase the overall elasticity. Thus, the elastic phase plays the principal role. The improvement of elasticity is also due to the formation of more hydrated products in the presence of nanoparticles.

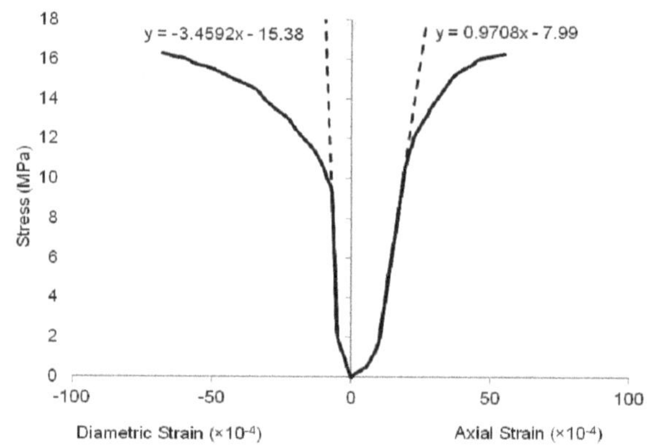

Figure 8: Stress-strain curve of the Flexstone cement

CONCLUSIONS

On the whole, the following conclusions can be deduced from the present study:

1. EX-RIPI reduces the compressive strength of cement in an appropriate range for casing support. Moreover, the addition of EX-RIPI to the cement slurries reduces the rate of compressive strength progression.
2. The Young's modulus of EX-RIPI cement is measured to be lower than the standard cements. This is due to the lower compressive strength.
3. EX-RIPI increases the axial and lateral strain of the cement when it is exposed to axial stress. But, the lateral strain increases more rapidly with respect to axial strain. Consequently, EX-RIPI increases the cement Poisson's ratio significantly.
4. The EX-RIPI-added cement is an elastic cement due to a low Young's modulus and a Poisson's ratio greater than 0.2.
5. EX-RIPI performs much better than the commercially available additives in the market.

ACKNOWLEDGMENT

The authors would like to thank Research Institute of Petroleum Industry (RIPI) for continuous help during conducting experiments and IOOC for the financial support of this work.

NOMENCLATURE

API	: American petroleum institute
BHST	: Bottom hole static temperature
BWOC	: By weight of cement
E	: Young's modulus (MPa)
FW	: Free water (cc)
HPHT	: High pressure high temperature
PV	: Plastic viscosity ($kg.m^{-1}.s^{-1}$)
TT	: Thickening time (s)
UCA	: Ultra-sonic cement analyzer
UCS	: Uniaxial compressive strength (MPa)
WCR	: Water cement ratio
YP	: Yield point (MPa)
v	: Poisson's ratio
ρ	: Density ($kg.m^{-3}$)

REFERENCES

[1] Mack J., Dillenbeck L. R., BJ Services Company, "Cement: How Tough is Tough Enough? A Laboratory and Field Study." SPE 78712-MS, SPE Eastern Regional Meeting, **2002,** Lexington, Kentucky, U.S.A., 23-26 October.

[2] Lullo G., Phil Rae, BJ Services Company. "Cements for Long Term Isolation Design Optimization by Computer Modeling and Prediction." SPE 62745-MS, IADC/SPE Asia Pacific Drilling Technology, Kuala Lumpur, Malaysia . **2000**, 11-13 September.

[3] Le Roy-Delage S., Baumgarte C., Thiercelin M., Vidick B., Schlumberger "New Cement Systems for Durable Zonal Isolation." *SPE 59132-MS, IADC/SPE Drilling Conference*, **2000,** New Orleans, Louisiana, U.S.A., 23-25 February.

[4] Mueller D. T., BJ Services Company, "Producing Stress-Resistant High-Temperature/High-Pressure Cement Formulations through Microstructural Optimization". SPE 84562-MS, SPE Annual Technical Conference and Exhibition, **2003**, Denver, Colorado, U.S.A., 5-8 October.

[5] Ravi K., Halliburton; Bosma M., Shell International E & P; Gastebled O., TNO Building and Construction Research, "Safe and Economic Gas Wells through Cement Design for Life of the Well." SPE 75700-MS, SPE Gas Technology Symposium, **2002** Calgary, Alberta, Canada, 30 April-2 May.

[6] Morris W., Criado M. A., Robles J., Bianchi G., San Antonio-Pride Int. "Design of High Toughness Cement for Effective Long Lasting Well Isolations." SPE 81001-MS, SPE Latin American and Caribbean Petroleum Engineering Conference, **2003,** Port-of-Spain, Trinidad and Tobago **.,** 27-30 April.

[7] Feder, J., "Casing and Cementing" 3rd Edition, *Petroleum Extension Service, University of Texas*, **2001**, p. 99.

[8] Boukhelifa L., Schlumberger; Moroni N., ENI SpA; James S. G., Le Roy-Delage S., Thiercelin M. J, Schlumberger, Lemaire G., INSA. Toulouse. "Evaluation of Cement Systems for Oil and Gas Well Zonal Isolation in a Full-Scale Annular Geometry." *SPE 87195-MS, IADC/SPE Drilling Conference*, **2004**, Dallas, Texas, U.S.A., 2-4 March.

[9] El-Hassan H., Sultan M., and Saeed A. A., ADCO, and Johnson C., SPE, Belmahi A., and Rishmani L., Schlumberger . "Using a Flexible, Expandable Sealant System to Prevent Microannulus Formation in a Gas Well: A Case History." *SPE 92361, SPE Middle East Oil & Gas conference*, **2005**, Bahrain., 12-15 March.

[10] Wray B., Cimarex Energy, Bedford D., Leotaud L., and Hunter B., Halliburton . "The Application of High-Density Elastic Cements to Solve HP/HT Challenges in South Texas: The Success Story." SPE 122762-MS, SPE Annual Technical Conference and Exhibition, **2009,** New Orleans, Louisiana, U.S.A, 4-7 October.

[11] The University of Texas at Austin-Petroleum Extension services, "Well Cementing (Oil and Gas production)", *The University of Texas at Austin-Petroleum Extension Service*, **1983**, p. 345.

[12] Montgomery C P., "Oil Well Cementing Practices Materials", *Halliburton Oil Well Cementing Co*, **1959**, p. 149.

[13] Marlow S R., "Effect of Temperature/ Pressure Cycling on Cement Bonding Characteristics in Gas Injection/Withdrawal Wells", *Southwest Research Institute*, **1987**, p. 202.

7

Lab and Field Scale Modeling of Near Miscible CO_2 Injection in Different Porous Mediums

Mohammad Parvazdavani*, Seyed Mahdia Motahhari, and Vahid Ansari

Reservoir Studies Research Division, Research Institute of Petroleum Industry (RIPI), Tehran, Iran

ABSTRACT

The main purpose of this investigation is to study the effect of near miscible CO_2 injection in different porous mediums on both lab and field scales. This effect can be traced by the change of two-phase gas-oil relative permeability curves. In this work, the experiments have been performed on three rock types (i.e. sandstone, dolomite, and artificial fractured sandstone) based on an incremental pressure algorithm approaching a near miscible condition. Lab-scale inverse modeling has been used to calculate relative permeability curves. Based on the experimental results, 85%of minimum miscibility pressure was defined as the near miscible pressure. Comprising the relative permeability curves in immiscible and near-miscible conditions, the results show that this change has become less significant from sandstone core type to artificial fractured. In other words, near miscible CO_2 injection would be recommended in rock types with a lower RQI. In addition, it was concluded that in the case of artificial fractured, simple conventional relative permeability methods obtain the same results as sophisticated inverse modeling method. Furthermore, in order to validate the lab scale results, the field scale modeling of the candidate reservoir was done using the 3D compositional reservoir simulator. 83% of minimum miscibility pressure was defined as near miscible pressure. Moreover, the simulation results confirmed lab-scale data regarding the recovery factor in different rock types. Additionally, the economic evaluation (NPV analysis) showed that use of near miscible CO_2 injection in lower RQI reservoirs was more economical rather than the other scenarios.

Keywords: Near-miscible CO_2 Injection, Lab and Field Scale, Inverse Modeling, Rock Types, Relative Permeability Curves, NPV Analysis

INTRODUCTION

Recently, carbon dioxide injection has been employed as a secondary or tertiary oil recovery method. Experimental studies showed that oil recovery factor after CO_2 injection improved up to 60% of original oil in place as the tertiary method, while this value was only 44% for water flooding as the secondary method [1].

***Corresponding author**

Mohammad Parvazdavani

Email: mohammadparvazdavani@yahoo.com

Dong et al. indicated that miscible CO_2 flooding was unfavorable for some reservoirs in southeast Saskatchewan due to high CO_2 minimum miscibility pressure or some available limitations on providing the required operational pressure. Their results showed that oil recovery was improved during near-miscible CO_2 injection and it could be inferred that the mechanism of oil mobilization in the CO_2 miscible and near-miscible injection was the same [2].

This increase in oil recovery is because of the changes in oil and gas relative permeability. Under miscible conditions, when the interfacial tensions between oil and gas phases decreases, the values of the relative permeability change; it is a function of the phase behavior of working fluids, their physical properties, and porous media characterization [3,4]. Al-Wahaibi et al. studied the two-phase drainage and imbibition relative permeability curves under different equilibrium conditions. They showed that when IFT decreased, the relative permeability curve of non-wetting phase changed more considerably than the wetting phase curve and hysteresis became no significant [3-5].

There are three major approaches toward the determination of two-phase relative permeability from unsteady state displacement experiments: analytical, semi analytical and history matching methods [6].

The possibility of overcoming the defects regarding the analytical and semi analytical methods encourages us to use the history matching method by using the inverse modeling technique to calculate the relative permeability values.

Very few experimental studies have been done to investigate the effect of near miscible CO_2 injection on oil recovery in various ranges of rock types. Trivedi and Babadagli provided an excellent review on practical recipes for CO_2 injection in fractured reservoirs. To make the experts capable of selecting the suitable rock type for applying the economic gas injection approach, an experimental study has been done with various rock types ranged from sandstone to dolomite to an artificial fractured rock from Iranian oil fields, which contains different levels of very low permeable to high permeable porous medias [7].

In this study, in the second section, lab scale experiments have been done to find the near miscible pressure and extract the production data. Then, in the third section, a lab scale inverse modeling has been performed to determine the relative permeability curves. By the analysis of the relative permeability curves, we aimed to determine the effect of near miscibility injection as an EOR method in different rock types. In the fourth section, a sector field scale modeling has been conducted to testify the accuracy of the lab test results. In the fifth section, an NPV analysis has been carried out to investigate the economical possibility of near miscible injection in different rock types. Finally, in the sixth section, the results have been discussed on both lab and field scales.

Lab-scale Experiments

Materials

A light oil sample is prepared from one of the west Iranian oil reservoirs and carbon dioxide as the injection gas is almost pure. The physical properties of oil and the lumped heavy component under the standard condition are shown in Table 1. The oil composition shown in Table 2 was determined by gas chromatography (GC) analysis [8].

Table 1: Fluid physical properties

Parameter (at 14.7 psi and 60 °F)	Value
API	41
Viscosity (μ_o) (cP)	3.25
Molecular Weigh (g/mol)	93.18
Density (ρ_l) (g/cc)	0.82
C_{12+} molecular weight (g/mole)	250
C_{12+} density (g/cc)	0.86
CO_2 Viscosity (μ_g) (cP)	0.014473
CO_2 density (ρ_v) (g/cc)	0.0018687

Table 2: Distribution of lab used oil components

Component	Molar percentage(%)
H_2S	0.88
CO_2	0.9
C_1	34.45
C_2	8.37
C_3	5.45
IC_4	1.66
NC_4	3
IC_5	1.92
NC_5	1.93
FC_6	3.05
FC_7	3.23
FC_8	4.16
FC_9	4.62
FC_{10}	2.38
FC_{11}	2.33
C_{12+}	21.67

MMP Determination

Slim-tube experiment was simulated using a 3D compositional simulator with the tuned Peng Robinsun (PR) EOS. The simulator was run and the ultimate oil recovery was calculated after 1.2 hydrocarbon pore volume (HCPV) in various incremental pressure steps. The plot of oil recovery versus pressure is shown in Figure 1. Minimum contact miscibility (MCM) pressure has been defined 950 psia as the refraction point (Figure 1). To investigate the pressure, the gas injection process was simulated by commercial software and total recovery factor at different stages of injection pressures were calculated. The pressure, at which the slope of the increase in the recovery factor was the highest and leveled off to zero afterwards, was called the near miscible pressure (85% of MMP (i.e. 800 psia)) (Figure 2). Thus 500 psia has been defined as the immiscible pressure and 800 psia is considered as the near miscible pressure [9, 10].

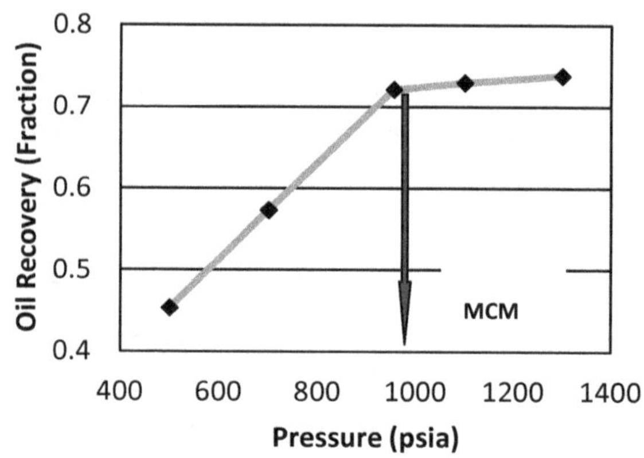

Figure 1: Oil recovery versus pressure in slim-tube simulation

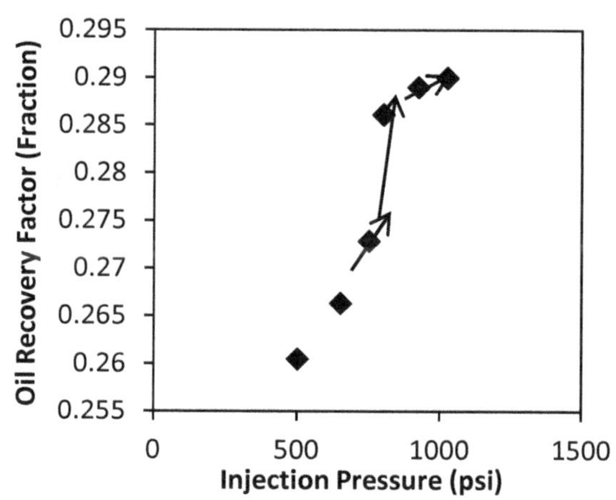

Figure 2: Oil recovery versus pressure in simulated gas injection through sandstone on lab scale

Core Properties

The displacement tests were performed on the low permeable (low reservoir quality index (RQI)) sand stone core sample and dolomite one (moderate RQI) from the south part of Iranian oil fields and their detailed information is shown in Table 3. To prepare the rock type with a high RQI value and investigate the effect of near miscibility on artificial fractured core types, fracturing process has been done on the sandstone core type. By this process, the obtained core had an absolute permeability of 459.3 mD, but its porosity was in range of the sandstone core sample.

Table 3: Sandstone/dolomite core sample Properties

Parameters		Dolomite	Sandstone
Length (L)	cm	5.37	6.59
Area(A)	cm^2	11.35	10. 95
Pore volume (PV)	cm^3	8.45	9.47
Porosity(φ)	Fraction	0.1386	0.1313
Permeability (K)	mD	16.45	1.31

Apparatus and Experimental Procedure

In this experimental work, the unsteady state displacement technique was used; the schematic diagram of relative permeability measurement apparatus is shown in Figure 3. By measuring the volume of oil and gas, and the corresponding pressure drop, the production data will be recorded dynamically. The relative permeability values have been calculated by analyzing the production data.

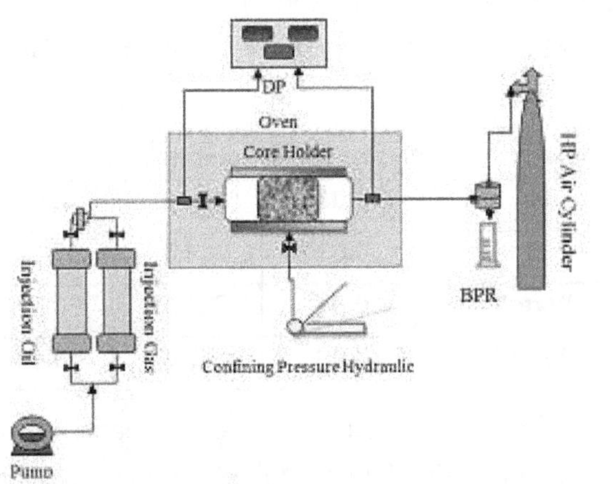

Figure 3: Plan of the experimental setup

First, the core is solvent cleaned, dried with hot nitrogen, and evacuated. After core preparation, the core was saturated by the saturator apparatus with the brine solution (10000 ppm). After the operational preparation (Table 4), absolute permeability was calculated by monophonic core flood test. Then, oil was injected into the core at the rate of 0.05 (cm^3/min) to reach irreducible water. Next, gas was injected at the rate of 0.05 (cm^3/min) up to 2 PV. The amount of the produced oil and gas was measured by camera tracker dynamically. The tests were done according to pressure steps from immiscible to near miscible condition and implemented by BPR.

Table 4: Core flooding set values

Parameter		Value
(pressure of back pressure regulator) BPR_P	psia	Two points 500 and 800
Overburden pressure	psia	BPR_P+650
Time interval for data recording	sec	10

Lab-scale Inverse Modeling

Inverse modeling (i.e. history matching technique) has been used to extract the relative permeability curves implicitly. In this approach, JBN method (1959) was firstly used to calculate the relative permeability as an initial guess, where it was used in compositional fluid flow simulation. Then, the production data from the simulator were compared with the experimental results. Finally, the difference between the experimental production results and simulation data was minimized by using a suitable optimization algorithm (classical conjugate gradient approach). The results of relative permeability obtained from the last step of the optimization process were defined as the implicit values [11].

JBN method extracts relative permeability curves individually based on Equation 1. The terms of W_i and I_r are determined through Equations 2 and 3 [11].

$$\frac{d(\frac{1}{W_i\, I_r})}{d(\frac{1}{W_i})} = \frac{f_o}{Kr_o} \quad (1)$$

$$W_i = \frac{1}{f'} = \frac{1}{\frac{df}{ds}} \quad (2)$$

$$I_r = \frac{\frac{u}{\Delta P}}{\frac{u_s}{\Delta P}} \quad (3)$$

$$Kr_g = \frac{1 - f_o}{f_o} \frac{\mu_g}{\mu_o} Kr_o \quad (4)$$

Relative permeability curves are drawn based on outlet gas saturation (Equation 5) and the value of $S_{g\,(ave)}$ in Equation 5 can be obtained by mass balance.

$$S_{g\,(2)} = S_{g\,(ave)} - W_i\,(f_o)_2 \quad (5)$$

In the second part (i.e. one dimensional, two-phase compositional simulation), where the values of initial relative permeabilities were embedded, the static model of simulator is firstly built in one dimensional (i.e. 11 grids in z-direction and 1 grid in the other two directions) based on the petro-physical data of core sample (Figure 4). Then, the dynamic behavior of fluid flow (i.e. gas injection process) is simulated fully implicit. The production results extracted from the simulation are history matched with the experimental results by using an optimizing process (conjugate gradient method) [12]. In this optimization process, the independent variable (*x*) is equivalent to K_r as the controlling parameter, and the objective function (*F*) is equivalent to the absolute difference of the production result (i.e. q_0 as the cumulative oil production) obtained by both experiments and simulation. By minimizing the objective function, the ultimate values of relative permeability were considered as the implicit values.

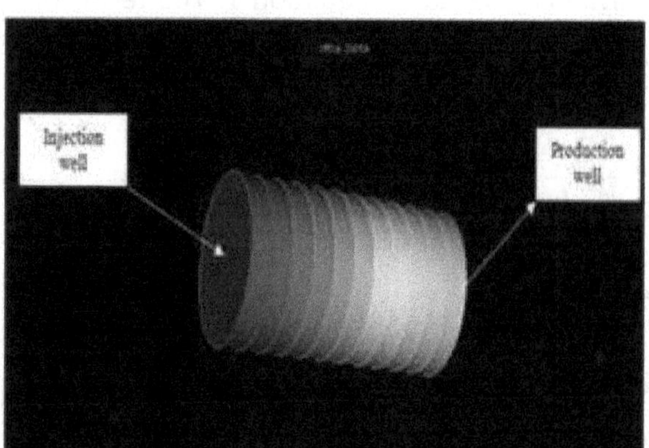

Figure 4: A schematic of core sample developed by the static model of compositional simulator

Field-scale Modeling

To investigate the near miscibility effect and to testify the obtained lab-scale results on field scale, a sector model has been constructed from the candidate reservoir. The sampled sector has eight injectors and one producer (Figure 5).

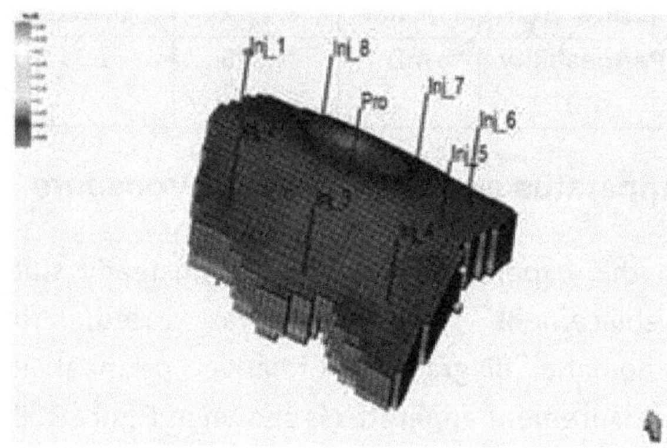

Figure 5: Constructed sector model

The total rock and fluid properties of the sector model have been shown in Table 5, 6, and 7. To use the miscible CO_2 injection on field scale, the screening criteria of this method has been analyzed based on the reservoir characteristics (Table 8).

Table 5: Properties of constructed sector model

Phi	0.01	0.26	0.19
PermX	0.01	214.14	56.67
PermY	0.01	214.14	56.67
PermZ	0.00	189.13	40.42

By running the slim-tube experiment using compositional simulation, multiple contact miscible pressure, 3100 psi, has been defined for the case of CO_2 injection into candidate reservoir fluid at reservoir temperature (T=260 °F). Different development strategies have been defined for the generation of successive increasing pressure algorithms as well as the laboratory set pressures. By running the strategies in case of candidate reservoir defined as lower RQI, (see the total information of strategies in Table 9), the pressure at which the oil saturation around the wellbore region of injection wells started to become less than the residual oil saturation would be called near miscible pressure. At this pressure, the IFT values between the injected gas and reservoir oil started to decrease and vaporizing mechanisms have been activated. To trace this pressure, like the laboratory phase, the recovery factor for each injection step has been plotted versus its injection pressure.

Table 6: Distribution of field used oil components

Parameter	Min	Max	Mean
Component	Residual Oil (mole %)	Associated Gas (mole %)	Reservoir Oil (mole %)
H_2S	0.00	0.00	0.00
N_2	0.00	0.83	0.36
CO_2	0.00	1.17	0.51
C_1	0.00	57.69	25.24
C_2	0.08	17.98	7.91
C_3	1.10	11.11	5.48
IC_4	0.59	1.70	1.07
NC_4	2.38	4.68	3.39
IC_5	1.52	1.31	1.42
NC_5	1.92	1.48	1.73
C_6	7.81	1.33	4.98
C_7^+	4.20	0.54	2.60
Molecular Weight of Residual Oil			243.0000
Molecular Weight of C_{12}^+ Fraction			325.0000
Molecular Weight of Reservoir Oil			149.0000
Sp. Gr. of C_{12}^+ Fraction (60/60 °F)			0.9567

Table 7: Fluid properties of field used oil

Parameter		Value
Bubble Point Pressure	Psia	1730
Solution Gas Oil Ratio	SCF/STB	430
Producing Gas Oil Ratio	SCF/STB	325.6
Oil FVF	bbl/STB	1.2398
Initial Reservoir Pressure	psia	1900
Oil Viscosity	cP	0.9520

The inflection point, 2600 psia, has been reported as the near miscible pressure (Figure 6). Then, the strategies have been repeated for the case of higher RQI (with higher permeability anisotropy and value) to trace the effect of near miscibility on the artificial fractured rock type on field scale.

Table 8: Screening criteria for CO_2 injection

Parameter		Value
Gravity	API	22-45 (Ave: 37)
Viscosity	cP	0-35 (Ave: 2.1)
Porosity	%	3-35 (Ave: 14.8)
Oil saturation	% (PV)	15-89 (Ave: 46)
Formation Type	-	Sandstone/Carbonate
Permeability	mD	1.5-4500 (Ave: 201.1)
Net thickness	ft	Wide range
Depth	ft	1500-13365 (Ave: 6171.2)
Temperature	°F	82-260 (Ave: 136.3)

Table 9: Total information on used strategies on field scale modeling

Parameter		Value
Minimum Bottom Hole Pressure for Production well	psia	1800
Maximum Bottom Hole Pressure for Injection well	psia	2100, 2600, 3100
Minimum GOR for Production well	SCF/STB	1.5
Production Time Interval	Years	40

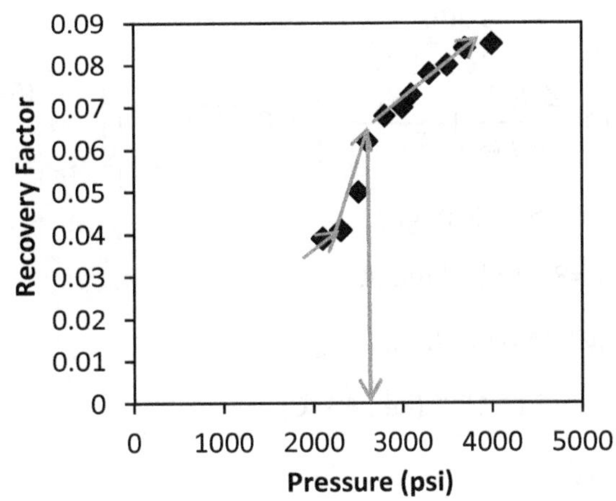

Figure 6: Oil recovery versus pressure in simulated gas injection into the reservoir on field scale

Economical Evaluation

To investigate the economical possibility of near miscible injection compared to the other ones, net present values (NPV) analysis has been done for three scenarios: miscible injection, 3100 psia; near miscible, 2600 psia; and immiscible, 2100 psia.

Since CO_2 has a low liquefied pressure at the reservoir temperature compared to other injection gases, two successive steps, namely compressor and pump, are necessary to increase the pressure. In each step, power calculation has first been performed to find the required power for the delivery of CO_2 into reservoir at the desired injection pressure. In the second step, capital and operational costs including the electrical power have been calculated. Finally, net present value has been calculated for each pressure.

To calculate the required power, the initial and final pressures of compressors have been defined 14.7 psia and 1085 psia (liquefied pressure at ambient temperature) respectively. Pumps transfer the flow from the final pressure of the compressor to set pressure in each scenario.

Compressor Power Calculations

The compression power requirement for each stage ($W_{s,i}$) is given by Equation 6 [13, 14].

$$W_{s,i} = \left(\frac{1000}{24 \times 3600}\right) \left(m \times \frac{Z_s RT_{in}}{M\eta\ is}\right)\left(\frac{k_S}{k_S - 1}\right)\left[(CR)^{k_S - \frac{1}{k_S}} - 1\right] \quad (6)$$

where, for all stages:

R=8.314 kJ/kmol.K

M=44.01 kg/kmol

T_{in}= 313.15 K (i.e., 40 °C)

η_{is} = 0.75

CR is the compression ratio for each stage and calculated by Equation 7.

$$CR = \left(\frac{P_{cut-off}}{P_{initial}}\right)^{\frac{1}{N_{stage}}} \quad (7)$$

where, N_{stage}=5

Values of each parameter for the calculation of compressor power are shown in Table 10 for all the stages. Finally, the total power requirement for the compressor has been defined by Equation 8.

$$W_{s-total} = (W_s)_1 + (W_s)_2 + (W_s)_3 + (W_s)_4 + (W_s)_5 \quad (8)$$

Table 10: Vital values for compressor power calculations

Stage No.	Z_s	K_s	Valid Pressure Range (MPa)	Valid Maximum Temperature (K)
1	0.995	1.277	0.1-0.24	356
2	0.985	1.286	0.24-0.56	356
3	0.970	1.309	0.56-1.32	356
4	0.935	1.379	1.32-3.12	356
5	0.845	1.704	3.12-7.38	356

Pump Power Calculations

The required power can be estimated by Equation 9.

$$W_p = \left(1000 \times \frac{10}{24 \times 36}\right) \times \left[\frac{m\,(P_{final} - P_{cut-off})}{\rho\eta_p}\right] \quad (9)$$

where, m is the CO_2 mass flow rate (tones/day), ρ=630 kg/m^3, and η_p=0.75.

As an example, power calculations of near miscible condition are shown in Figure 7.

Capital Cost Calculations

Capital costs of compressor and pump have been defined by Equations 10 to 13.

$$m_{train} = \frac{(100 \times m)}{(24 \times 3600 \times N_{train})} \quad (10)$$

$$C_{comp} = m_{train} \times N_{train} \left[\begin{array}{l}(0.13 \times 10^6)(m_{train})^{-0.71} + \\ (1.40 \times 10^6)(m_{train})^{-0.6} \ln\left(\frac{P_{Cut-off}}{P_{initial}}\right)\end{array}\right] \quad (11)$$

$$C_{pump} = \left\{(1.11 \times 10^6) \times \left(\frac{W_p}{1000}\right)\right\} + 0.07 \times 10^6 \quad (12)$$

$$C_{total} = C_{comp} + C_{pump} \quad (13)$$

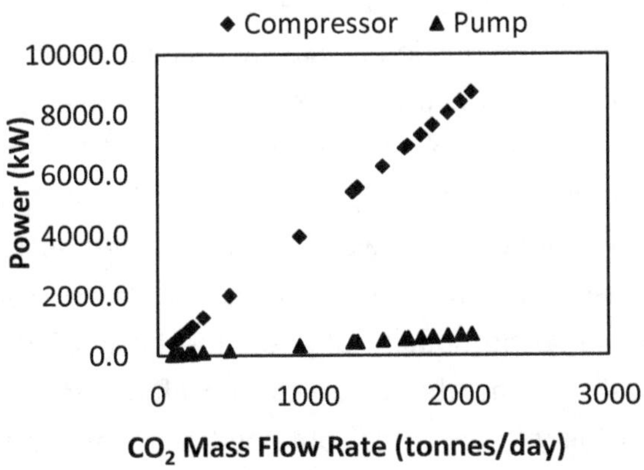

Figure 7: Sampled power calculations for compressor and pump in near miscible condition

The capital cost can be annualized by applying capital recovery factor (CRF) of 0.15.

$$C_{annual} = C_{total} \times CRF \quad \text{(where CF=0.80)}. \quad (14)$$

The total amount of CO_2 that must be compressed every year is found by applying a capacity factor (CF) of 0.80.

$$m_{year} = m \times 365 \times CF \quad \text{(where CF=0.80)} \quad (15)$$

The levelized capital costs (C_{lev}) can be defined by Equation 16.

$$C_{lev} = \frac{C_{annual}}{m_{year}} \quad (16)$$

Operational and Maintenance Costs Calculations (O&M)

Annual O&M can be found by applying an O&M factor of 0.04 to the total capital cost.

$$O\&M_{annual} = C_{total} \times O\&M_{factor} \quad \text{(where } O\&M_{factor} = 0.04) \quad (17)$$

The levelized O&M costs ($O\&M_{lev}$) can be defined by Equation 18.

$$O\&M_{lev} = \frac{O\&M_{annual}}{m_{year}} \quad (18)$$

Total Electric Power Costs of the Compressor (E_{comp}) and Pump (E_{pump})

This factor is calculated by multiplying the total power requirement by the capacity factor (CF) of 0.08 and price of electricity (P_c). It can be assumed that the electricity price is 0.065 $ per kWh [15].

$$E_{annual} = E_{comp} + E_{pump} = p_c \times (W_{s\text{-}total} + W_p) \times (CF \times 24 \times 365) \quad (19)$$

The levelized power costs (E_{lev}) can be defined by Equation 20.

$$E_{lev} = \frac{E_{annual}}{m_{year}} \quad (20)$$

Finally, the total annual and levelized costs of CO_2 for compression and pumping have been calculated by Equation 21.

$$\textit{Total Annual Cost } [\$/yr] = C_{annual} + O\&M_{annual} + E_{annual}$$

$$\textit{Total Levelized Cost } [\$/tonne\ CO_2] = C_{lEV} + O\&M_{lev} + E_{lev} \quad (21)$$

The last cost item is the price of consumed CO_2, which can be defined based on 5000 $ per 1 MMscf of injection CO_2.

As an example, all of the levelized costs including the total levelized costs for near miscible condition are shown in Figure 8.

RESULTS AND DISCUSSIONS

Six unsteady state displacement experiments have been performed and their production data in the drainage process are used to extract the relative permeability curves using an inverse modeling method. These experiments were carried out at an injection pressure of 500 psi as the immiscible condition and at 800 psi as the near miscible condition for three rock types of core samples with different RQI (sandstone (the lowest RQI), dolomite (moderate RQI), and artificial fractured sandstone (the highest RQI)). In addition, the length of the core in this study is not enough to establish the situation for near miscibility state. To cover this drawback, the soaking time is considered between the injected gas and crude oil before the effluent fluids are collected. This process facilitates the approach to near miscible condition during the experiments as Asghari and Torabi reported [16].

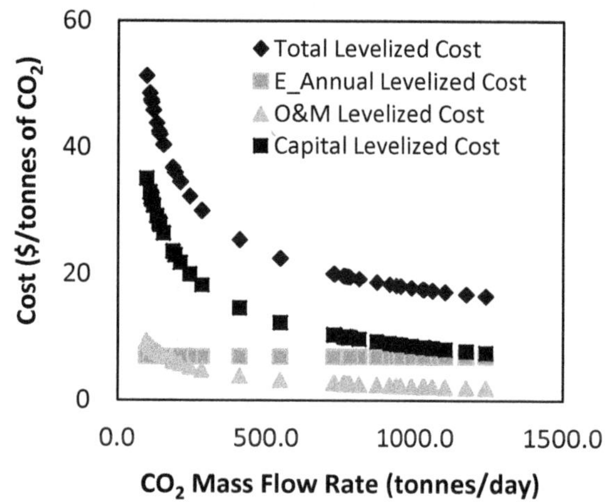

Figure 8: Levelized costs calculations in near miscible condition

To be certain that the near miscible condition was achieved, BSTEN keyword was used on lab scale compositional simulator data file to plot the trend of IFT values around the injection well (Figure 9). This figure shows the decreasing trend of IFT dimensionless value up to 50% of the initial amount as the injection time increased.

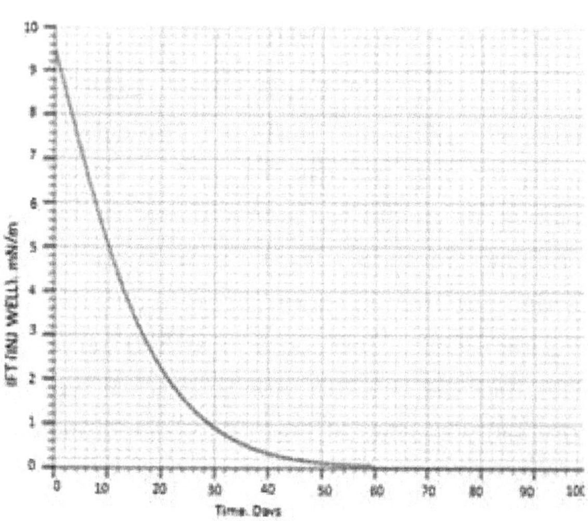

Figure 9: Trend of the change of IFT around the injection well based on elapsed injection time

After the achievement of near miscible condition, the direct production results including cumulative oil production and pressure drop along the core samples were measured (see Figures 10.a and 10.b for sandstone); then, implicit relative permeability curves were extracted (Figures 11, 12, and 13) by using the lab scale inverse modeling method. Finally, on the basis of the relative permeability curves, the effect of near miscibility on the obtained curves was investigated.

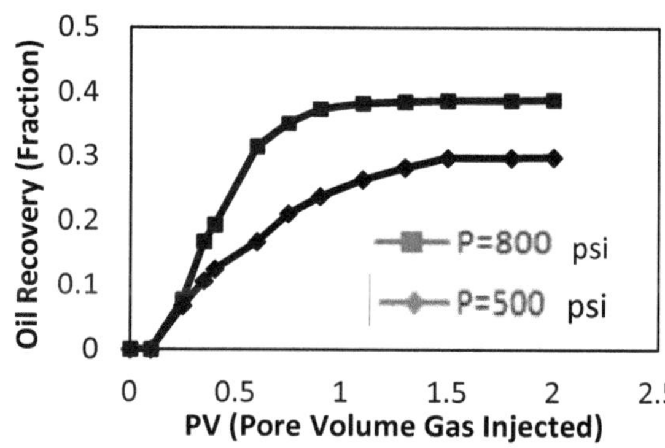

Figure 10.a: Oil recovery comparison in both immiscible and near miscible conditions for sandstone sample

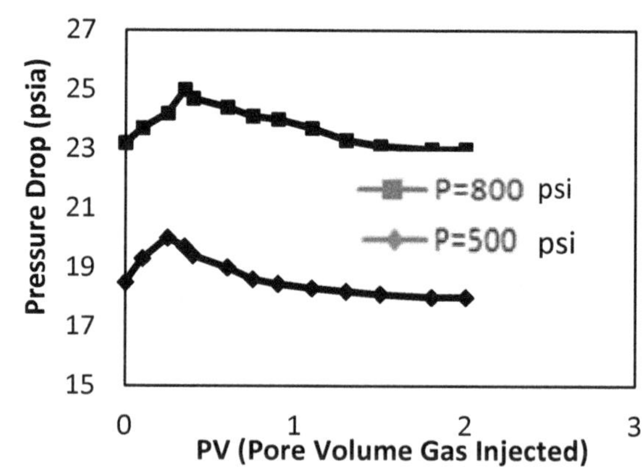

Figure 10.b: Pressure drop comparison in both immiscible and near miscible conditions for sandstone sample

To compare the results of implicit relative permeability curves in cases of immiscible and miscible conditions, the relative permeability curves were plotted based on normalized gas saturation (*Sg* (norm)), which was calculated by Equation 22.

$$S_{g(norm)} = \frac{S_g - S_{g(\min)}}{S_{g(\max)} - S_{g(\min)}} \quad (22)$$

where, $S_{g(\min)}$ and $S_{g(\max)}$ refer to the minimum and maximum outlet gas saturation respectively.

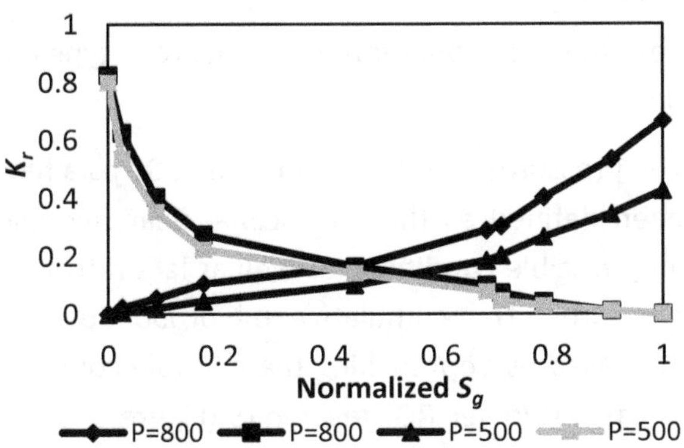

Figure 11: K_r comparison in both immiscible and near miscible conditions for sandstone sample

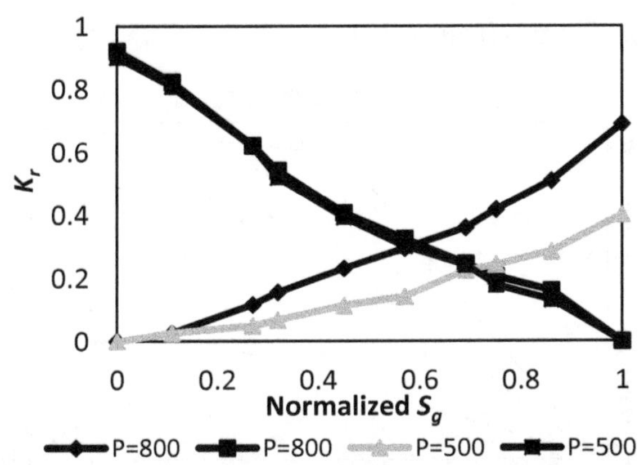

Figure 12: *Kr* comparison in both immiscible and near miscible conditions for dolomite sample

As it can be observed, the change of relative permeability is more dominant for non-wetting gas phase, when the near miscible condition approaches [5]. The main reason is the decrease in oil saturation along the core in the near miscible condition compared to immiscible. This decrease causes more space for gas to pass through itself, which leads to an increase in non-wetting relative permeability curve.

On the other hand, the resistance between oil and gas phases decreases and this change affects the relative permeability of the phase occupied the larger pores (i.e. the non-wetting phase). When there is a high absolute permeability ratio of artificial fractured sample to sandstone sample (K_f/K_s), the gas phase is not tended to pass the matrix part and it passes through fractures. This phenomenon not only causes a decrease in ultimate oil recovery, but also it stabilizes oil saturation along the core in both immiscible and near miscible conditions. This unchanging saturation profile is the reflective of no substantial change of relative permeability curves as shown in Figure 13. In this experiment, this ratio (K_f/K_s) is nearly more than 350 and it can be inferred that, in artificial fractured core samples with a high ratio (K_f/K_s), the effect of near miscibility on changing relative permeability curves is ignorable and the conventional relative permeability methods such as JBN show the same results compared to the sophisticated inverse modeling method (Figure 14).

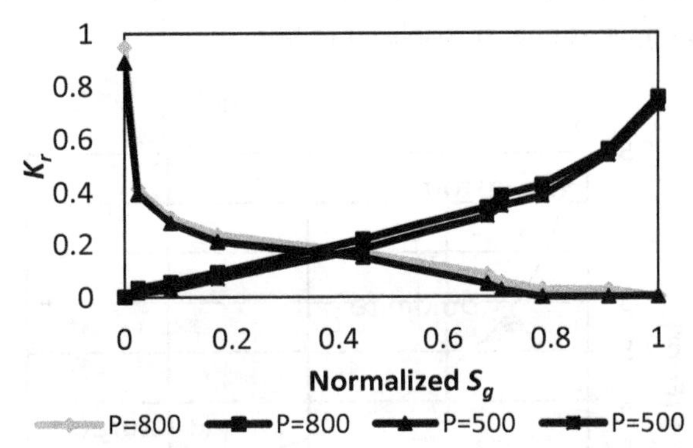

Figure 13: K_r comparison in both immiscible and near miscible conditions for artificial fractured sample

Due to lack of the activation of near miscible mechanisms and successively lack of decrease in ultimate oil saturation (S_{or}), this effect has become less important and not been observed in cases of dolomite and artificial core types respectively (Figures 12 and 13) compared to sandstone result (Figure 11).

This effect on the change of relative permeability curves has been computed by

Equation 23. To have a correlation for this change based on RQI, parameter E, calculated by Equation 23, has been plotted versus RQI (Equation 24) as can be seen in Figure 15.

$$E(Error) = \frac{K_{r(o,g)imp} - K_{r(o,g)JBN}}{K\_r(o,g)imp} \tag{23}$$

$$RQI = \sqrt{\frac{K}{\emptyset}} \tag{24}$$

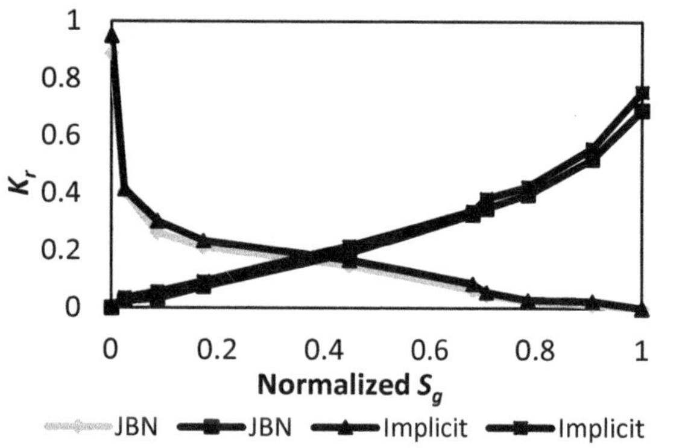

Figure 14: *Kr* comparison in both JBN and implicit inverse modeling method for artificial fractured sample at near miscible condition

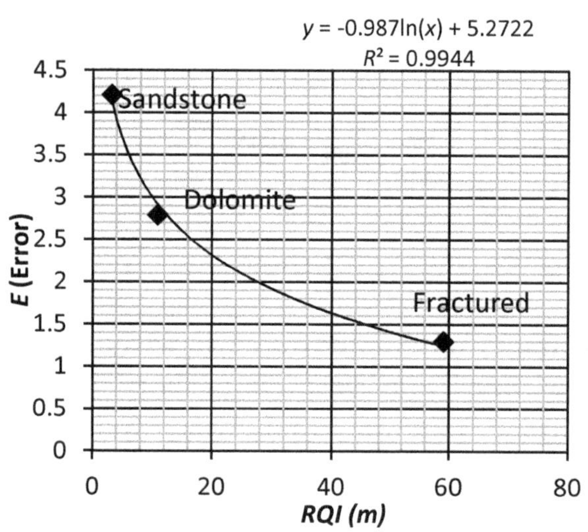

Figure 15: Change of relative permeability values (*E*) versus RQI

To validate the lab scale results, field scale modeling has been investigated using a constructed sector model. The obtained results considers two main issues; it firstly takes into account comprising the daily and cumulative injection and production results in cases of immiscible, near miscible, and miscible conditions as well as economic evaluations for rock types with lower RQI. Secondly, the same procedures have been applied in case of higher RQI (higher x-direction permeability leading to more anisotropic reservoir) to investigate the both economic and technical effectiveness of near miscible condition in various rock types on field scale.

The pressures of 2100, 2600, and 3100 psia have been defined as the immiscible, near miscible, and miscible conditions as well as laboratory set pressures. The cumulative oil production was increased by approaching the miscible condition in case of lower RQI reservoirs (Figure 16). The increase of the slope of recovery factor decreased after the near miscible pressure. This trend can be traced from the daily production profile too (Figure 17).

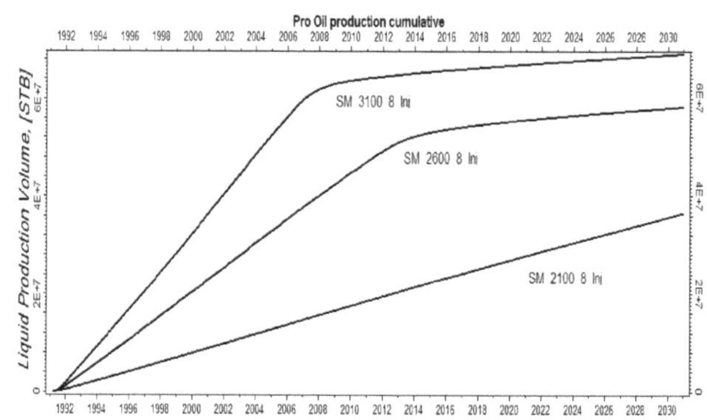

Figure 16: Comparison of cumulative oil production in immiscible, near miscible, and miscible conditions in lower RQI

This phenomenon has happened due to the production of by-passed oil remained in the process of immiscible gas injection. This volume was produced mainly based on the activation of vaporizing mechanism and successively IFT reduction. In this condition, the reduction of IFT between oil and gas phases moved by-passed oil from pore throats to the main oil stream. This condition can be traced by shifting the non-wetting phase saturation into lower value and decrease the curvature of relative permeability in lab scale results, too (Figures 11).

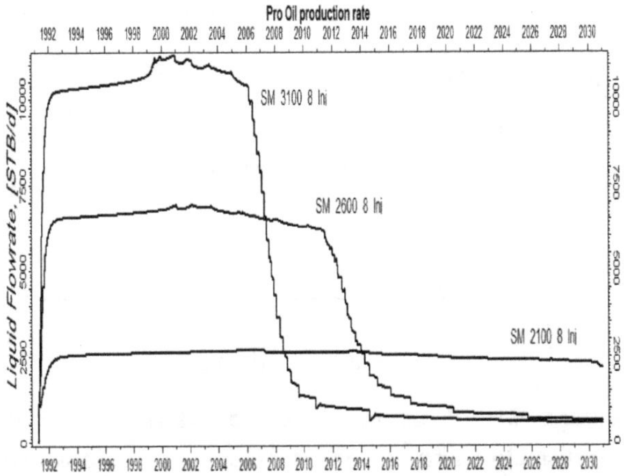

Figure 17: Comparison of oil production rate in immiscible, near miscible, and miscible conditions in lower RQI

To investigate the activation of near miscible mechanisms such as vaporizing in field scale modeling, the oil saturation profile are shown in cases near miscible and miscible injections (Figure 18.a and 18.b). As it can be observed, the oil saturation has been started to decrease to a lesser extent than the residual oil saturation value (0.46) when approaching the miscible condition, which indicates the start of IFT reduction and the activation of near miscibility condition. The corresponding oil saturation values of near miscible and miscible conditions are tabulated in Table 11.

Table 11: Oil saturation profile of specified grid cell in near miscible and miscible conditions in case of lower RQI

Injection Condition	Oil saturation (Fraction)
Near miscible	0.32
Miscible	0.064

To investigate the near miscibility behavior in case of high RQI reservoirs on field scale, a sector with higher *x*-direction permeability (10 times higher) has been constructed to generate the desired anisotropy. The final oil recovery factor was increased by approaching the near miscible condition; however, this increasing ratio, 1.04, was smaller in case of higher RQI than the case of lower RQI, i.e.1.6 (Figure 19).

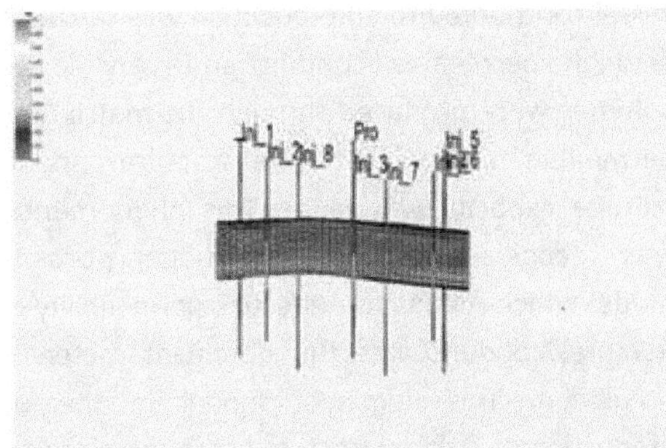

Figure 18.a: Oil saturation profile of the sector model in near miscible condition (P=2600 psia)

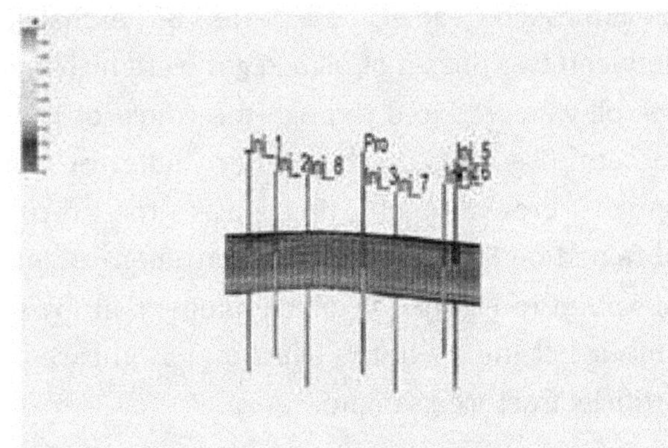

Figure 18.b: Oil saturation profile of sector model in miscible condition (P=3100 psia)

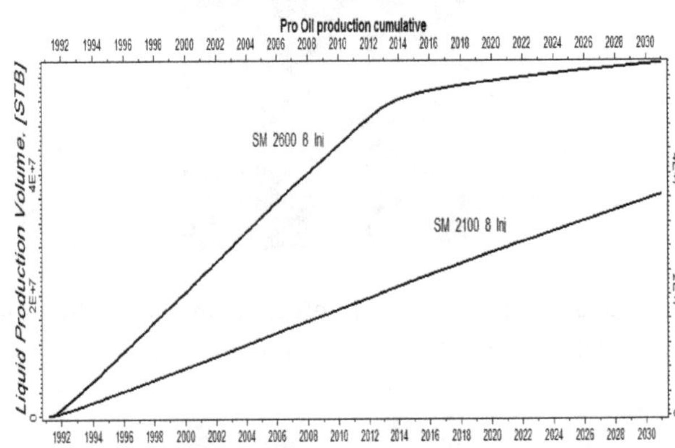

Figure 19: Comparison of cumulative oil production in immiscible and near miscible conditions in higher RQI

In the case of higher RQI reservoirs, oil production did not increase much in near miscible condition compared to the immiscible one. The main reason was due to early breakthrough time and inadequate time for the activation of vaporizing mechanism. This mechanism caused the low rate of mass transfer between two phases. The main vessel contributed to oil production was fractures or high permeable conduit and very lower volumes were produced through the matrix/low permeable section (i.e. the fracture/conduit storage capacity was high). This phenomenon was recognized as mono-perm-dual porosity model which means the effective permeability of fractures/conduits was the dominant factor in simulation. This dominant model in the oil production of artificial fractured core samples was schematically shown in Figure 20.a. In the case of lower RQI reservoirs, as the gas did not breakthrough earlier, the rate of exchange between two phases of oil and gas were high and the oil was produced through the whole of pore throats (Figure 20.b). Another indicator for earlier breakthrough time was the result obtained on lab scale, i.e. the sharp peak points observed in Figure 21, which happens in lower amount of the PV of the injection gas in case of artificial fractured sample.

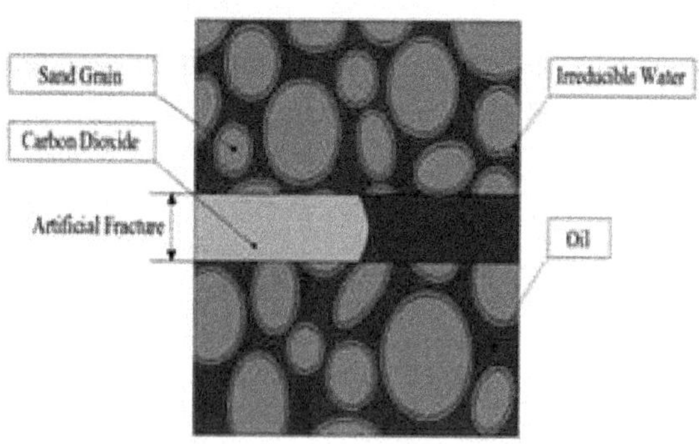

Figure 20.a: A schematic diagram of oil production model in an artificial fractured core type

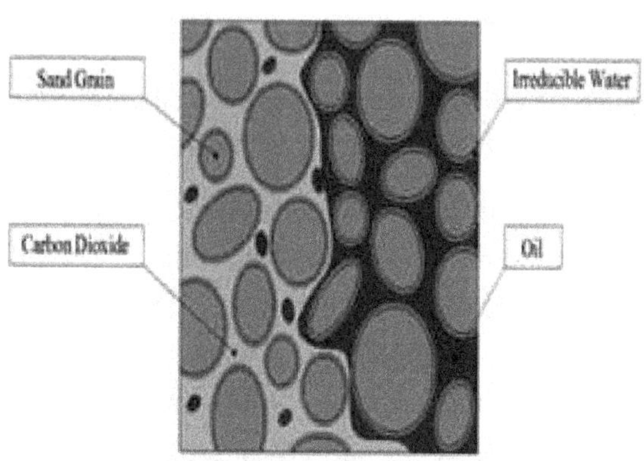

Figure 20.b: A schematic diagram of oil production model in non-fractured core type

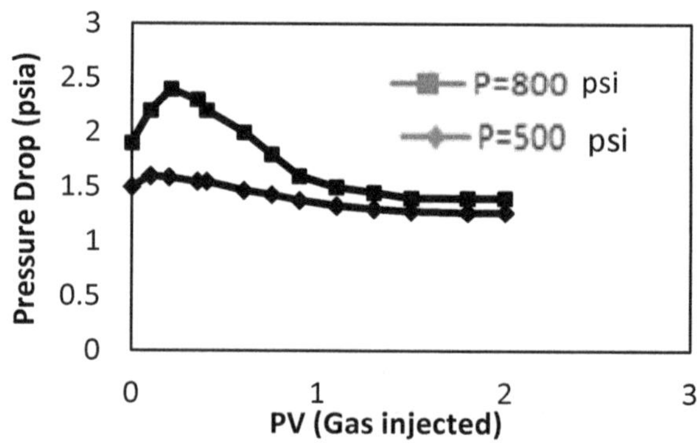

Figure 21: Pressure drop comparison in both immiscible and near miscible conditions for dolomite sample on lab scale

To investigate the economical possibility of near miscible injection compared to the other conditions, NPV analysis has been performed in case of lower RQI (Table 12). To have a better analysis, NPV values have been plotted versus injection pressures in case of lower RQI (Figure 22). As it can be observed, the slope of the increase of NPV decreased and this means that use of near miscible injection is the most economical scenario in case of lower RQI reservoirs. On the other hand, this trend can be traced by the NPV ratio values, which are the ratio of the NPV of each injection stage to the previous one (Table 12). This trend showed a decrease in the ratio, which means gas injection at higher pressures does not increase NPV at the same speed compared to pressures lower than the near miscible pressure.

Table 12: NPV analysis results in case of lower RQI

Pressure (Psia)	Total Cost ($)	Total Income ($)	NPV	NVP ratio
2100	9.869E+07	9.214E+08	8.227E+08	2.354E+00
2600	1.411E+08	2.078E+09	1.937E+09	1.413E+00
3100	1.616E+08	2.899E+09	2.737E+09	1.144E+00
3400	1.537E+08	3.284E+09	3.131E+09	1.111E+00
3700	1.607E+08	3.639E+09	3.479E+09	1.090E+00
4000	1.675E+08	3.958E+09	3.791E+09	1.000E+00

Moreover, NPV and recovery factor values were plotted simultaneously to define the near miscible pressure more certainly. As it can be observed, the value pointed by the cursor was near miscible pressure at which both NPV and RF curves started to bend and their slopes decreased (Figure 23).

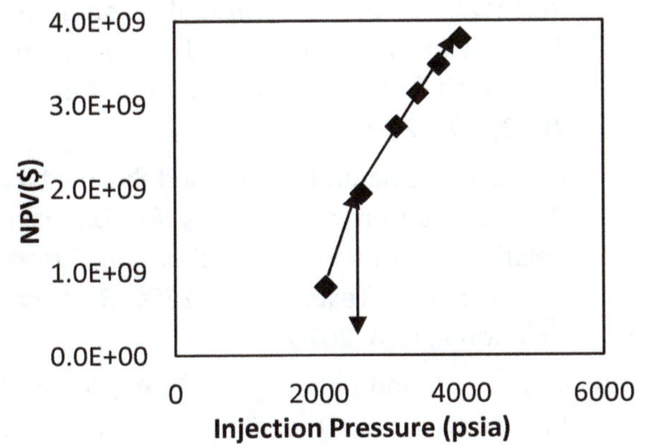

Figure 22: NPV analysis results for injection stages including immiscible, near miscible, and miscible conditions in lower RQI

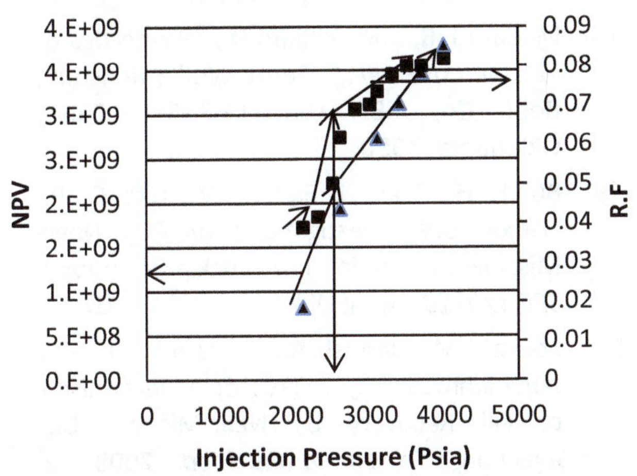

Figure 23: Simultaneous NPV and RF results for injection stages including immiscible, near miscible, and miscible conditions in lower RQI

CONCLUSIONS

To observe the effect of near miscibility on lab scale in different rock types (artificial fractured core, dolomite, and sandstone samples), relative permeability curves have been extracted by lab scale inverse modeling and the results were testified by field sector modeling. The following conclusions can be drawn based on the experiments and modeling results:

- 85% of MMP, obtained by the slim-tube experimental simulation, was reported as the near miscible pressure on lab scale and defined as the inflection point of the curve;
- The effect of near miscibility on two-phase oil-gas relative permeability curves was more significant, moderate, and ignorable in rock types with lower, moderate, and higher RQI respectively;
- In case of artificial fractured core type (higher RQI), the conventional relative permeability methods such as JBN had the same results compared to the sophisticated inverse modeling method;
- Field sector modeling results satisfied lab data regarding the recovery factor in different rock types. 83% of MMP was reported as the near miscible pressure at which mechanisms such as vaporizing activated and oil recovery factor of near miscible injection was near immiscible one in case of higher RQI reservoirs;
- Use of near miscible gas injection in lower RQI reservoirs is more economical rather than the other scenarios.

ACKNOWLEDGMENTS

Authors would like to thank Mr. Safari and Habibi for their useful experience shared.

NOMENCLATURES

f, s	Fracture, sandstone
s	Start of injection
I_r	Injection capacity
	Viscosity
Ø	Porosity
K	Absolute permeability
BPR_P	Pressure of Back Pressure Regulator

Z_s	average CO_2 compressibility for each individual stage [-]
M	Molecular weight of CO_2 [kg/kmol]
K_s	(C_p/C_v) = average ratio of specific heats of CO_2 for each individual stage [-]
m_{train}	CO_2 mass flow rate through each compressor train [kg/s]
1,2	Inlet, outlet of the core
o, g	Oil, gas phase
W_i	Cumulative injection
f	Fractional flow
L	Core length [cm]
A	Cross section area [cm^2]
ρ	Density [kg/m^3]
T_{in}	CO_2 temperature at compressor [K]
η_{is}	isentropic efficiency of compressor inlet [K]
η_p	Efficiency of pump [-]
CF	Capacity factor

REFERENCES

[1] Wang X. and Gu Y., "Oil Recovery and Permeability Reduction of a Tight Sandstone Reservoir in Immiscible and Miscible CO_2 Flooding Processes," *Industrial and Engineering Chemistry Research*, **2011**, *50*(4), 2388-2399.

[2] Dong M., Huang S. S., and Srivastava R., "Laboratory Study on Near-miscible CO_2 Injection in Steelman Reservoir," *Journal of Canadian Petroleum Technology*, **2001**, *40*(2), 53-61.

[3] Al-Wahaibi, Y. M. and A. K. Al-Hadrami, "The Influence of High Permeability Lenses on Immiscible, First- and Multi-contact Miscible Gas Injection," *Journal of Petroleum Science and Engineering 77*, **2011**, *77*, 313-325.

[4] Al-Wahaibi Y. M., Grattoni C. A., and Muggeridge A. H., "Drainage and Imbibition Relative Permeabilities at Near Miscible Conditions," *Journal of Petroleum Science and Engineering*, **2006**, *53*, 239-253.

[5] Parvazdavani M., Masihi M. and Ghazanfari M. H., "Gas-Oil Relative Permeability at near Miscible Condition: An Experimental and Modeling Approach," *Journal of Scientia Iranica*, **2012**, *20*, 626-636.

[6] Hussain F., Cinarand F., and Bedrikrikovetsky P., "Comparison of Methods for Drainage Relative Permeability Estimation from Displacement Tests," *SPE 129678*, Tulsa, Oklahoma, USA, **2010**.

[7] Trivedi J. J. and Babadagli T., "Experimental Investigations on the Flow Dynamics and Abandonment Pressure for CO_2 Sequestration and Oil Recovery in Artificially Fractured Cores," *Journal of Canadian Petroleum Technology*, **2010**, *49*(3), 22-27.

[8] McCain J. R. and William D., *"Properties of Petroleum Fluids,"* Penn Well publishing Book Co, ISBN 0-87814-335-1, Tulsa, Oklahama, **1990**.

[9] Bui L. H., Tsau, J. S., and Willhite G. P., "Laboratory Investigations of CO_2 Near-Miscible Application in Arbuckle Reservoir," *SPE 129710*, Tulsa, Oklahoma, USA, **2010**.

[10] Sohrabi M., Danesh A., Tehrani D. H., and Jamiolahmady M., "Microscopic Mechanisms of Oil Recovery by Near-Miscible Gas Injection," *Transp Porous Med.*, **2008**, *72*, 351–367.

[11] Johnson E. F., Bossler D. P., and Naumann V. O., "Calculation of Relative Permeability from Displacement Experiments," *Trans. AIME*, **1959**, *216*, 370-372.

[12] Fletcher R., "Conjugate Direction Methods," Chapter 5 in Numerical Methods for Unconstrained Optimization, edited by W. Murray, Academic Press, New York, **1972**.

[13] Mohitpour M., Golshan H., and Murray A., "Pipeline Design & Construction: A Practical approach," *The American Society of Mechanical Engineers*, New York, **2000**.

[14] IEA Greenhouse Gas R&D Programme, Transmission of CO_2 and Energy, Report No. PH4/6, **2002**.

[15] Kreutz T., Williams R., Consonni S., and Chiesa P., "Co-production of Hydrogen, Electricity and CO_2 from Coal with Commercially Ready Technology Part B: Economic Analysis," *International Journal of Hydrogen Energy*, **2005**, *30*, 769-784.

[16] Asghari K. and Torabi F., "Laboratory Experimental Results of Huff 'n' Puff CO_2 Flooding in a Fractured Core System," Proceeding of the *SPE Annual Technical Conference and Exhibition* held in Anaheim California, US, **2007**.

8

Performance of Air Foam Flooding under Low Frequency Vibration

Pu Chunsheng[1,2], Zheng Liming*[1], Liu Jing[1], and Xu Jiaxiang[1]

[1] School of Petroleum Engineering, China University of Petroleum (Eastern China), Qingdao, China
[2] State Key Laboratory of Heavy Oil, Qingdao, China

ABSTRACT

Foam injection is widely applied in amounts of fields to drilling, production, and formation protection. Sometimes, the application result is disappointing, which is caused by the failure of bubble generation in foam flooding. Therefore, it is necessary to seek ways for improving the performance of foam injection. An increased disturbance to the stratum, like the vibration caused by a seismic oil recovery technique, would be helpful. In the current work, the seepage of air foam in porous media under low frequency (LF) vibration is analyzed with experiments and an investigation of bubble creation/destruction rate change is carried out using mathematical modeling. The resistance factor of foam flooding under indoor vibration increases by 1.5 times and the valid time is obviously extended compared with when no vibration is used. The optimal vibrating acceleration and frequency of 0.7 m/s^2 and the natural frequency of the cores-nearby of 18 Hz are achieved in the indoor experiments. Under vibration, the bubble generation rate increases, while bubble break rate by internal expansion or by gas diffusion and transfer decreases. An interesting phenomenon is also observed, which might develop a power level formula between the initially defined dimensionless MRF (maximum foam flooding resistance factor) and dimensionless DMRF (duration of maximum foam flooding resistance factor). The power product and sum of the power exponents of the above formula both equal approximately to 1. With the assistance of LF vibration, the increase of security, adaptability, and efficiency in foam injection may improve the reservoir recovery and extend its application.

Keywords: Air Foam Flooding, Low Frequency Vibration, Resistance Factor, Foam Stabilization, Low Permeable Reservoir

INTRODUCTION

Air foam flooding [1-2] and low frequency vibration oil extraction technology [3-5] (also known as the artificial seismic oil recovery technology) are widely applied to low permeability reservoirs. For their wide adaptability and better performance in dead oil displacement inside the fine pores or throats, the development efficiency could be improved. The gas source in air foam flooding is free and convenient to be obtained. The low frequency (LF) vibration is of a large action radius due to its weak attenuation when transmitting in porous media, and more wells could share the running cost. The low cost of air foam flooding and LF vibration oil extraction technology is attractive for application in oilfield.

***Corresponding author**

Zheng Liming

Email: upczlm@sina.cn

However, low efficiency and disappointing economy effect were sometimes caused by the failure of bubble generation in foam flooding, which had caused a block in pilot tests. Furthermore, the safety concerns might be a key factor, even though foam injection had been widely applied and is relatively inexpensive. It is necessary to improve the performance of foam injection. Therefore, an increased disturbance to the stratum, like the vibration caused by LF vibration oil extraction technology, would be helpful to increase the bubble generation.

LF vibration oil extraction technology improves oil recovery through the following mechanisms [6-8]: (a) the absolute permeability is increased, for the capillary forces inside matrix and the particle blocking inner pore throats are reduced; (b) the seepage rate is improved, with the decrease in the viscosity of crude oil, the increase in hydrophilicity on rock surface, and the additional driving force added by vibrating; (c) the residual oil saturation is reduced and oil recovery is enhanced due to the increase in the detachment of crude oil in die pores and around the rock surface.

Air foam flooding mechanisms for enhancing oil recovery mainly include the following [9-11]: (a) the sweep efficiency is expanded, for the high permeable layers or seepage channels are plugged to achieve oil displacement in low permeability layers or bypassed throats; (b) the water-oil mobility ratio is increased on account of the higher foam viscosity; (c) the reservoir energy is increased and a driving force in the upper layer is generated, with the injection of gas; (d) the additional effect of chemical flooding is achieved, with the main components of foam system being surfactants as well.

Researches on foam under vibration were conducted mainly on static foam performance without porous media and wave propagation in foam systems [12-15], indicating an improvement in foam behavior with the impact of the wave. The static foam performance includes foaming volume, the half-time of deposition, interfacial tension, compressibility, strain rate, etc. Wave propagation in foam systems includes flooding model, wave velocity, energy absorption, etc. However, there was less consideration in the real porous flow of foam with three phase coexistence. Therefore, what and how the foam performance change under low frequency vibration are investigated in wave loading experiments. Moreover, the variation in bubble creation/ destruction rate is analyzed in terms of the foam flooding model and elastic wave theory.

MATERIALS AND METHODS

Apparatus and Materials

As Figure 1 shows, the experiment apparatus is conducted on the horizontal vibration table (Figure 2) designed by the Physical & Eco-Chemical Technology and Engineering Center (PECTEC) of China University of Petroleum. The unconsolidated sand pack (25 cm length) is fixed on the horizontal vibration table with the foam liquid and gas injected into the inlet. A hand pump and a backpressure valve are used to establish the backpressure. The electronic balance, 2PB00C-type constant flux pump, gas flow meter, intermediate container, and electronic pressure gauge (0.001 MPa) are also used in the foam flooding.

The materials include distilled water, sodium chloride, calcium chloride, magnesium chloride, kerosene, AES (fatty alcohol-polyoxyethylene ether sodium sulfate), and hydrolyzed polyacrylamide (molecular weight of 10 million). The simulated formation water used in this study is $CaCl_2$ type and with a salinity of 40000 mg/L, according to the formation water of certain oil-field in Ordos basin. The sand is quartz sand with a uniform size, 80-mesh, and good abrasion. After filling and compaction in the unconsolidated sand pack, the sand is low permeable for about $30\times10^{-3}\ \mu m^2$.

Figure 1: Indoor experimental device for low frequency vibration

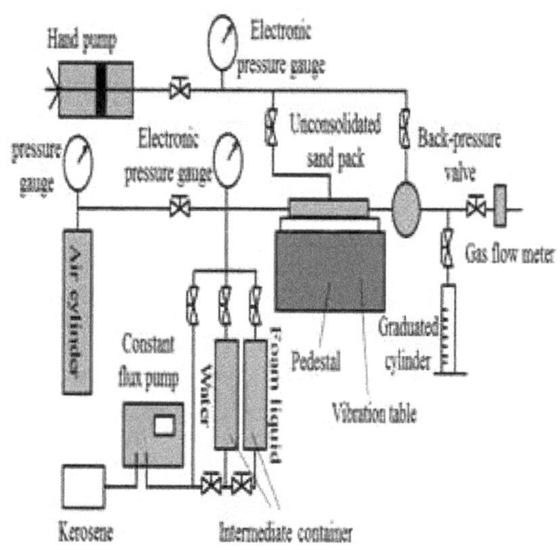

Figure 2: A schematic of the experimental workflow and apparatus

Experiment Procedure

Foam liquid, with an agent concentration of 0.5%, and air are injected into the unconsolidated sand pack repeatedly for three times. The total amount of foam liquid is 0.15 pore volume (PV) and gas-liquid volume ratio under the entrance pressure is 3:1. The back pressure is 1.0 MPa. Once the foam liquid is injected, the vibration table begins to work with a sinusoidal motion in the horizontal direction. The flowing fluid is along the same direction with generated propagating wave.

The pressure drop of water drive before or after air foam injection, expressed as $P_1$and P_2 respectively, are measured successively at 30 °C and the same flow rate. The resistance factor *K*, defined as the ratio of pressure drops above ($K = P_2 / P_1$), is an important indicator to evaluate foam mobility in porous media [16-17]. Through the analysis of *K* versus time under different vibrating parameters, the performance change in the water plugging of air foam is revealed.

The evaluation of dynamic foam performance under vibration includes the influence of vibrating time, vibrating frequency, and acceleration. To explore the influence of vibrating time on the resistance factors of air foam flooding, vibrating time is changed for 1, 2, and 3 hrs. During the process containing foam injection and water displacement afterwards, vibration continues until the set time. The vibrating frequency and acceleration are kept 18 Hz and 0.4 m/s^2 respectively. Resembling the above, four samples were prepared with a vibrating frequency of 5, 10, 18, and 25 Hz to explore the influence of vibrating frequency, while the vibrating time and acceleration were kept 1 hr and 0.4 m/s^2 respectively. The influence of vibrating acceleration on foam flooding was conducted by setting the vibrating acceleration to 0.2, 0.4, 0.7, and 1.0 m/s^2, while the vibrating time and frequency were kept constant at 1 hr and 18 Hz respectively. In contrast, a sand sample was also foam flooded without vibration.

RESULTS AND DISCUSSION

Vibrating Time

Variation tendencies of resistance factor at different vibrating times are similar as shown in Figure 3. It is characterized by three stages, namely a sharply increasing stage, a stable stage, and a slowly decreasing stage. The first stage lasts for about 30-100 minutes before the resistance factor increases to a maximum value; the second stage remains for about 80-150 minutes in our experiments and then the resistance factor decreases gradually.

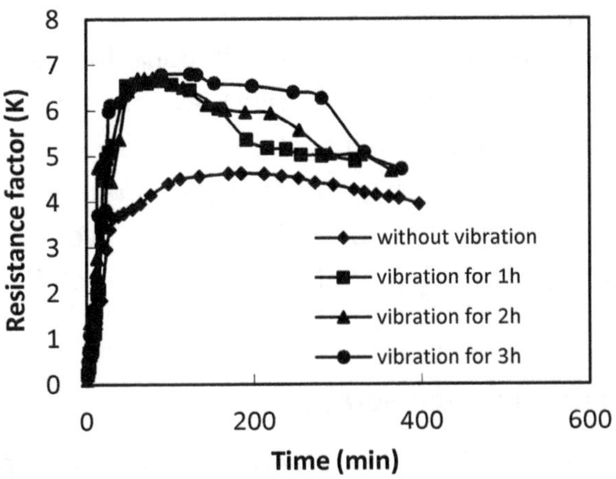

Figure 3: Influence of vibration time on resistance factor

At different vibrating times, the maximum resistance factor (MRF) is found approximately the same, which indicates that MRF and instant foaming speed are scarcely affected by the change of the whole vibrating time. However, the durations of maximum resistance factor (DMRF) in three experiments are different. Because the resistance factor is always changing over time and the maximum is an isolated point, DMRF is defined as the period in which resistance factor value is not less than a certain percentage of the MRF. The percentage chosen herein is 85%, because the value in the range of most samples is close to the MRF. The analysis of DMRF versus vibrating time is shown in Figure 4.

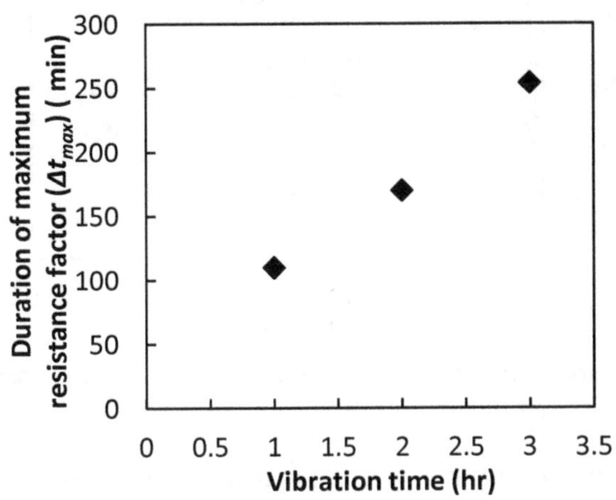

Figure 4: Duration of maximum resistance factor versus vibration time

The DMRF becomes larger as the vibrating time is prolonged and the data in Figure 4 is linearly fitted by Equation 1:

$$\Delta t_{\max} = 72t_v + 34 \quad (1)$$

where, Δt_{max} is the duration of maximum resistance factor in minute; t_v is the LF vibrating time.

In the foam drive process, as the foam liquid is consumed, foaming speed would gradually decrease. When the foaming speed is higher than the defoaming speed, the bubble number will instantly increase and so will the resistance factor. When the foaming speed is lower than the defoaming speed, the bubble number and the resistance factor will obviously decline. Hence the disturbance added by vibration is the reason causing the enhancement in foam plugging effect. Furthermore, the bubble size is decreased and even more distributed [18], leading to stronger foam strength and extended defoaming process, although the maximum amount of foam is reckoned to be constant as shown in Figure 3. Another key factor promoting the foam flooding stability is probably the increased refoaming ability of in situ foam liquid. After some bubble break, new bubbles are soon generated under vibration. In addition, the decreased surfactant consumption for absorption or retention due to LF vibrating might also play a role in extending the duration [19]. Therefore, vibrating time should be prolonged as much as possible when the foam drive is applied in combination with LF vibration technology in field tests.

Vibrating Frequency

It could be noticed in Figure 5 that the resistance factor has shown different tendency for various vibration frequencies. With the increase of vibrating frequency, the MRF becomes larger firstly and then decreases' an optimum value is observed nearby 10 Hz, which indicates a relatively better foam behavior. For a while after foam injection, increased foaming speed because of a higher frequency shock has improved the foam strength and the plugging efficiency, resulting in a larger value in MRF. However, the faster the oscillation is, the quicker the bubble quantity is decomposed and displaced by subsequent water per unit time. A rapid reduction of foam liquid amount in situ and a smaller resistance factor at the third stage of foam behavior appear under faster oscillation. Nevertheless, an opposite tendency is observed with the DMRF (Figure 6).

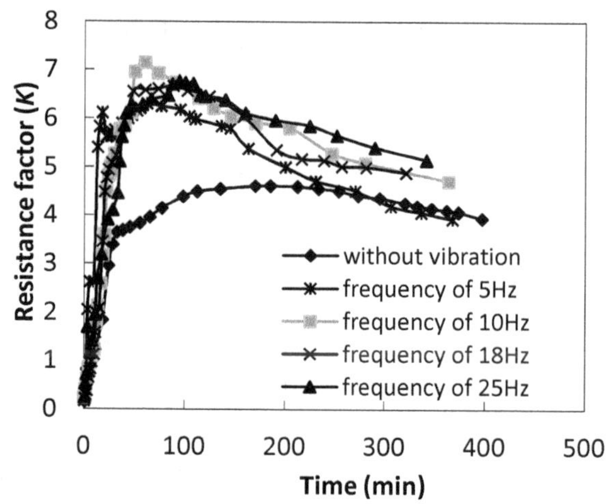

Figure 5: Influence of vibration frequency on foam flooding resistance factor

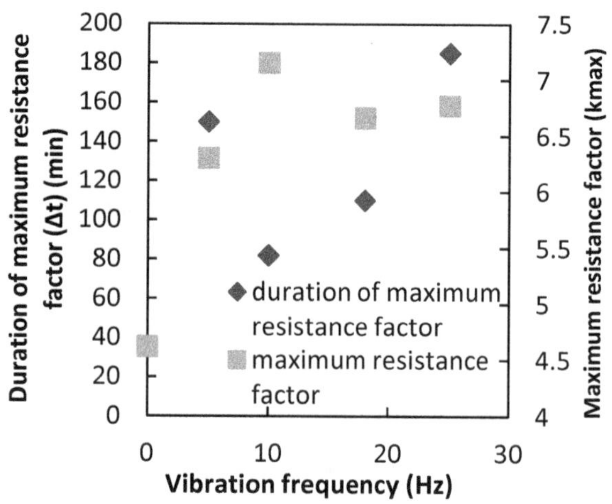

Figure 6: Influence of vibration frequency on the maximum resistance factor and its duration

A larger foaming speed is obtained at a higher vibrating frequency. Correspondingly, the foam is not only of a poor foaming capacity but it also has a low breaking rate under LF vibration. Therefore, a larger value of DMRF is achieved in relatively higher or lower vibrating frequency ranges. In a comprehensive view of MRF and DMRF versus vibrating frequency, an optimal vibrating frequency (10 Hz in the current work) should be considered to improve the comprehensive foam flooding effect in oil field tests.

Vibrating Acceleration

The alterations of resistance factor over time at different vibrating accelerations are not identical (Figure 7); the MRF and its duration also follow similar trends. As shown in Figure 8, MRF increases at first and then decreases as vibrating acceleration increased in the range of 0 to 1.0 m/s^2, with a peak achieved at 0.4 m/s^2. It indicates that at the above optimal vibration the largest foaming speed is obtained and the critical amplitude leads to the largest apparent relative velocity of solid-foam phase and solid-water phase. Following above, a rapid decline in resistance factor, as well as relatively reduced foam stability, appears due to the increase of foam liquid consumption.

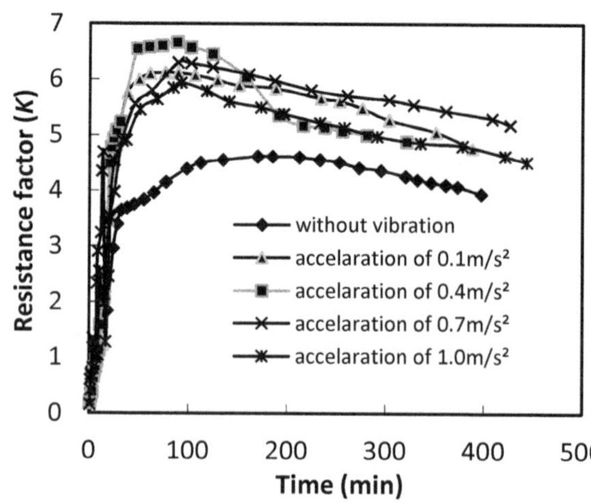

Figure 7: Influence of vibration acceleration on foam flooding resistance factor

Considering the overall performance in foaming and stability, the optimal water plugging effect is observed at acceleration of 0.7 m/s^2. The MRF at acceleration of 0.7 m/s^2 is slightly smaller than the one at acceleration of 0.4 m/s^2, but the DMRF of the former acceleration is far more than that of the latter. The experiments have shown that proper vibrating acceleration should be seriously selected in oilfield tests for vibration coupling foam injection in order to further enhance the foam displacement efficiency and oil recovery.

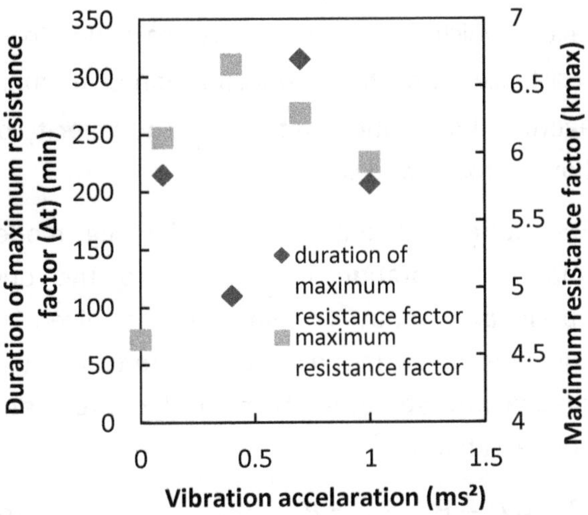

Figure 8: Influence of vibration acceleration on the maximum resistance factor and its duration

FOAM FLOODING MODELING

Foam Flooding Model under LF Vibration

Different from the usual model of foam flooding, the motion equation of foam flooding model under LF vibration needs to take the influence of FSI and the mass conservation equation into consideration.

In the foam flooding model under LF vibration, the following assumptions are made:

1. multiphase and multiple components are included;
2. the gas dissolved in water is not taken into account;
3. only the water phase contains the surfactant, and the influence of surfactant on the density and viscosity of each phase is omitted;
4. the injected fluid percolating in porous media obeys the generalized Darcy's law;
5. the wave attenuation is negligible;
6. the heat transfer and change in the foam flooding process are not considered.

Based on the theory of elastic wave propagation in elastic porous media containing multiphase fluid, the motion equation under the LF vibration of phases as matrix, foam liquid without bubbles (water phase), oil, free gas, and the foam (gas+liquid) is defined by Equation 2 [20-21]:

$$\begin{cases} \nabla \tau_{ij,i} = \rho \ddot{\mathbf{u}}_i + \sum \rho_m \ddot{\mathbf{v}}_{m,i} \\ P_{m,i} = \rho_m \left(\ddot{\mathbf{u}}_i + \ddot{\mathbf{v}}_{m,i} / \phi S_m \right) + \eta_m \ddot{\mathbf{v}}_{m,i} / \kappa_m \end{cases} \tag{2}$$

where τ_{ij} is the total stress and expressed as $\tau_{ij} = \lambda e \delta_{ij} + 2\mu\varepsilon_{ij} - \alpha\delta_{ij} P$ in unsaturated porous media; $P_{m,i}$ is the fluid pressure of component i in phase m; $\mathbf{u}_i$, $\overline{\mathbf{v}}_{w,i}, \overline{\mathbf{v}}_{o,i}, \overline{\mathbf{v}}_{g,i}, \overline{\mathbf{v}}_{f,i}$ is the displacement of the matrix phase, water phase, oil phase, free gas phase, and foam phase respectively; the relative displacement of mobile phase is;

$$\frac{\partial \mathbf{v}_{m,i}}{\partial t} = \phi S_m \left(\frac{\partial \overline{\mathbf{v}}_{m,i}}{\partial t} - \frac{\partial \mathbf{u}_i}{\partial t} \right) \tag{3}$$

S_m represents the saturation of the mobile phase; ρ, ρ_m, and ρ_s are the densities of porous media, fluid, and skeleton respectively. η_m stands for the viscosity of fluid and ϕ is the porosity;

$$\rho = \phi \sum S_m \rho_m + (1-\phi)\rho_s \tag{4}$$

Based on the relationship of the strain, porosity and pore pressure in Biot theory, the constitutive equation is defined as follows:

$$d\phi = \alpha de + \frac{dP}{Q} \tag{5}$$

where e is the volume strain of the rock; P is the confining pressure on solid particles, and is expressed as $P = \sum S_m P_m$ with the average volume method; $\frac{1}{Q}$ is the positive coefficient representing the fluid compressibility under fluid-solid interaction; α is an equivalent representing the action of shear stress on porous media.

The mass conservation equation of component i in phase m is given by Equation 6 [16, 22-24]:

$$\sum_{m=1}^{Np}\nabla\left[\rho_m x_{I,m}\mathbf{v}_m\right]+\phi\sum_{m=1}^{Np}\nabla\left(D_{I,m}\rho_m\nabla x_{I,m}\right)+ \sum_{m=1}^{Np}q_m x_{I,m}+\phi\sum_{K=1}^{Nr}\left(r_{g,K}-r_{c,K}\right)= \frac{\partial}{\partial t}\left(\phi\sum_{m=1}^{Np}\rho_m S_m x_{I,m}\right)+ \frac{\partial}{\partial t}\left[(1-\phi)\rho_r C_{I,r}\right], I=1,2,\cdots,N \tag{6}$$

where $\mathbf{v}_m$ stands for the velocity of mobile phase and expressed in Darcy motion equation as;

$$\mathbf{v}_m=\frac{kk_{rm}}{\mu_m}\left(\nabla P_m-\nabla P_{cm}-\gamma_m Z\right) \tag{7}$$

k is the absolute permeability of reservoir and k_{rm} represents the relative permeability; $x_{I,m}$ means the mole fraction of component I in phase m; μ_m, P_m, P_{cm}, γ_m, q_m, and Z represent the viscosity, reservoir pressure, capillary pressure, specific gravity of overlying rock, source term, and reservoir vertical depth respectively. $r_{g,K}$ and $r_{c,K}$ mean the generation rate and dissipation rate of component I in the physical or chemical reaction K respectively, which is mainly used for the foam phase and could be expressed by the semi-quantitative formula or experimental model formula; Np is the number of phases.

The parameters like viscosity or relative permeability in the foam phase could be expressed as a function of the ones in the gas phase as given below, which is a common approach in foam injection modeling:

$$\mu_f=\mu_g+\frac{\alpha\rho_{ff}}{v_f^{1/3}} \tag{8}$$

$$k_{rf}=k_{rg}\left(S_w\right)\cdot F_M \tag{9}$$

where ρ_{ff} is the density or bubble number per unit volume of gas in flowing foam, not including the foam captured; F_M stands for a dimensionless factor which is related to the maximum foam resistance factor $K_{\max}$, concentration of foaming agent, oil saturation, saturation and velocity of flowing foam, and so on.

The state equation of foam flooding model under LF vibration is identical to the one obtained without LF vibration as given in Equations 10-12, including capillary force formula, saturation normalization, and component normalization.

$$P_{cwo}=P_o-P_w,\ P_{cwg}=P_g-P_w \tag{10}$$

$$S_w+S_o+S_g+S_f=1 \tag{11}$$

$$\sum_{I=1}^{Nc}x_{I,m}=1, m=1,2,\cdots,N_p \tag{12}$$

Firstly, one may substitute the physical equation, state equation, and the fluid relative displacement of the momentum conservation equation into the continuity equation; then, by removing the item of $d\phi$ from the continuity equation and differentiating with respect to time, the seepage continuity equation of pore fluid is obtained. Based on the solution of simultaneous equations containing the seepage continuity equation, momentum conservation equation, and stress-strain equation in unsaturated porous media, the displacement and seepage rate of multicomponent systems would be obtained.

Bubble Creation and Destruction under LF Vibration

To demonstrate the influence of LF vibration,

the change of bubble generation rate and bubble break rate is analyzed. According to the experimental static performance of foam under LF vibration [12] and expressions about bubble generation [25-26], the foaming speed r_g increases with higher relative velocities of solid-foam phase and solid-water phase v_w and v_f under LF vibration.

$$r_g = k_1^0 v_w v_f^{1/3}\left[1-\left(\frac{n_f}{n^*}\right)^w\right] \quad (13)$$

where r_g is the bubble generation rate; k_1^0 is a constant for bubble generation; v_w, v_f, n^*, and w, are water phase velocity, velocity of flowing foam, an upper limit value of foam density, and a constant determined by experiment respectively. Vibration is known to increase fluid seepage velocity; therefore v_w and v_f will increase under vibration and so will the bubble generation rate.

The reasons causing bubble break include gas expansion and gas diffusion, etc. With the analysis of experimental foam performance and expression about bubble break rate (Equation 14) [27-28], the break rate decreases and the foam half-life decomposition extends under LF vibration.

With the impact of shock, the foam size is smaller and evenly distributed; therefore, the sum of liquid membrane surface area and boundary surface area has decreased; the bubble internal break rate (left side of the formula below) also drops, which is related to bubble break due to gas expansion.

$$\left(\frac{d\varepsilon_F}{d\tau}\right)=\frac{1}{V_F}\frac{dA}{d\tau}=\frac{1}{V_F}\left(\frac{dA_f}{d\tau}+\frac{dA_b}{d\tau}\right) \quad (14)$$

where ε_F is the specific area of the bubble; τ and V_F represent time and foam volume respectively; A is the sum of liquid membrane surface area (A_f) and boundary surface area (A_b), i.e. $A=A_f+A_b$.

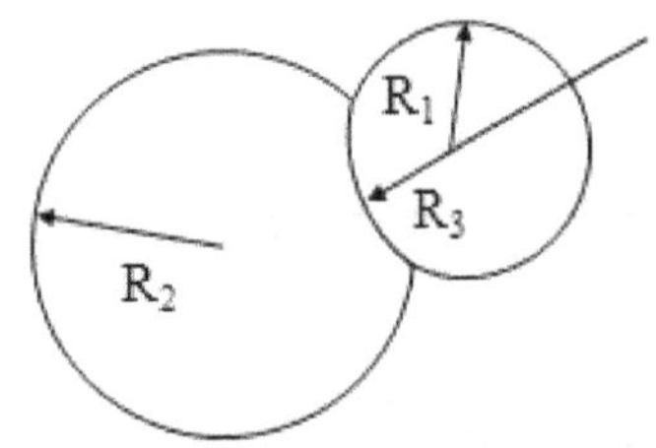

Figure 9: Membrane schematic diagram between two contacted polyhedral bubbles of different radius

The more evenly distributed bubbles (a smaller difference between diameters $|R_2-R_1|$ of the adjacent bubbles shown in Figure 9) and a decreased contact angle on surfactant solution membrane will cause the curvature radius R_3 of sharing parts between adjacent membranes to increase under LF vibration. Expression about the radius of the sharing membrane between two bubbles of different diameters is detailed as given below. An increased membrane radius R_3 and a decreased specific area ε_F of polyhedral bubble will result in a decrease in the differential pressure P_G' between adjacent bubbles. Gas diffusion, transfer between bubbles, and the value of gas diffusion coefficient (D) all decline. We assume that the parameters (such as degree of bubble number decline due to gas transfer, ambient temperature, the solubility of gas in foam liquid, interfacial tension, and external pressure) remain constant under vibration or without vibration. According to model of De Vrise [27], the bubble break rate caused by gas diffusion and transfer will decrease and the time for generating a certain number of new bubbles will extend as gas diffusion coefficient decreases under LF vibration.

$$R_3=\frac{2R_1R_2}{R_2-R_1}\cos\theta \quad (15)$$

$$P_G'=P_G-P_o=\frac{2}{3}\sigma\varepsilon_F\frac{n}{n-1}+\frac{\rho g z}{n} \quad (16)$$

$$\frac{N_t}{N_0} = 1 + K_D^{'}t = 1 + \frac{4RTD\beta\sigma}{P_0 h}t \quad (17)$$

where R_1 and R_2 are the radius of two adjacent bubbles; θ, R_3, $P_G^{'}$, σ, and n are contact angel on surfactant solution membrane, the curvature radius of sharing parts between adjacent bubbles, the differential pressure between adjacent bubbles, the interfacial tension of solution, and the dimensions of polyhedron respectively; D, R, $\frac{N_t}{N_0}$, P_0, T, and β represent gas diffusion coefficient, the universal gas constant, the degree of bubble number decline due to gas transfer, the external pressure, the ambient temperature, and the solubility of gas in foam liquid respectively.

In summary, the foaming speed increases, but the bubble break rate caused by internal expansion and gas diffusion decreases under LF vibration.

Potential for Application in Low Permeability Reservoirs

Through the research on air foam performance in dynamic displacement under LF vibration, the potential for the application of LF vibration coupling foam drive in low permeability reservoirs is embodied in the following aspects:

(1) Increasing the plugging efficiency of air foam flooding: air foam performance for water plugging under the conditions of vibration has been obviously enhanced. The MRF is increased by about 1.5 times in this paper. The MRF is influenced by the vibrating frequency and acceleration instead of the vibrating time.

(2) Extending the valid time of air foam flooding: the entire process for bubble break and foam liquid consumption, as well as the valid time of air foam flooding, is expanded under vibration; this has already been proved by the experimental results for static foam half-life change under a resonance wave. Because the actual formation is far greater than indoor experimental model, the property for the extended valid time of foam flooding will be more obvious.

(3) To a certain extent, improving the safety in air injection. With the extended valid time, more gas retention and consumption will be achieved, when the foam is displaced in the underground reservoir. Particularly, when air is injected as a source of foam generation in low-temperature shallow reservoirs, low-temperature oxidation between crude oil and O_2 will be improved; this is indeed good news for providing a way to increase the safety and adaptability of air injection techniques [29-31]. It is a new, albeit important, research direction for improving the development efficiency in air foam flooding and controlling the oxygen concentration from the production well lower than the explosion limit through optimization with the injection parameters or changing the injection technology.

It should be noted that, although the test is carried out with air as a gas source, it does not mean that a specific effect barely exists in air foam flooding under LF vibration and it is the potential for foam flooding with other kinds of gases as the gas source.

CONCLUSIONS

Under LF vibration, air foam performance in dynamic displacement experiments was studied and it was found out that resistance factor under vibration increased by 1.5 times and the valid time was obviously extended. Optimal vibrating frequency and acceleration, resulting in the best foam plugging efficiency, were obtained.

With the analysis of bubble generation and burst rate, foaming speed increased, but foam break rate caused by internal expansion or by gas diffusion and transfer decreased under LF vibration. Dimensionless maximum resistance factor and dimensionless duration of maximum resistance factor revealed a stronger influence of maximum foaming speed than in situ foam stability.

LF vibration oil extraction technology could enhance the plugging efficiency, extend the valid time, and improve the safety and adaptability of air foam flooding to a certain extent.

ACKNOWLEDGEMENTS

We thank the National Natural Science Foundation of China (No. 51274229) for supporting our research on LF vibration oil extraction technology in low permeability reservoirs. We are also grateful to China University of Petroleum (East China) and Yanchang Petroleum Group Co. for their continuous support of this work.

REFERENCES

[1] Zhu Y. Y., Hou Q. F., and Weng R., "Recent Progress and Effect Analysis of Foam Flooding Field Tests in China," *SPE 165211-MS*, **2013**.

[2] Farajzadeh R., Andrianov A., and Krastev R., "Foam-oil Interaction in Porous Media: Implications for Foam Assisted Enhanced Oil Recovery," *Advances in Colloid and Interface Science*, **2012**, *183-184*, 1-13.

[3] Liao J. H., Sun L., and Jiang S. X., "Research and Application of Seismic Shooting Oil Recovery," *Drilling & Production Technology*, **2003**, *5*, 56-58.

[4] Miao X. M., Zheng L. G., and Chen G., "Discussion of Feasibility of Vibration Stimulation Technology by Low-frequency Wave in the Low Permeability Oilfield," *SINO-GLOBAL ENERGY*, **2010**, *12*, 57-59.

[5] Kurawle I., Kaul M., and Mahalle N., "Seismic EOR- The Optimization of Aging Water Flood Reservoirs," *SPE 123304*, **2009**.

[6] Ariadji T., "Effect of Vibration on Rock and Fluid Properties: on Seeking the Vibroseismic Technology Mechanisms," *SPE 93112*, **2005**.

[7] Sun R. Y. and Cheng G. X., "Effect of Artificial Vibration on Liquids Flow through Porous Media," *Journal of Hydrodynamics*, **2004**, *19*(4), 552-557.

[8] Ma J. G., Jin Y. H., and Zhou S. P., "Experiments on the Effects of Mechanical Vibration on Core Permeability," *Journal of Xi'an Shiyou University (Natural Science Edition)*, **1996**, *11*(5), 8-15.

[9] Teng L., Zhaoming Li., and Jing Li., "A Mathematical Model of Foam Flooding Based on Foam Microscopic Seepage Characteristics," *Chinese Journal of Computational Physics*, **2012**, *29*(4), 519-524.

[10] Solbakken J. S., Skauge A., and Aarra M. G., "Supercritical CO_2 Foam - The Importance of CO_2 Density on Foams Performance," *SPE 165296-MS*, **2013**.

[11] Sheng Q., Guo P., and Chen J., "Research on Visual Characteristics of Foam in Porous Media," *Special Oil and Gas Reservoirs*, **2012**, *19*(4), 122-125.

[12] Zhang Lei., "Low-frequency Resonant Wave N_2 Foam Flooding Effect of the Influence of Experimental Study," Xi'an: Xi'an Shi You University, **2012**.

[13] Ashoori E., Marchesin D., and Rossen W. R., "Roles of Transient and Local Equilibrium Foam Behavior in Porous Media: Traveling Wave Original Research Article," *Colloids and Surfaces A: Physicochemical and Engineering Aspects*, **2011**, *377*(1–3), 228-242.

[14] Gong L., Kyriakides S., and Triantafyllidis N., "On the Stability of Kelvin Cell Foams under Compressive Loads Original Research Article," *Journal of the Mechanics and Physics of Solids*, **2005**, *53*(4), 771-794.

[15] Petel O. E., Ouellet S., and Higgins A. J., "The Elastic–plastic Behavior of Foam under Shock Loading," *Shock Waves*, **2013**, *23*(1), 55-67.

[16] Yu H. M., Ren S. R., and Zuo J. L., "Experiment of Improved Oil Recovery by Air Foam Injection Low Temperature

Oxidation Process," *Journal of China University of Petroleum (Edition of Natural Science)*, **2009**, *33*(4), 94-98.

[17] Zhao J. S., Zhang M., and Li T. T., "Study on the Influential Factor of Foam Filtrational Resistance Factor Based on Uniform Design Method," *DRILLING & Production Technology*, **2009**, *32*(4), 74-76.

[18] Bouix R., Viot P., and Lataillade J. L., "Polypropylene Foam Behavior under Dynamic loadings: Strain Rate, Density and Microstructure Effects," *International Journal of Impact Engineering*, **2009**, *36*, 329–342.

[19] Liu J., Pu C. S., and Zheng L. M., "Research on Oil Recovery Technology of Surfactant Flooding Assisted by Low Frequency Resonance Wave," *Oil Drilling & Production Technology*, **2012**, *34*(5), 87-94.

[20] Berryman J. G., Thigpen L., and Chin R., "Bulk Elastic Wave Propagation in Partially Saturated Porous Solids," *Acoust. Soc. Amer*, **1988**, *84*, 360-373.

[21] Cai Y. Q., Li B. Z., and Xu C. J., "Analysis of Elastic Wave Propagation in Sandstone Saturated by Two Immiscible Fluids," *Chinese Journal of Rock Mechanics and Engineering*, **2006**, *25*(10), 2009-2016.

[22] Kam S. I., Nguyen Q. P., and Rossen W. R., "Dynamic Simulations with an Improved Model for Foam Generation," *SPE Journal*, **2007**, *3*, 35-48.

[23] Li S. Y., Li Z. M., and Li B. F., "Experimental Study and Application on Profile Control Using High Temperature Foam," *Journal of Petroleum Science and Engineering*, **2011**, *78*(3-4), 567-574.

[24] Li S. Y., Li Z. M., and Li B. F., "Experimental Study of Effect of Permeability on Foam Diversion," *Petroleum Sci. and Tech.*, **2012**, *30*(18), 1907-1919.

[25] Falls A. H., Gauglitz P. A., and Hirasaki G. J., "Development of a Mechanistic Foam Simulator: The Population Balance and Generation by Snap-off," *SPE 14961*, **1986**.

[26] Yu H. M., Ren S. R., and Zuo J. L., "A Mathematical Model and Numerical Simulation Method for Air-foam Flooding," *Acta Petrolei Sinica*, **2012**, *4*, 653-657.

[27] Exerowa D. and Kruglyakov P. M., "Foam and Foam Films: Theory, Experiment, Application," Elsevier, **1998**.

[28] Pinazo A. P., Infante M. R., and Frances E.I., "Relation of Foam Stability to Solution and Surface Properties of Gemini Cationic Surfactants Derived from Arginine," *Colloids and Surfaces A*, **2001**, *189*, 225-235.

[29] Alvarez J. M., Rivas H. J., and Rossen M. R., "Unified Model for Steady Sate Foam Behavior at High and Low Foam Qualities," *SPE 56825*, **1999**.

[30] Dong X., Liu H., and Sun P., "Air-foam-injection Process: An Improved-oil-recovery Technique for Waterflooded Light-oil Reservoirs," *SPE Reservoir Evaluation & Engineering*, **2012**, *15*(4): 436-444.

[31] Dong X. H., Liu H. Q., and Pang Z. X., "Study on Reaction Kinetics of Low Temperature Oxidation of Air Injection in Light Crude Oil," *Applied Chemical Industry*, **2012**, *41*, 1135-1139.

9

Catalytic and Non-catalytic Conversion of Methane to C_2 Hydrocarbons in a Low Temperature Plasma

Hamid Reza Bozorgzadeh

Research Institute of Petroleum Industry, West Blvd. of Azadi Sports Complex, Tehran, Iran

ABSTRACT

The direct conversion of methane to C_2 hydrocarbons, in a quartz tube reactor enforced by a DC corona discharge, was investigated at atmospheric pressure. The process was carried out in the presence of metal oxide catalysts of $Mn/W/SiO_2$, $Mn/W/SiO_2$ (tetraethyl orthosilicate, TEOS), and Mn/W/CNT (supported on carbon nanotubes). The total yield to C_2 hydrocarbons in the presence of metal oxide catalysts in plasma environment was in the order of $Mn/W/SiO_2$> $Mn/W/SiO_2$ / TEOS> Plasma only> Mn/W/CNT. The order changes to $Mn/W/SiO_2$>Mn/W/CNT>Plasma only> $Mn/W/SiO_2$/ TEOS, when the selectivity and yield of ethylene is considered. The highest yield to C_2 hydrocarbons was 15.8%, which was obtained by using $Mn/W/SiO_2$ in combination with gas discharge plasma without external heating; it was lower when the same feed composition was tested over this catalyst at 825 °C. The catalyst $Na_2WO_4/Mn_2O_3/SiO_2$ –b1, which produces the least carbon oxides, gives rise to the highest production of higher hydrocarbons and ethylene. Catalysts $Na_2WO_4/Mn_2O_3/SiO_2$ /TEOS-b2 and Na_2WO_4/Mn_2O_3/CNT-b3, due to their high selectivity toward carbon oxides, show low efficiency in producing more valuable hydrocarbons.

Keywords: Methane Conversion, Non-thermal Plasma, Catalysis, Metal Oxide, Hydrocarbons

INTRODUCTION

Although methane is an excellent raw material for the production of fuels and chemicals, its main use is still restricted to fuel for power generation for domestic and industrial use. In many respects, methane is an ideal fuel for these purposes because of its availability in most populated centres, its ease of purification to remove sulphur compounds, and the fact that among the hydrocarbons, it has the largest heat of combustion relative to the amount of CO_2 formed. On the other hand, methane is a greatly underutilized resource for chemicals and liquid fuels [1]. Methane can be converted directly to C_2 hydrocarbons by pyrolysis or thermal coupling. The reaction is highly endothermic and the heat must be supplied at high temperatures. Ethane, ethylene, benzene, and hydrogen are the main products. Excessive carbon formation can be avoided using short reaction times and low partial pressures of methane preferably by hydrogen dilution of the feed. More than 90% selectivity of C_2 hydrocarbons may be obtained from methane. High yields of ethane (>85%) are obtainable at extreme conditions of temperatures (> 2000 K) and short reaction times ($<10^{-2}$ s) [2, 3].

Methane can be converted to chemicals and

*Corresponding author
Hamid Reza Bozorgzadeh
Email: bozorgzadehhr@ripi.ir

fuels in two ways, either via synthesis gas or directly into C_2 hydrocarbons or methanol. Almost all commercial processes for large scale natural gas conversion involve synthesis gas. Steam reforming is the dominant process for the production of synthesis gas [4, 5] at high input energy as shown in Equation 1.

$$CH_4 + H_2O \leftrightarrow CO + 3H_2$$
$$\Delta H^o_{298K} = 206 \frac{kJ}{mol} \quad (1)$$

The synthesis gas ($CO + H_2$) is converted to higher hydrocarbons or fuels via Fischer-Tropsch (FT) synthesis. The catalysts employed are based mainly on cobalt and iron at pressures as high as 22 bars and temperatures of higher than 560 K [6, 7].

A direct method for the conversion of methane includes the partial oxidation of methane to methanol [8, 9] or the oxidative coupling of methane (OCM) to C_2 hydrocarbons. In the oxidative coupling reaction, CH_4 and O_2 react over a catalyst at temperatures higher than 700 °C to form C_2H_6, C_2H_4, and CO_2. Unfortunately, both the C_2H_6 and the C_2H_4 may be converted to CO_2, and the single-pass combined yield of C_2H_4 and C_2H_6 (C_2 products) is limited to about 25% [1]. Over better catalysts, including SrO/La_2O_3 [10] and $Mn/Na_2WO_4/SiO_2$ [11], a C_2 selectivity of about 80% can be achieved at a methane conversion of 20%. About half of the C_2 hydrocarbons obtained is C_2H_4 and the rest is C_2H_6, although the C_2H_4/C_2H_6 ratio can be enhanced by using a second catalyst. A high C_2 selectivity is almost always achieved under oxygen limiting conditions; thus the specific activity of the catalyst is not a factor. Because the overall reaction is exothermic, a zone within the catalyst bed may be 150-300 °C hotter than the external temperature [12]. Heat management in this process, therefore, is a serious engineering problem.

All the mentioned reactions need high intensive energy input to establish high temperature or pressure.

To reduce energy costs, it is possible to use electric power in a plasma process instead of thermal processing. In most of gas discharge plasma, free molecular radicals generated by excitation, dissociation, and the ionization of gas molecules are essential for the subsequent free radical reactions. The control of electron energy by the suitable design of discharge reactor and the mode of gas discharge may lead to favorable products. Non-thermal plasmas at ambient temperature and pressure are recently being investigated as the promising alternative to convert methane to C_2 hydrocarbons [13-22]. Corona discharge, in which the geometry of electrodes is quite different like a needle and plate electrodes, is widely used for CH_4 conversion on laboratory scale [14-17]. Dielectric barrier discharge (DBD) is the most common used method of atmospheric pressure non-thermal plasma to convert methane to mainly oxygenate and methanol [17-21]. Recently pulsed power supplies of nanosecond rise time of voltage and current have been employed to improve the energy efficiency of plasma techniques for methane conversion [22-24].

The relative advantages of plasma techniques over conventional routes for the production of more valued products from methane is that, in gas discharge plasma, the main electrical energy is transferred to energetic electrons and active radical species rather than simple gas heating. As electrons process minor mass compared to heavy ions, they gain much higher speed in an electric field, and thereby a higher temperature. The electrons in a gas discharge collide with gas molecules and impart the whole or a portion of their kinetic energy to exit the molecules to a higher energy states. Ultimately, the molecule of the gas dissociates to radicals and other species and gives rise to the synthesis of new products [13].

In this study, methane conversion using atmospheric corona discharge was carried out in two parallel circular discs and different OCM catalyst prepared by impregnation method were investigated. The effect of input power, feed flow rate, and catalysts on the methane and oxygen conversion, the selectivity of products, and the yield of C_2 hydrocarbons were also studied.

Experiments

The experimental set-up is shown in Figure 1. The corona discharge is maintained in the gap between a tungsten wire electrode with a thickness of 2 mm and a circular stainless steel disc with a diameter of 6.6 mm. The electrodes were housed in a quartz tube with an outer diameter (O.D.) of 7 mm and a wall thickness of 0.2 mm. A few holes were drilled in the circular electrode to allow the passage of gases from the plasma zone. The circular electrode in all the experiments was kept at the potential of zero, i.e. it was grounded; the wire electrode had different positive potentials. The wire electrode was perpendicular to the circular plate electrode at the centre at a distance of 10 mm from the disc electrode. The triggering voltage for maintaining gas discharge was supplied with a regulated DC power supply, which could supply voltages up to 15 KV. The measuring system was composed of high voltage probe (Tektronix P6015), an oscilloscope (Tektronix, TDS 2024B), a galvanometer, and a gas chromatograph (GC). The feed gases, namely methane (99.999%), argon (99.999%), and oxygen (99.99%), were introduced into the discharge system via three mass flow controllers (Brooks, model 5850 TR).

The feed gases were mixed before entering the reactor by a static mixture, which was a 20 cm long stainless steel tube with a diameter of 3 cm filled with glass spheres. The feed gases (methane, oxygen, and argon) and the products were analyzed by an online gas chromatograph (Agilent 6890N) equipped with a flame Ionization detector (FID) linked to alumina plot column and a thermal conductivity detector (TCD). This thermal detector was connected to HYSEP Q and N and molecular sieve 139 (MS-139) packed columns. The column temperature was programmable between 60 and 120 °C and the carrier gas was He and N_2. The flow rate of carrier gas was 25 ml/min. At higher input power applied to the reactor, there were some liquid products collected in condenser, which were kept away without analysis.

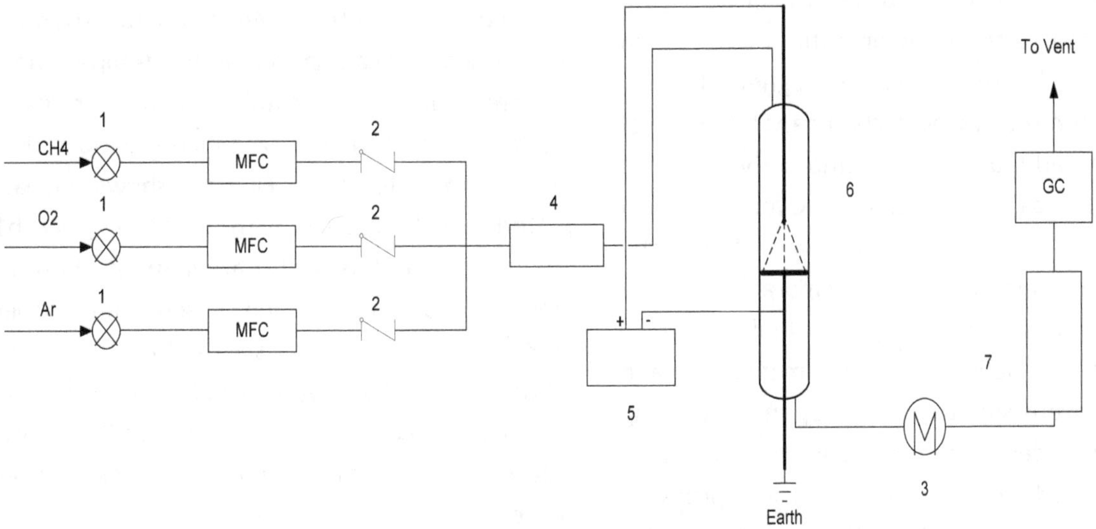

Figure 1: A schematic diagram of corona discharge reactor setup; (1) pressure regulator, (2) check valves, (3) condenser, (4) static mixer, (5) power supply, (6) reactor, and (7) silica gel.

Catalyst Preparation

The first catalyst was prepared by sol-gel procedures; first pure sodium silicate solution (Merck) with a pH of 11-11.5 and 98% sulfuric acid (Merck) were slowly added together. The resulting solution was stirred at 80 °C for 2 hours until a pH of 8±0.1 was achieved. When the precipitation was completed, the obtained gel was filtered out, washed with warm de-ionized water, and dried at 120 °C followed by calcinations in air at 550 °C for 4 hours. The active components Na_2WO_4 (5 wt.%) and Mn_2O_3 (2 wt.%) was loaded to SiO_2 support by an incipient wetness impregnations procedure. The salts $Mn(NO_3)_24H_2O$ (Merck) 98.5% and $Na_2WO_42H_2O$ (Merck) 99% were dissolved in water, the amorphous SiO_2 precursor was gradually added to the solution, and the sample was dried by a rotary evaporator at 80 °C. The sample was dried at 120 °C overnight and calcined up to 800 °C for 16 hours under a flow of air at a heating rate of 3 °C per minute. The resulting powder catalyst $Na_2WO_4/Mn_2O_3/SiO_2$ were mixed thoroughly, pressed into tablets, crushed, and sieved to 20-40 mesh before use.

To investigate the effect of different sources on the catalyst precursor, tetraethyl orthosilicate (TEOS) 99.999% (Sigma Aldrich) was used as the SiO_2 source. 0.11 mol of tetraethyl orthosilicate was diluted with a molar ratio of 1:1 with ethanol, while stirring with a magnetic heater; 1.000 gr of multi-walled carbon nanotubes (CNT) (RIPI) was added to the solution; the resulted suspension was well mixed at 80 °C for 2 hours. In next step, water was removed from the suspension with a rotary evaporator at low pressure followed by drying the sample at 120 °C overnight. The active metals, i.e. manganese and tungsten, were supported on SiO_2 support taking the same procedure as for catalyst b1. The water was removed from the bulk in a low pressure rotary evaporator, dried at 120 °C and the sample was calcined at 850 °C in a flow of oxygen for 16 hours consequently the CNT was removed from the sample in the form of CO_2. The solid obtained $Na_2WO_4/Mn_2O_3/SiO_2/$ (TEOS) was designated as catalyst b2.

Catalysts Characterization

Standard BET method was used to measure the specific surface areas of the prepared catalyst. The calculated values are shown in Table 1.

Table1: Results of specific surface area of different supported catalysts

Unit cell composition	Specific surface area (m^2/gr)
SiO_2	155
$Na_2WO_4/Mn_2O_3/$ SiO_2–b1	6
$Na_2WO_4/Mn_2O_3/$ SiO_2 (TEOS) –b2	2
Na_2WO_4/Mn_2O_3 /CNT– b3	110

The X-ray diffraction patterns of the catalysts were obtained by Philips PW 1840 diffractometer using Cu-Ka radiation. As it is clear from Figure 2, the XRD patterns of the SiO_2 as the support of the prepared catalysts investigated in this study show no specific line intensity, which indicates that it is an amorphous material. By supporting the active elements of Mn/W on the supports and subsequent calcinations at high temperatures, the resultant solid catalyst demonstrates a crystalline structure and exhibits the characteristic of α-crystoballite phase. Figure 3 shows the XRD patterns of the catalyst $Na_2WO_4/Mn_2O_3/SiO_2$-b1, when active metals Mn/W are supported on the amorphous SiO_2 support and subsequently calcined at 850 °C. The two catalysts $Na_2WO_4/Mn_2O_3/SiO_2$ (TEOS)-b2 and Na_2WO_4/Mn_2O_3/CNT-b3 also show XRD patterns closely resembling Figure 3, but with weaker characteristic line intensities, which implies possessing lower crystallinity.

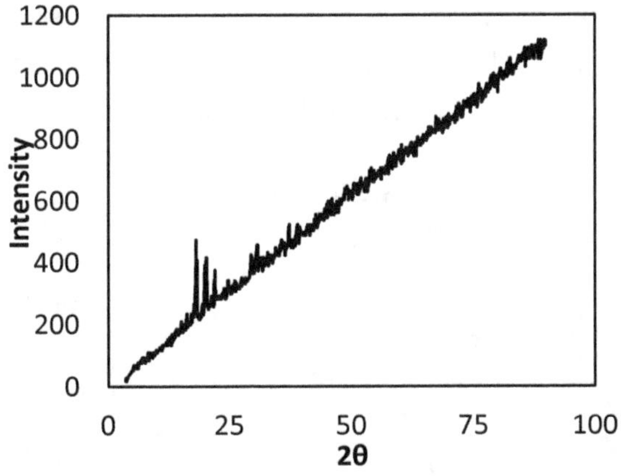

Figure 2: XRD patterns of SiO_2 prepared with sol- gel method

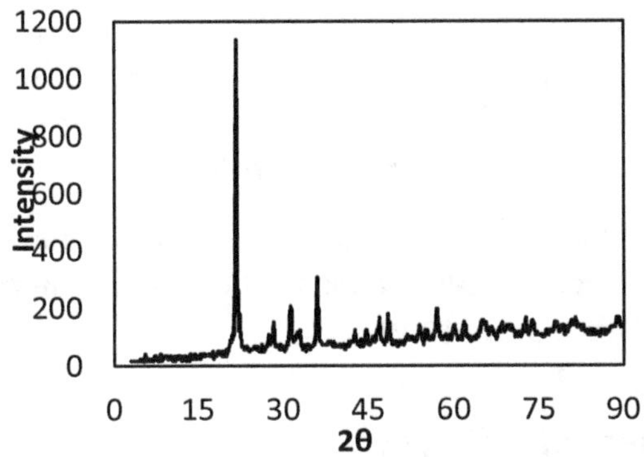

Figure 3: XRD patterns of $Mn_2O_3/Na_2WO_4/SiO_2$-b1

RESULTS AND DISCUSSION

The following relations were used to determine the conversion of reactant, selectivity and the yield of products.

CH_4 conversion = (moles of CH_4 consumed/moles of CH_4 introduced) × 100

O_2 conversion = (moles of O_2 consumed/moles of O_2 introduced) × 100

Selectivity of C_2H_6= 2×(moles of C_2H_6 formed/moles of CH_4 consumed) × 100

Selectivity of C_2H_4= 2×(moles of C_2H_4 formed/moles of CH_4 consumed) × 100

Selectivity of C_2H_2= 2×(moles of C_2H_2 formed/moles of CH_4 consumed) × 100

Yields of C_2 hydrocarbons = CH_4 conversion × ∑ (Selectivity's of C_2H_6, C_2H_4, C_2H_2)

Effect of Input Power

All the experiments were carried out at atmospheric pressure and no external heating was used to heat the reactor. The first set of experiments were conducted using a corona discharge plasma at various feed flow rates and a fixed methane to oxygen ratio of 4:1 (volume basis). To enhance the stability of gas discharge, the feed was diluted with argon (60% of volume).

The effect of the input power and the total feed flow rates on methane and oxygen conversion is illustrated in Figure 4. The conversions of methane and oxygen are increased at higher levels of input power to the reactor. Increasing the input power to the reactor gives rise to an enhancement in the intensity of the internal electric field developed across the discharge gap; consequently, the density of high energy electrons in the generated gas discharge plasma will also increase. The high energy electrons upon interactions with methane and oxygen molecules will increase the probability of breaking the bonds between hydrogen and carbon in methane molecules and hence the methane conversion increases at higher levels of input power supplied to the reactor. Figure 4 also confirms that increasing the total flow rates to the reactor reduces the conversion of both methane and oxygen. It can be inferred that the energy levels of the electrons produced in gas discharge plasma are not equal but obey the Boltzmann's energy distribution function; hence only a limited number of electrons which have sufficiently high energy can participate in the formation of active species on collision with methane or oxygen molecules. When the residence time of the reaction decreases, the probability that each methane or oxygen molecule will successfully interact with any of the sufficiently high energy electrons is decreased, and thereby reducing the rate of methane and oxygen conversion.

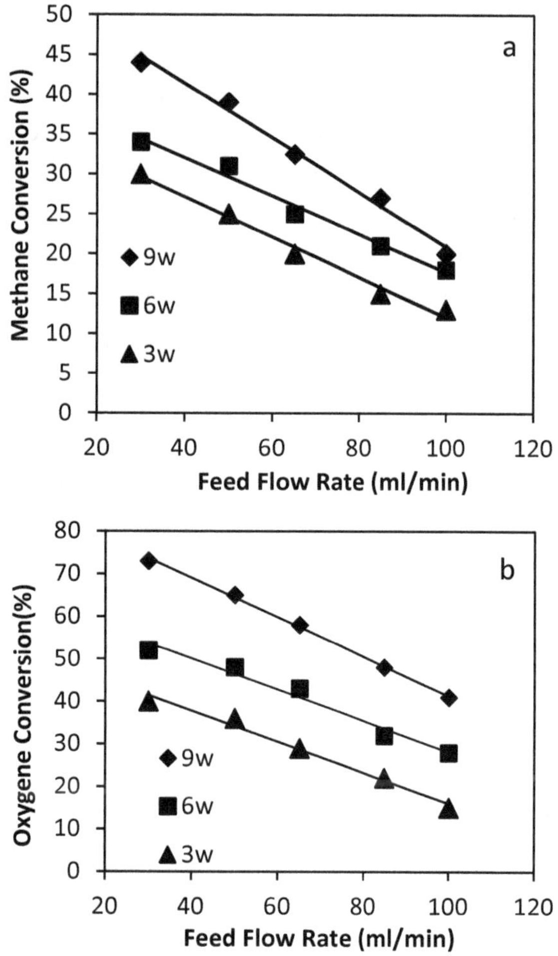

Figure 4: Effect of feed flow rates on conversion of methane and oxygen at various input powers to the reactor; CH_4/O_2 ratio: 4:1; a) methane conversion and b) oxygen conversion.

The distribution of product selectivities at various input powers to the reactor and a fixed flow rate of 65 ml/min is shown in Figure 5. The formation of acetylene below 400°C at an ordinary temperature in plasma environment is very interesting, as its formation does not takes place in OCM reactions, which usually occur at temperatures above 600 °C. Carbon dioxide was not detected in these series of experiments at the measured temperatures up to 222 °C, which is not favorable for the formation of this gas. At higher levels of input power to the reactor, a fraction of higher hydrocarbon molecules dissociates to their radical spices and hence the selectivities of C_2 products decreases, while the selectivity of carbon monoxide is increased. As the bond dissociation energy of carbon monoxide is relatively high [26], at a higher field intensity, it remains stable and therefore its selectivity is improved.

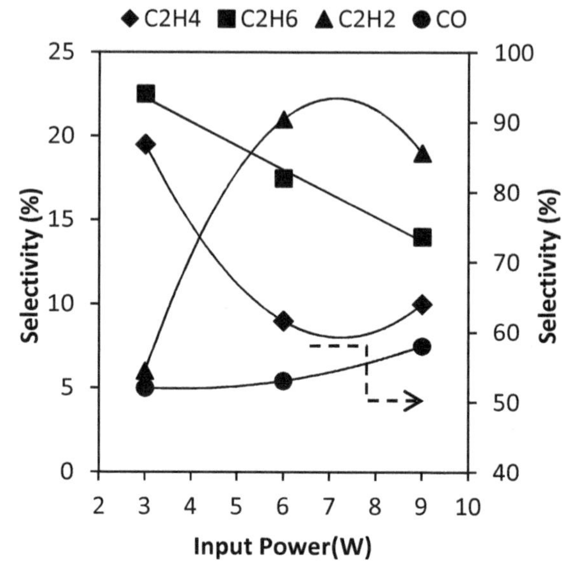

Figure 5: Selectivity of products vs. input power to the reactor; CH_4/O_2 ratio: 4:1, flow rate: 65 ml/min, and input power: 3 W.

Effect of Catalyst

To investigate the effect of the prepared catalysts on methane and oxygen conversion and product selectivities, 250 mg of the catalyst was introduced in the gap between the two electrodes. After each experiment, the tested catalyst was removed from the reactor, the reactor was cleaned, and finally fresh catalyst was loaded. To minimize the rate of coke formation, the new series of experiments was performed at an applied power of 3 Watt. Figure 6 demonstrates the methane and oxygen conversions at various feed flow rates to the reactor. The highest methane conversion (43%) belongs to the $Na_2WO_4/Mn_2O_3/SiO_2$ (TEOS) catalyst. The catalyst Na_2WO_4/Mn_2O_3/CNT-b3 in plasma environment has improved the CH_4 conversion from 30%, when only gas discharge was available to 37%. The catalyst $Na_2WO_4/Mn_2O_3/SiO_2$-b1, however, demonstrates a slight raise in methane conversion; on the other hand, oxygen converion showes a considrable enhancment.

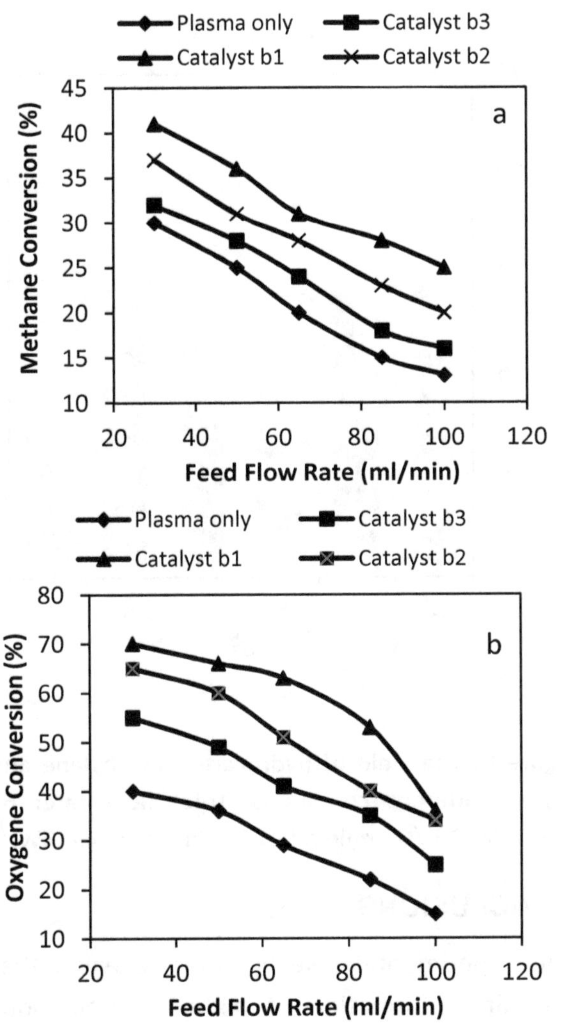

Figure 6: Effect of feed flow rates on conversion of feed constituents in corona discharge plasma and over the catalysts b1, b2, and b3. Total flow rate: 65 cm^3/min, CH_4/O_2 ratio: 4:1, input power: 3W; a) methane conversion and b) oxygen conversion.

Figure 6 also demonstrates that by increasing the feed flow rates to the reactor the total conversion for oxygen and methane generally decreases. The rate of oxygen conversion in the combined plasma-catalyst system shows a high conversion at shorter residence times. For instance, when the catalyst b2 is present in plasma environment, it shows oxygen conversion up to 70%, while it is 40% in the absence of this catalyst.

Figure 7 demonstrates the changes in the selectivity of products at a fixed input power of 3 Watt and a total feed flow rate to the reactor of 65 ml/min. The formation of acetylene at the measured temperature of up to 225 °C in a new set of experiments is an important phenomenon. All the three catalysts have increased the selectivity of ethylene. The stable gas discharge effect was observed over $Na_2WO_4/Mn_2O_3/SiO_2$-(b1) and the process of coke formation was considerably delayed. For this reason, this catalyst is preferred to the other tested catalysts as it reduces the charges of maintenance and need for the frequent overhaul of large scale industrial units. The catalyst b1 also reduces the rate of formation of carbon monoxide and at the same time the selectivity to C_2 hydrocarbons is the highest compared to the other tested catalysts and gas discharge alone. The selectivity of the products over $Na_2WO_4/Mn_2O_3/SiO_2$ (TEOS)-b2 is interesting; this catalyst exhibits a high selectivity of carbon dioxide formations in OCM reactions in the temperature range of 770-850 °C. In the same way, this catalyst also shows a high carbon dioxide selectivity of about 15% in the plasma environment, which happens at a much lower temperature than its OCM reactions. Therefore, the overall effect of high CO_x selectivity is a reduction in the selectivity of C_2 hydrocarbons. This catalyst also shows the highest acetylene selectivity.

In the presence of Na_2WO_4/Mn_2O_3/CNT- b3 catalyst, ethylene selectivity improves in plasma environment.

Figure 8 also compares the total yield of C_2H_4 and C_2 hydrocarbons in discharge plasma environment, plasma combined catalysts at the measured temperature of 300 °C, and OCM reaction over b1 and b2 catalysts at 825 °C. The b3 catalyst was not tested in high-temperature OCM reaction, as it would eventually destroy and oxidize to carbon dioxide. Moreover, Figure 8 shows that the yield of ethylene is increased from 3.9%, when only plasma is the forcing agent, to 8.4%, when catalyst b1 is also available in discharge zone.

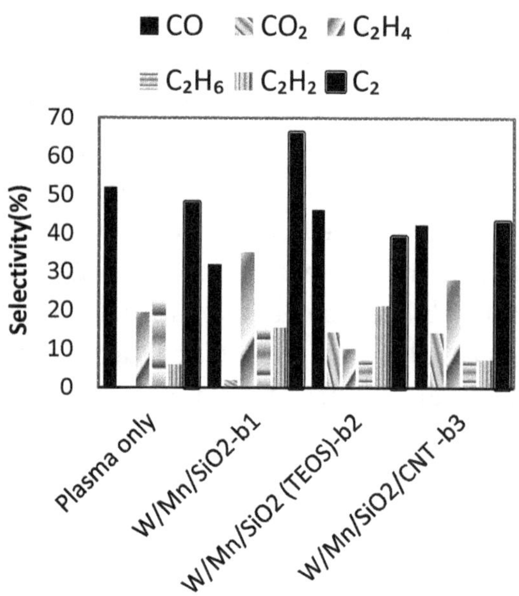

Figure 7: Product selectivity of corona discharge plasma over the catalysts b1, b2, and b3; total flow rate: 65 cm^3/min; CH_4/O_2 mole ratio: 4:1; input power: 3W.

The other two catalysts also show significance gain in the yield of C_2 hydrocarbons in plasma environment.

Although the methane conversion in the combined plasma-catalytic conversion over this catalyst was higher than their corresponding OCM reactions, due to their high selectivity to carbon monoxide, their yield to higher hydrocarbons is still lower than their corresponding OCM reactions. According to Figure 8, the thermal catalytic methane conversions over catalyst b1 are 21.8% and 14% for C_2 hydrocarbons and ethylene yield respectively, while they are respectively reduced to 15.8% and 8.4% for the same catalyst in plasma environment.

An important point about W/Mn catalysts is that the catalysts show no sign of catalytic activity for methane conversion at temperatures below 650 °C. The enhancement of C_2 productions when they are placed in plasma environment indicates that there is a strong interaction between the free electrons of plasma and the active site of the three tested catalysts.

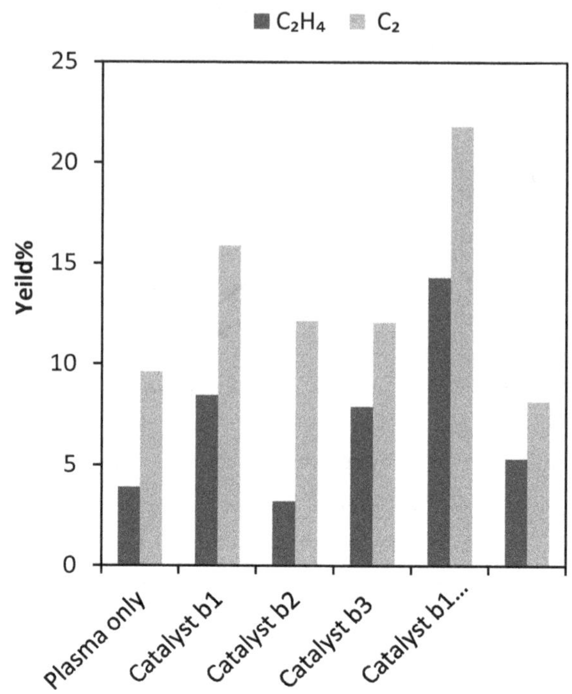

Figure 8: The yield of hydrocarbons (ethylene and C_2) in different conditions; total flow rate: 65 cm^3/min; CH_4/O_2 mole ratio: 4:1; input power: 3W.

CONCLUSIONS

The experimental investigations revealed that the direct formation of higher hydrocarbons from CH_4 could be demonstrated in a catalytic and non-catalytic DC corona discharge. The two catalysts, namely Na2WO4/Mn_2O_3/SiO_2 –b1 and $Na_2WO_4/Mn_2O_3/SiO_2$/TEOS-b2, which were mainly used in OCM reactions, and the catalyst Na_2WO_4/Mn_2O_3/CNT-b3 enhanced the CH_4 conversion combined with gas discharge compared with gas discharge in the absence of catalysts. It was shown that the catalysts tested in the presence of gas discharge plasma improved the total yield to C_2 hydrocarbons. The catalyst $Na_2WO_4/Mn_2O_3/SiO_2$ –b1, which produced the least carbon oxides, gave rise to the highest productions of higher hydrocarbons and ethylene. The catalysts Na_2WO_4/Mn_2O_3/ SiO_2/TEOS-b2 and Na_2WO_4/Mn_2O_3/CNT-b3, due to their high selectivity toward carbon oxides, did not yield more valuable hydrocarbons. The catalyst b3 was different from the other catalysts in the sense that the support of the catalyst b3 was multi-walled carbon nanotubes, which was electrically conductive. This catalyst conducted the corona discharge current and the overall discharge effect was the formation of a stable corona induced plasma and the inhibition of the formation of cokes. The ratio of yield of C_2H_4 to C_2H_6 for this catalyst was 3.7, which was the highest value compared to 2.3 and 1.3 for b1 and b2 catalysts respectively.

ACKNOWLEDGEMENTS

The financial support by the Research and Technology Directorate of NIOC and the provision of the experimental facilities by Research Institute of Petroleum Industry (RIPI) are gratefully acknowledged.

REFFERENCE

[1] Lunsford J. H., "Catalytic Conversion of Methane to More Useful Chemicals and Fuels," *Catal. Today*, **2000**, *63*, 165-173.

[2] Holeman A., Olsvik O., and Rosktad O. A., "Pyrolysis of Natural Gas: Chemistry and Process Concepts," *Fuel Processing Technology*, **1995**, *42*, 249-260.

[3] Serrano D. P., Botas J. A., and Guil-Lopeza R., "H_2 Production from Methane Pyrolysis over Commercial Carbon Catalysts Kinetic and Deactivation Study," *International Journal of Hydrogen Energy*, **2009**, *34*, 44-88.

[4] Holeman A., "Direct Conversion of Methane to Fuels and Chemicals," *Catal. Today*, **2009**, *142*, 2-8.

[5] Rostrup-Nielsen J., "Production of Synthesis Gas," *Catalysis Today*, **1993**, *18*, 305–324.

[6] Schulz H., "Short History and Present Trends of Fischer-Tropsch Synthesis," *Applied Catalysis A*, **1999**, *186*, 3-12.

[7] Nakhaei Pour A., Housaindokht M. A., Tayyari S F., Zarkesh J. et al., "Deactivation Studies of Fischer-Tropsch Synthesis on Nanostructured Iron Catalyst," *Journal of Molecular Catalysis A*, **2010**, *330*,112-120.

[8] Qijian Z., Dehua H., and Qiming Z., "Direct Conversion of Methane into Oxygenates Role of Catalyst Location," *Natural Gas Chemistry*, **2008**, *17*, 24-28.

[9] Otsuka K. and Wang Y., "Direct Conversion of Methane into Oxygenates," *Applied Catalysis A*, **2001**, *222*, 145-161.

[10] Mimoun H., Robine A., Bonnaudet S., and Cameron C. J., "Recent Advances in the Oxtoattve Coupling of Methane," *Appl. Catal. A: Gen.*, **1990**, *58*, 269-280.

[11] Fang X. and Li S., "Role of Sodium in the Oxidative Coupling of Methane over NA-W-MN/SIO_2 Catalysts," **1992**, 427.

[12] Schweer D., Mleczko L., and Baerns M., "OCM in a Fixed-bed Reactor: Limits and Perspectives," *Catal. Today*, **1994**, *21*, 357-369.

[13] Liu C., Marafee A., Hill B., Xu G., et al., "Oxidative Coupling of Methane with AC and DC Corona Discharges," *Ind. Eng. Chem. Res.*, **1996**, *35*, 3295-3301.

[14] Liu C., Mallinson R., and Lobban L., "Hydrocarbons in a Corona Discharge over Metal Oxide Catalysts with OH Groups," *Applied Catalysis*, **1997**, *164*, 21-33.

[15] Liu C., Mallinson R., and Lobban L., "Non-oxidative Methane Conversion to Acetylene over Zeolite in a Low Temperature Plasma," *Journal of Catalysis*, **1998**, *179*, 326-334.

[16] Marafee A., Liu C., Xu G., Mallinson R. et al., "Methane Conversion to Higher Hydrocarbons in a Corona Discharge over Metal Oxide Catalysts with OH Groups," *Ind. Eng. Chem. Res.*, **1997**, *36*, 632.

[17] Zhou L. M., Xue B., Kogelschatz U., and Eliasson B., "Partial Oxidation of Methane to Methanol with Oxygen or Air in a Non-equilibrium Discharge Plasma," *Plasma Chemistry and Plasma Processing*, **1998**, *18*, 375-393.

[18] Liu Changjun., Mallinson R., and Lobban L., "Comparative Investigations on Plasma Catalytic Methane Conversion to Higher Hydrocarbons over Zeolites," *Applied Catalysis*, **1999**, *178*, 17-27.

[19] Eliasson B., Jun Liu C., and Kogelschatz U., "Effect of Catalysts in Carbon Dioxide Reforming," *Ind. Eng. Chem. Res.*, **2000**, *39*, 1221-1227.

[20] Aghamir F. M., Jalili A. H., Esfarayeni M. H., Khodagholi M .A. et al., "Conversion of Methane to Methanol in an AC Dielectric Barrier Discharge," *Plasma Sources Science and Technology*, **2004**, *13*, 707-711.

[21] Haji Tarverdi M. S., Mortazavi Y., Khodadadi A. A., and Mohajerzadeh SH., "Conversion of Methane to C_{2+} Hydrocarbons in a Dielectric Barrier Discharge Reactor," *Chemistry and Chemical Engineering*, **2005**, *24*(4), 63–71.

[22] Yao S. L., Suzuki E., Meng N., and Nakayama A., "Effect of Voltage Waveform on Dielectric Barrier Discharge," *Energy and Fuels*, **2001**, *15*, 1300-1303.

[23] Shuiliang Y., Suzuki E., and Nakayama A., "A Novel Pulsed Plasma for Chemical Conversion," *Thin Solid Films*, **2001**, *390*, 165-169.

[24] Ghorbanzadeh A. M. and Matin N. S., "Methane Conversion to Hydrogen and Higher Hydrocarbons by Double Pulsed Glow Discharge," *Plasma Chemistry and Plasma Processing*, **2005**, *25*, 19-29.

[25] Hongfei L., Yanying W., Jürgen C., and Haihui W., "Oxidative Coupling of Methane with High C_2 Yield by using Chlorinated Perovskite $Ba_{0.5}Sr_{0.5}Fe_{0.2}Co_{0.8}O_{3-\delta}$ as Catalyst and N_2O as Oxidant," *Chem. Cat. Chem.*, **2010**, *2*, 1539-1542.

[26] Darwent B., "Bond Dissociation Energies in Simple Molecules," *U.S. National Bureau of Standards*, **1970**, 23.

10

The Gene Cloning, Overexpression, Purification, and Characterization of Dibenzothiophene Monooxygenase and Desulfinase from Gordonia Alkanivorans Ripi90a

Mahmoud Shavandi[1], Majid Soheili[1]*, Shekufeh Zareian[2], Neda Akbari[2], and Khosro Khajeh[2]

[1] Microbiology and Biotechnology Group, Environment and Biotechnology Research Division, Research Institute of Petroleum Industry, Tehran, Iran

[2] Biochemistry Department, Faculty of Biological Science, Tarbiat Modares University, Tehran, Iran

ABSTRACT

The biodesulfurization (BDS) of sulfur compounds in fossil fuels is a process to reduce sulfur dioxide emissions that cause environmental pollution. *Gordonia alkanivorans* RIPI90A is able to convert dibenzothiophene, an organic sulfur compound in petroleum, to 2-hydroxybiphenyl (2-HBP) in 4S pathway. In this study, (DBT), DszA and DszB, DBT sulfone monooxygenase, and desulfinase were respectively isolated from *G. alkanivorans* RIPI90A. PCR amplified fragments were obtained by using primers designed based on known sequences from *G. alkanivorans* RIPI90A. They are identified as *dszA* and *dszB* and have shown high similarity compared to *Rhodococcus erythropolis* IGTS8 (88% for *dszA* and 88% for *dszB*). Subsequently, *dszA* and *dszB* genes were expressed under the control of T7 promoter in *Escherichia coli*. The recombinant proteins were purified to achieve homogeneity using Ni-agarose column chromatography. The molecular mass of the purified DszA and DszB were determined to be 51.9 and 39.2 kDa respectively by using SDS-polyacrylamide gel electrophoresis. DszA showed a K_m of 0.14±0.005 mM and a maximal velocity of 0.004±0.0004 mM/min. DszB showed a wide substrate range in a way that all aromatic sulfonates compounds acted as its substrate; as it seemed the active site was suitable for the sulfonated aromatic rings. The K_m and V_{max} values of DszB were calculated to be 1.81±0.02 mM and 6.55 ± 0.005 μM/min respectively using 4-Amino-3-hydroxy-naphthalene-sulfonic acid as a substrate.

Keywords: DszA, DszB, Purification, *Gordonia alkanivorans* RIPI90A, Biodesulfurization

INTRODUCTION

The presence of sulfur in petroleum products is a serious concern for the petroleum industry. The combustion of sulfur-containing fuels leads to the atmospheric emission of sulfuroxides (SO_x), which is the major cause of acid rain and other health and environmental problems [1, 2]. Hydrodesulphurization (HDS) is the conventional method for removing non-volatile organic sulfur compounds from petroleum fractions. However, HDS is operated under high temperature and pressure and is not effective for the removal of DBT and alkylated DBT's, which constitute the main organosulfur compounds in fossil fuels [3, 4]. Microorganisms are considered to be

***Corresponding author**

Majid Soheili

Email: soheilym@ripi.ir

potential catalysts for removing sulfur from the refractory organic sulfur compounds, and biodesulfurization (BDS) has attracted much attention as an alternative and complementary method. The desulfurization pathway of DBT has extensively been studied with *Rhodococcus erythropolis* strain IGTS8 [1]. The pathway, also called 4S pathway, proceeds via two cytoplasmic monooxygenases (DszC and DszA) supported by flavin reductase (DszD) and a desulfinase (DszB). DBT monooxygenase (DszC) catalyzes the sequential oxidation of DBT to DBT-sulfoxide and DBT-sulfone. DBT-sulfone converts to 2-(2′-hydroxyphenyl) benzene sulfinate (HPBS) through the oxidative cleavage of one of the DBT-sulfone carbon-sulfur bonds by second enzyme, DBT-5, 5′-dioxide monooxygenase (DszA). The final enzyme in the pathway, HPBS desulfinase, catalyzes the desulfination of HPBS to produce 2-hydroxybiphenyl (2-HBP) and sulfite. The oxygen atom incorporated at each step of this desulfurization pathway is derived from molecular oxygen [5,6]. DszC and DszA do not use NADH directly, but use $FMNH_2$ from a FMN:NADPH oxidoreductase (DszD) [7]. DszD couples the oxidation of NADH with substrate oxygenation by DszA and DszC. The DszA enzyme is widely studied and has been purified from different strains, namely *R. erythropolis* IGTS8, *R. erythropolis* D-1, and *Paenibacillus* sp. Strain A11-2 [8,9]. It is a homodimer with a subunit molecular weight of 50 kDa and a reaction rate of 5-10-fold higher than DszC [7]. A great deal of genetic studies have been done on the *dsz* operon but only a few reports are available on the purification and characterization of Dsz enzymes. A DBT-desulfurizing bacterial strain has been previously isolated and identified as *Gordonia alkanivorans* RIPI90A based on the 16S rDNA gene sequence (GeneBank accession no. DQ321498) [10]. In this study, we have isolated, cloned, and sequenced the DBT monooxygenase and desulfinase genes from *G. alkanivorans* RIPI90A. Subsequently, they were expressed in *E. coli* as a heterologous host. The enzymes were purified to near homogeneity and the kinetic parameters were determined.

MATERIALS AND METHODS

Chemicals

DBT sulfone was obtained from Aldrich (Milwaukee, USA). As stock solutions for culture media and reaction mixtures, DBT sulfone was dissolved in acetonitrile. Restriction endonucleases, *Pfu* DNA polymerases, T4 DNA ligase, and isopropyl-beta-D-thiogalacto pyranoside (IPTG) were obtained from MBI Fermentas. NADH, NADPH, and FMN were from Fluka, as were all sulfonated substrates.

Bacterial Strains, Plasmids, and Culture Conditions

Gordonia alkanivorans RIPI90A was used as the source of genomic DNA for cloning of *dszA* and *dszB* genes. *E. coli* DH5α was used as a host for general cloning and *E. coli* BL21 cells were used as a host for gene expression. Plasmid pUC19 (Fermentas) was used as a cloning vector and plasmid pET21a (Novagen) as an expression vector. *G. alkanivorans* RIPI90A was cultivated on Luria-Bertani (LB) medium at 30 °C.

General Recombinant DNA Techniques

Recombinant DNA techniques were performed as described by Sambrook and Russell [11] or according to manufacturer's protocols. Total DNA of *Gordonia alkanivorans* RIPI90A was prepared as previously described [12]. Plasmid DNA was prepared using high pure plasmid isolation kit (Roche Applied Sciences, Mannheim, Germany). Plasmids were introduced into *E. coli* by electroporation under standard conditions using Gene pulser X-cell electroporator (BioRad). All sequencing was carried out in both directions to eliminate reading errors by Applied Biosystems 3130XL Genetic Analyzer instrument (USA).

Gene Cloning

To isolate the entire *dszA* and *dszB* open reading frames from *G. alkanivorans* RIPI90A genomic DNA, primers annealed to the 5′ and 3′ ends of the coding regions of *dszA* and *dszB* gene were designed based on the known sequences of *G. alkanivorans* RIPI90A dsz operon (GenBank accession number EU364831). PCR amplification was carried out using programmable termal-cycler (eppendorf). The reaction mixture contained 50 μl of 20 mM Tris-HC1 (with a pH of 8.8 at 25 °C), 10 mM of $(NH_4)_2SO_4$, 10 of mM KC1, 0.1% (v/v) Triton X-100, 0.1 mg/ml of BSA, 2 mM of $MgSO_4$, 0.2 mM of (each) deoxynucleotide triphosphate, 50 μmol of each primer, 0.5 μg of RIPI90A genomic DNA, and 1.25 U of *pfu* DNA polymerase. The reaction was performed at 95 °C for 1 min, then at 61 °C for 30 s, and finally at 72 °C for 3 min for a total of 35 cycles within a 5-min initial denaturation step at 95 °C in the first cycle and a 5-min extension at 72 °C in the last cycle. The forward and reverse primers designed for *dszA* and *dszB* genes have the sequences of AF: 5′-AAGGATCCATGGCTCAACGGCGACAAC-3′ and AR: 5′-AAAAAGCTTGTGTGTCGAGGATGCC-GGTAT-3′, BF: 5′-AAAGAATTCATGGCAGGCCGCC-TCAG-3′ and BR: 5′-AAAAAGCTTTCGGTGACGG-TTGAGGCTG-3′ respectively. These primers were designed with *Bam*HI and *Hind*III recognition sites for *dszA* gene and *Eco*RI and *Hind*III recognition sites for *dszB* gene for the convenient in-frame cloning of the genes into the multiple cloning site of the pET21a vector. This placed the *dszA* and *dszB* genes under the control of strong bacteriophage T7 transcription and translation signals. The cloned genes were fully sequenced to confirm the fidelity of PCR amplification and the accuracy of reading frame.

Bacterial Growth and Induction of DszA and DszB Protein

Recombinant *E. coli* cells were cultivated at 37 °C in an LB medium containing 50 μg of ampicillin per milliliter. For the induction of the *dszA* and *dszB* genes under the control of the T7 promoter, 1 mM isopropyl-beta-D-thiogalacto-pyranoside (IPTG) was added to the LB medium and the recombinant cells were cultivated at 25 °C and 200 rpm for 6 hrs. The estimation of inducted DszA and DszB protein was carried out by SDS-PAGE according to the protocol described by Laemmli [13].

Protein Purification

The cells were harvested by centrifugation at 5000 rpm for 20 min, resuspended in a lysis buffer (50 mM of Tris with a pH of 8.0, 300 mM of NaCl, 2 mM of imidazole, 1 mM PMSF, 0.2 mg/mL of lysozyme, 2% Triton X-100, 0.02% NaN_3, and 10% glycerol), and finally sonicated on ice for 10 min for 10 times with an interval of 1 min. The total lysate was centrifuged at 12000 rpm and 4 °C for 20 min. Because of a high level of expression, large amounts of the proteins would appear in the pellet, which were not supernatant. Thus some additional processes were employed for purification; The pellet was dissolved in a denaturation buffer containing 50 mM Tris with a pH of 8.0, 300 mM of NaCl, 8M of urea, 30 mM of imidazole, and 10% glycerol (5 mL/1gr wet weight), and was then centrifuged at 12000 rpm for 20 min. The supernatant was shaken with 2 mL of Ni-agarose resin for 2 hours at room temperature. As the produced proteins were appended to a His-domain at the N-terminus, binding to the Ni-agarose was facilitated. A 1×10 cm column was packed with this protein-tagged resin. In the next step, the column was washed with eight volume of washing buffer containing a gradient of urea from 4 to 0 M (4 M of urea, 50 mM of Tris with a pH of 8.0, 300 mM of NaCl, 30 mM of imidazole, and 10% glycerol). The His-tagged proteins were eluted with four volumes of elution buffer containing 50 mM of Tris with a pH of 8.0, 300 mM of NaCl, 250 mM of imidazole. The eluted protein was assayed for purity by SDS-PAGE and used for enzyme characterization.

Analytical Technique

For DszA activity measurement, a reaction mixture (with a total volume of 250 μL) containing 100 mM of potassium phosphate buffer (pH=7.0), 10 μM of FMN, 6 mM of NADH, 0.25 mM of DBT sulfone (in acetonitrile), 2 unit/mL of FMN-NAD(P)H oxidoreductase, and DszA was prepared. After 30 min of incubation at 35 °C and 175 rpm, the reaction was terminated by the addition of 12 μL of 12 N HC1 followed by the extraction with 200 μL of ethyl acetate. After centrifugation at 13000 rpm for 20 min, the supernatant was injected into the HPLC for measuring the amount of residual DBT sulfone in the organic layer. 1 unit of enzyme activity was defined as the amount which produces 1 μmol of HBPS per minute.

The HPLC was carried out with a C18-S5ODS1 column (250 mm×4.6 mm). A spectrophotometric UV detection was used (λ= 280 nm). The mobile phase contained methanol and 20 mM of KH_2PO_4, which was adjusted to have a pH of 2.5 by using H_3PO_4. The ratios of potassium phosphate solution to the methanol were 40:60 for the measurement of DszA activity (the retention times of 2'-hydroxybiphenyl-2-sulfinate (HBPSi) and DBT sulfone were 3.4 and 4.5 min respectively) at a flow rate of 1 mL/min. DszB activity was assayed with Ellman's reagent (5, 5′-dithio bis (2-nitrobenzoic acid)) which produced a bright yellow color due to the production of disulfide and TNB (2-nitro-5-thiobenzoic acid) upon reaction with sulfite produced by enzyme reaction with sulfonated substrates [14, 15].

TNB was the colored species produced in this reaction and had the molar extinction coefficient of 14150 $M^{-1}cm^{-1}$ at 412 nm. The specific substrate of DszB in the desulfurization pathway is HBPSi which converts to HBP and sulfite upon reaction. In this study, we measured the kinetic parameters of the enzyme by measuring the amount of sulfite produced from different sulfonated substrates such as 4-amino-3-hydroxy-naphthalene-sulfonic acid, 3-(N-morpholino) propanesulfonic acid (MOPS), 4-(2-hydroxy-ethyl)-1-piperazine-ethanesulfonic acid (HEPES), and 2-morpholino-ethane-sulfonic-acid. The reaction mixture (with a total volume of 200 μl) containing a 100 mM potassium phosphate buffer (with a pH of 7.0), 500 μM of sulfonated substrate, and an appropriate amount of DszB was prepared. The reaction was initiated by the addition of enzyme to the reaction mixture. After 30 min, at 35 °C and 175 rpm, the reaction was terminated by the addition of Ellman's reagent; colorimetric reaction was allowed to continue at room temperature for 2 min. Absorbance was measured at 412 nm. One unit of enzyme activity was defined as the amount which produced 1 μmol of sulfite per minute.

RESULTS AND DISCUSSION

Cloning and Sequencing of *dszA* and *dszB* Genes

Polymerase chain reaction (PCR) experiments were performed with oligonucleotides based on the known nucleotide sequences of *G. alkanivorans* RIPI90A dsz operon (GenBank accession number EU364831) and the total DNA from RIPI90A as template. The fragments containing *dszA* and *dszB* genes with lengths of 1.4 kb and 1.1 kb respectively were PCR-amplified and cloned into pET system and the nucleotide sequence of these products were determined. The overall nucleotide sequence of *dszA* and *dszB* genes of *G. alkanivorans* showed 83.1% homology to the corresponding genes of *Rhodococcus* sp. IGTS8 and 83.2% to those of *G. nitida*.

Expression and Purification of DszA and DszB Proteins

DszA and DszB were overproduced in *E. coli* BL21 (DE3) cells by using IPTG as an inducer. Cells were harvested after 6 hr induction, and then the supernatant and debris of sonicated cells were collected. A small amount of soluble enzyme detected in the supernatant. SDS-PAGE analysis revealed that overproduced proteins were present in the insoluble fraction indicating that the major part of the recombinant protein accumulated as inclusion bodies (Figure 1a and 1b). To decrease the extent of protein aggregation, the expression conditions such as induction time, temperature, and IPTG concentration were optimized and a very little activity of either enzyme was detected in the cell-free extracts (data are not presented). However, the main part of the recombinant protein still remained in the insoluble fraction. To refold the inactive proteins, a gradient of urea in the elution buffer was applied during the purification process. The purification of His6-N-tagged DszA and DszB proteins in a single-step affinity chromatography using Ni-agarose column resulted in almost homogenous protein preparations. The effecttiveness of purification process was evaluated by SDS–PAGE analysis. The results of the electrophoretic analysis of DszA and DszB overexpression and purification are shown in Figure 1a and 1b respectively. The molecular mass of the DszA and DszB was 51.9 kDa and 39.2 kDa respectively and no extra bands were observed

Determination of Kinetic Parameters of DszA and DszB

Kinetic parameters of the purified refolded DszA were obtained at room temperature using different concentrations of DBT-sulfone for the depiction of standard curve. The Michaelis constant was 0.14±0.005 mM and the maximal velocity was 0.004±0.0004 mM/min. The Eadie-Hofstee plot is shown in Figure 2. Enzyme activity for DszB was studied in the presence of different sulfonated substrates (Table 1) and finally kinetic parameters, namely K_m and V_{max}, were determined using different concentrations of 4-amino-3-hydroxy-naphtalen-sulfonic acid. Each value represented the mean of at least three independent sets of experiments. The Eadie-Hofstee plot is shown in Figure 3. The Michaelis constant was 1.81±0.02 mM and maximal velocity was 6.55±0.005 µM/min.

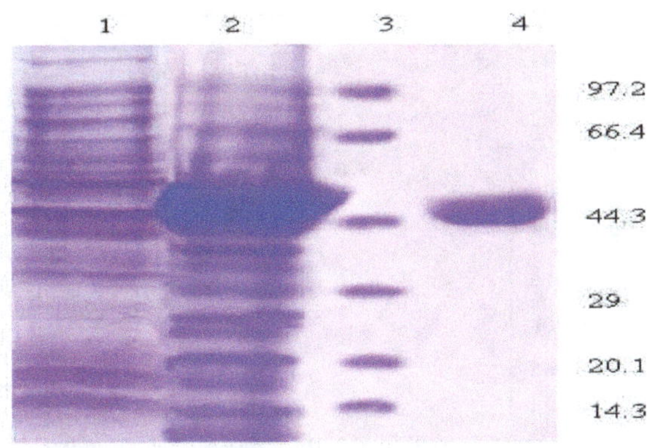

Figure 1a: SDS-PAGE analysis (12.5% polyacrylamide gel) of overexpression and purification. Proteins were detected by coomassie brilliant blue; lane 1: crude extract before induction with IPTG, lane 2: crude extract after induction, lane 3: molecular size markers (kDa) 97.2, 66.4, 44.3, 29.0, 20.1, 14.3, lane 4: purified enzyme (DszA).

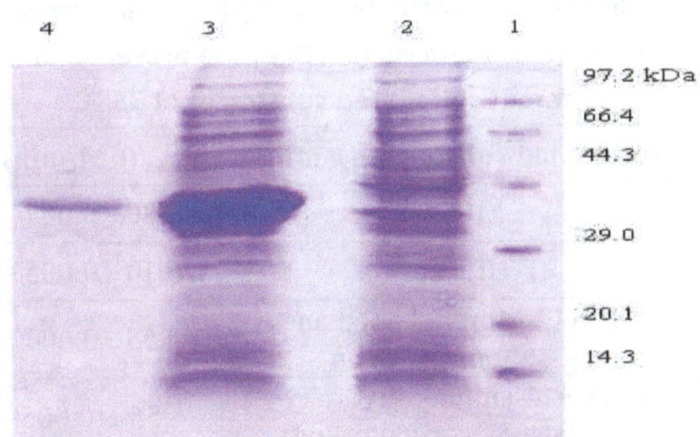

Figure 1b: lane 1: molecular size markers, lane 2: crude extract before induction with IPTG, lane 3: crude extract after induction, lane 4: purified enzyme (DszB).

With reference to activity, the refolding yield of enzymes is often estimated. Our results indicated that the refolding yield under refolding conditions was 37% and 40% for DszA and DszB, respectively (Table 2).

In 2007, Alves et al. isolated new desulfurization genes from *G. alkanivorans* strain 1B and successfully expressed into *E. coli* in the presence of inorganic sulfur. Therefore, it was strongly suggested that alternative culture media containing cheaper carbon sources, like those obtained from lignocellulosic or agro-industrial by-products should further be used [16].

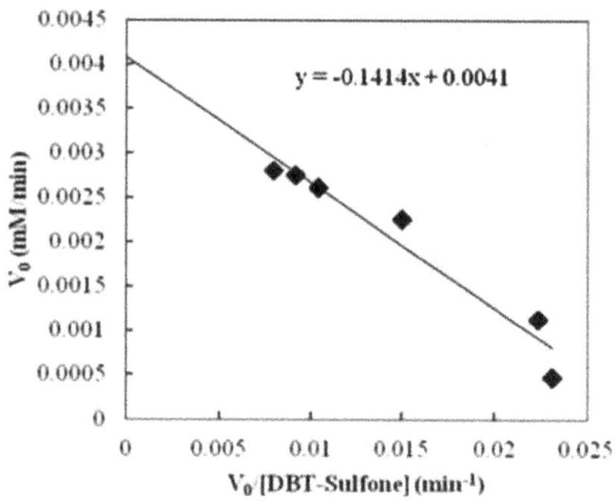

Figure 2: The Michaelis–Menten constant (K_m) and maximum velocity (V_{max}) of DszA calculated by Eadie-Hofstee plot using DBT-sulfone as the substrate

Table 1: Enzyme activity of 2′-hydroxybiphenyl-2-sulfinate desulfinase (DszB) in the presence of different sulfonated substrates at 35 °C

Sulfonated substartes (1 mM)	V_{max} (μM/min)
MOPS	5.80± 0.004
HEPES	5.10± 0.005
MES (2-Morpholinoethane sulfonic acid)	2.55± 0.004
AHNS (4-amino-3-hydroxy-naphtalene-1-sulfonic acid)	6.55 ± 0.005

Table 2: Comparison of kinetic parameters between soluble and refolded forms of DszA and DszB

Protein	K_m (mM)	K_{cat} (V_{max}/μM)	Refolding Yield
Soluble DszA	0.14	35	100%
Refolded DszA	0.14	13	37%
Soluble DszB	1.83	3	100%
Refolded DszB	1.81	1.1	40%

G. alkanivorans RIPI90A has been previously isolated and reported as an efficient DBT-desulfurizing bacterium with unique advantages for the BDS process; however, as other natural desulfurizing bacteria, the expression of Dsz enzymes are strongly repressed in the presence of sulfate; nevertheless, we have successfully expressed the methionine and cysteine [10, 17]. Therefore, it is impossible to obtain high desulfurization activity in the presence of sulfur compounds in culture media. Accordingly, we decided to clone the dsz genes under the control of a sulfate non-repressible promoter.

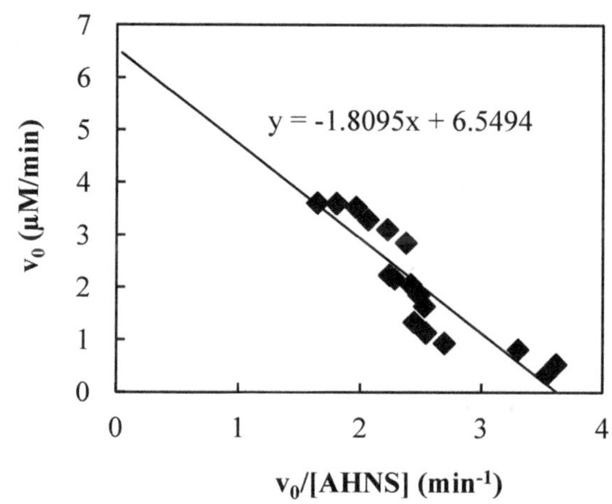

Figure 3: The Michaelis–Menten constant (K_m) and maximum velocity (v_{max}) of DszB calculated by Eadie-Hofstee plot using 4-amino-3-hydroxy-naphthalene-1-sulfonic acid as the substrate

In the previous study, we successfully expressed the *G. alkanivorans* RIPI90A dsz operon under the control of an inducible lac promoter. In this study, the T7 promoter was selected for the strong expression of recombinant *dszA* and *dszB* genes separately to obtain a high level of active enzyme. Subsequently, recombinant construct was made to place the *dsz* genes under the control of T7 promoter and transferred to *E. coli* BL21 as heterologous host, but the expression of the Dsz enzymes in recombinant strain resulted in a great amount of insoluble enzymes even after lowering the induction temperature and IPTG concentration. Therefore, the development of a strategy for the efficient refolding of the proteins from inclusion bodies has been considered as an important issue in the recovery of soluble active proteins for functional studies. In the present work, the refolding of insoluble proteins with the gradient of urea during the purification process resulted in a marked increase in the amount of active enzymes. The efficiency of refolding was obtained, with reference to activity, to be 37% and 40% for DszA and DszB respectively. In order to obtain the maximum refolding yields, the refolding conditions were optimized by statistical methods.

In this study, the activity of DszB was studied in presence of different sulfonated substrates, and according to the results, AHNS was selected to determine kinetic parameters and DBT-sulfone for DszA. DszA, the sulfone monooxygenase, is the second enzyme in microbial DBT desulfurization metabolism and has a great significance in DBT desulfurization metabolism, because it cleaves one of the two C-S bonds of the DBT molecule. DszB is a unique enzyme capable of cleaving carbon–sulfur of 2-hydroxybiphenyl-2-sulfinate (HBPSi) in the final step in the microbial DBT desulfurization and produce 2-HBP [18].

CONCLUSIONS

In conclusion, we reported the expression, purification and refolding of DszA and DszB of Dsz enzymes of *Gordonia alkanivorans* RIPI90A. It was shown that *E. coli*-derived enzymes could effectively act on sulfonated substrates in vitro. The effective strategies for DszA and DszB production could be suggestive for a further study on the structure and function of the enzymes.

NOMENCLATURE

2-HBP : 2-hydroxybiphenyl

BDS : Biodesulfurization

DBT : Dibenzothiphene

DszA : DBT-5, 5′-dioxide monooxygenase

dszA : Gene encoding DszA

DszB : 2-(2′-hydroxyphenyl) benzene sulfinate desulfinase

dszB : Gene encoding DszB

DszC : DBT monooxygenase

dszC : Gene encoding DszC

FMN : Flavin mononucleotide

HDS : Hydrodesulphurization

NADH : Nicotinamide adenine dinucleotide

NADPH : Nicotinamide adenine dinucleotide phosphate

REFERENCES

[1] Konishi, J., Ishii, Y., Onaka, T., Maruhashi, K., "Purification and Characterization of the Monooxygenase Catalyzing Sulfur-atom Specific Oxidation of Dibenzothiophene and Benzothiophene from the Thermophilic Bacterium Paenibacillus sp. strain A11–2," *App. Microbiol. Biotechnol.*, **2002**, *60*, 128-33.

[2] Guerinik, K., Al-Mutawah, Q., "Isolation and Characterization of Oil-desulphurizing Bacteria," *World. J. Microbiol. Biotechnol.*, **2003**, *19*, 941-945.

[3] Finnerty, W. R., Robinson, M., "Microbial Desulfurization of Fossil Fuels: A Review," Biotechnol" *Bioengin. Symp.*, **1986**, *16*, 205-221.

[4] Gallardo, M. E., Ferrández, A., De Lorenzo, V., García, J. L., Díaz, E., "Designing Recombinant Pseudomonas Strains to Enhance Biodesulfurization," *J. Bacteriol.*, **1997**, *179*, 7156-7160.

[5] Konishi, J., Ishii, Y., Onaka, T., Ohta, T., Suzuki, M., Maruhashi, K., "Purification and Characterization of Dibenzothiophene Sulfone Monooxygenase and FMN-Dependent NADH Oxidoreductase from the Thermophilic Bacterium Paenibacillus sp. strain A11-2.," *J. Bioscience.* Bioengin., **2000**, *90*, 607-13.

[6] Izumi, Y., Ohshiro, T., "Purification and Characterization of Enzymes Involved in Desulfurization of Dibezothiophene in Fossil Fuels," *J. Mol. Catal B: Enzym.*, **2001**, *11*, 1061-1064.

[7] Gray, K. A., Pogrebinsky, O. S., Mrachko, G. T., Xi, L., Monticello, D. J., Squires, C. H., "Molecular Mechanisms of Biocatalytic Desulfurization of Fossil Fuels," Nature" *Biotechnol.*, **1996**, *14*, 1705-1709.

[8] Matsubara, T., Ohshiro, T., Nishina, Y., Izumi, Y., "Purification, Characterization, and Overexpression of Flavin Reductase Involved in Dibenzothiophene Desulfurization by Rhodococcus erythropolis D-1," *Appl. Environ. Microbiol.*, **2001**, *67*, 1179-1184.

[9] Hasan, N., Nester E. W., "Purification and Characterization of NADPH-dependent Flavin Reductase: An Enzyme Required for the Activation of Chorismate Synthase in Bacillus Subtilis," *J. Biol. Chem.*, **1978**, *14*, 4987-4992.

[10] Mohebali, G., Ball, A. S., Rasekh, B., Kaytash, A., "Biodesulfurization Potential of a Newly Isolated Bacterium, Gordonia Alkanivorans RIPI90A," *Enzyme. Microb. Technol.*, **2007**, *40*, 578-584.

[11] Sambrook, J., Russell, D. W., *Molecular Cloning: A Laboratory Manual*, 3rd Ed, Cold Spring Harbor Laboratory Press, New York, **2001**.

[12] Banh, Q., Arenskötter, M., Steinbüchel, A., "Establishment of Tn5096-based Transposon Mutagenesis in Gordonia Polyiso-Prenivorans," *Appl. Environ. Microbiol.*, **2005**, *71*, 5077-84.

[13] Laemmli, U. K., "Cleavage of Structural Proteins during the Assembly of the Head of Bacteriophage T4," *Nature*, **1970**, *227*, 680-685.

[14] Johnston J. B., Murray, K., Cain, R. B., "Microbial Metabolism of Aryl Sulphonates a Re-assessment of Colorimetric Methods for the Determination of Sulphite and their Use in Measuring Desulphonation of Aryl and Alkylbenzene Sulphonates," *Antonie Van Leeuwenhoek*, **1975**, *41*, 493-511.

[15] Riddles, P. W., Blakeley, R. L., Zerner, B., Reassessment of Ellman's Reagent," *Methods. Enzym.*, **1983**, *91*, 49-60.

[16] Alves, L., Melo, M., Mendonca, D., Simoes, F., Matos, J., Tenreiro, R., Girio, F. M., "Sequencing, Cloning and Expression of the dsz Genes Required for Dibenzothiophene Sulfone Desulfurization from Gordonia Alkanivorans Strain 1B," *Enzym. Microb. Technol.*, **2007**, *40*, 1598-1603.

[17] Shavandi, M., Sadeghizadeh, M., Zomorodipour, A., Khajeh, KH., "Biodesulfurization of Dibenzothiophene by Recombinant Gordonia Alkanivorans RIPI90A," *Biores. Technol.*, **2009**, *100*, 475-479.

[18] Nakayama, N., Matsubara, T., Ohshiro, T., Moroto, Y., Kawata, Y., Koizumi, K., Hirakawa, Y., Suzuki, M., Maruhashi, K., Izumi, Y., Kurane, K., "A Novel Enzyme, 2'-hydroxybiphenyl-2-sulfinate Desulfinase (DszB), from a Dibenzothiophene-desulfurizing Bacterium Rhodococcus Erythropolis KA2-5-1: Gene Overexpression and Enzyme Characterization," *Biochim. Biophys. Acta.*, **2002**, *1598*, 122-130.

11

Development of a Pellet Scale Model for Trickle Bed Reactor Using CFD Techniques

Mansoor Bazmi[1], Sayed Hasan Hashemabadi[2]*, and Mahmood Bayat[1]

[1] Research Institute of Petroleum Industry (RIPI)

[2] Computational Fluid Dynamics (CFD) Research Laboratory, School of Chemical Engineering, Iran University of Science and Technology, Tehran, 16846, Iran

ABSTRACT

In this study, a pellet scale model was developed for trickle bed reactor utilizing CFD techniques. Drag coefficients were calculated numerically at different velocities and bulk porosities in the case of single phase flow through the dry bed. The simulation results were then compared with the prediction of Kozeny-Carman (K-C) equation. The results indicated that drag coefficients calculated from the square arrangement of cylindrical particles in the pellet scale model were in good agreement with Kozeny-Carman equation prediction; however, triangular arrangement had over prediction comparing with Kozeny-Carman equation. Afterward, the pellet scale model with square arrangement was developed for fully pre-wetted particles which were enveloped with a liquid film. The VOF model was used to investigate the boundary condition on the surface of the static liquid layer. The results of CFD simulation in various gas velocities indicated that, at the adjacent of the particle walls, the no-slip boundary condition was acceptable. This pellet scale model was also in good agreement with the Kozeny-Carman equation.

Keywords: Trickle Bed Reactor, Cylindrical Particle, Drag Coefficient, Pressure Drop, Particles Arrangement, CFD Simulation.

INTRODUCTION

Trickle bed reactors (TBR's) are three phase reactors in which gas and liquid phases simultaneously flow downward to a fixed bed of catalyst particles; they are widely used in various petroleum industries such as hydro-cracking, hydro-desulphurization, hydridenitro-genation, catalytic de-waxing as well as chemical industries including reactive amination, liquid phase oxidation, and wastewater treatment.

Due to the importance of pressure drop and liquid hold up predictions in the design and control of TBR's, a number of studies have been carried out to understand and quantify the hydrodynamics of TBR's [1-5]. One of the most important terms in the hydrodynamic modeling of a TBR is the drag force as the inter-phase momentum exchange term [6], which is usually calculated with Ergun equation as the sum of viscous and inertial terms. However, various models have been proposed to calculate the drag force such as relative permeability [7, 8], slit [9, 10] and two-fluid models [11, 12] which

*Corresponding author

Sayed Hasan Hashemabadi

Email: hashemabadi@iust.ac.ir

are commonly used in the simulation of trickle bed reactors. Furthermore, during the last decade, the pressure drop and flow drag coefficient through the TBR's were estimated by using CFD techniques [13-15]. Various parameters such as the size and shape of particles, operating conditions, fluid properties, and bed tortuosity affect the drag force applied on the particles and on the hydrodynamics of TBR. Unfortunately, although during the last decade many authors have investigated the hydrodynamic behavior of trickle bed rectors, the studies on particle shape and bed tortuosity are scarce and the previous calculations are usually based on particles with a spherical shape. Similarly, for other particle shapes, the equivalent diameter is considered in the literature as a shape effect. Lakoda et al. [7] studied the effect of particle shape (spherical, cylindrical, and rings) on pressure drop values and found that particle shape affected the relative permeability constant. Also, Nemec and Levec [16] investigated the effect of particle shape on the pressure drop of a single phase flow through the TBR's. Based on their experimental results, the original Ergun constants can accurately predict the pressure drop for only single phase flow over spherical particles, but these constants should be modified for other particle shapes. In another study, Nemec and Levec [8] concluded that particle shape did not have any effect on the liquid hold up values. Trivizadakis et al. [17] studied the effects of spherical and cylindrical particles on the hydrodynamics of a two-phase flow through a TBR and it was found out that particle shape had a significant effect on pressure drop and liquid holdup. On the other hand, the pressure drop and liquid hold up values of a packed bed loaded with cylindrical particles were greater than those of a bed with spherical particles of the same size.

The tortuosity of the bed is another parameter that characterizes the packing structure. It depends on the factors such as particle arrangement, media homogeneity, and channel shape. Investigations on tortuosity are usually restricted to beds with spherical particles. Lanfreg et al. [18] presented a theoretical model for particles which was indicated with a parking structure factor the tortuosity of a fixed bed randomly packed with identical particles. They found that tortuosity depended on bed voidage and the sphericity of the particles.

Hellstrom and Lundstrom [19] developed a micromechanical model with the square arrangement of cylindrical particles to study the real mechanisms of flow in a porous media using a CFD approach. Also, Lopes et al. [20] used triangular arrangement of spherical particles to model TBR's using a CFD approach. In their computational domain, the catalyst particles did not contact each other. To facilitate the grid generation, they considered 3% of the sphere diameter as a distance between two particles. Gunjal et al. [21] investigated the fluid flow through a bed packed with different arrangements of sphere particles (simple cubical, 1-D rhombohedra, 3-D rhombohedra, and face-centered cubical geometries) using a CFD approach.

In this study, a pellet scale model is proposed to calculate the drag force of dry and wet particles with cylindrical shape. The effect of particle arrangement on the drag force is also investigated based on the slit model and CFD techniques. The proposed model can be used for a regime in which liquid film flows over the particles.

MATHEMATICAL FORMULATION

Modeling Single Gas Phase Flow Using CFD Techniques

Continuity and momentum equations for gas flow through the bed are expressed according to Equations 1 and 2 as given below:

$$\frac{d\rho_g}{dt}+\nabla.\left(\rho_g u_g\right)=0 \tag{1}$$

$$\frac{d\left(\rho_g y_g\right)}{dt}+\nabla.\left(\rho_g u_g^2\right)=\nabla P_g+ \nabla\left(\mu_g \nabla u_g\right)+\rho_g g \tag{2}$$

These two above equations are solved simultaneously to calculate velocity and pressure field. According to Newtonian viscosity law, for incompressible fluid, the stress tensor is related to velocity gradient tensor as follows [22]:

$$\tau=-\mu\left(\nabla u+\nabla u^t\right) \tag{3}$$

$$C_D=\frac{\tau_w}{\frac{1}{2}\rho v^2} \tag{4}$$

On the other hand, the drag coefficient can be obtained from the two-fluid model.

$$F_D=\frac{E_1(1-\varepsilon)^2\mu_g U_g}{\varepsilon^3 d_p^2}+\frac{E_2(1-\varepsilon)\rho_g U_g^2}{\varepsilon^3 d_p} \tag{5}$$

$$C_D=\frac{F_D}{\frac{1}{2}\rho A U^2} \tag{6}$$

Thus, these two results are compared with each other. The two-fluid model [11, 12] is based on this assumption that trickle flow can be idealized as a stratified flow in which the gas and liquid phases are completely separated by a smooth interface. Therefore, according to Kozeny-Carman (K-C) equation, each fluid behaves like as continuous phase. Accordingly, the particle-gas drag force per unit volume of space and drag coefficient can be calculated respectively as follows [1]:(5) and (6) where, E1 and E2 are Ergun constants that might be changed due to trickle bed parameters (such as particle size and shape, operating conditions, fluids properties, particles arrangement, and bed tortuosity). The original Ergun constants E_1 and E_2 were proposed to be 150 and 1.75 respectively; but, Macdonald et al. [23] recommended 180 and 1.8 for these constants. Similarly, in this study the same values are considered to implement in the proposed model.

In case of a fully wetted bed with the static liquid hold-up, due to the presence of a thin layer of liquid on the surface of particles and consequently the reduction of porosity, the drag force values are higher than when the bed is loaded with dry particles. In this case, the K-C equation is modified as follows [1]:

$$F_D=\alpha\left[E_1\frac{(1-\alpha\varepsilon)^2\mu_g U_g}{\alpha^3\varepsilon^3 d_p^2}\left(\frac{1-\varepsilon}{1-\alpha\varepsilon}\right)^{\frac{2}{3}}+ E_2\frac{(1-\alpha\varepsilon)\rho_g U_g^2}{\alpha^3\varepsilon^3 d_p}\left(\frac{1-\varepsilon}{1-\alpha\varepsilon}\right)^{\frac{1}{3}}\right] \tag{7}$$

Using CFD techniques, the governing equations on flow field (Equations 1 and 2) can be solved in the proposed geometries. Therefore, the drag force imposed on particles can be calculated from Equation 4 if their arrangement is specified. The appropriate representative geometry for the bed can be determined by a suitable pellet scale model. Afterward, the obtained results are compared with those of K-C equations (5-7).

Pellet Scale Modeling Using CFD

In this study, the slit model [9, 10] was used to interpret the pellet scale model for a bed loaded with cylindrical particles. By obtaining the void volume versus the surface area of the bed particles, the slit between two particles through which fluid passes can be calculated (Figure 1). The parameters, w (slit half void thickness) and S (slit half wall thickness) can be determined from Equations 8 and 9. Therefore, the distance between two cylindrical particles is 2w [9].

$$S = \frac{1-\varepsilon}{a} \tag{8}$$

$$\frac{w}{S} = \frac{\varepsilon}{1-\varepsilon} \tag{9}$$

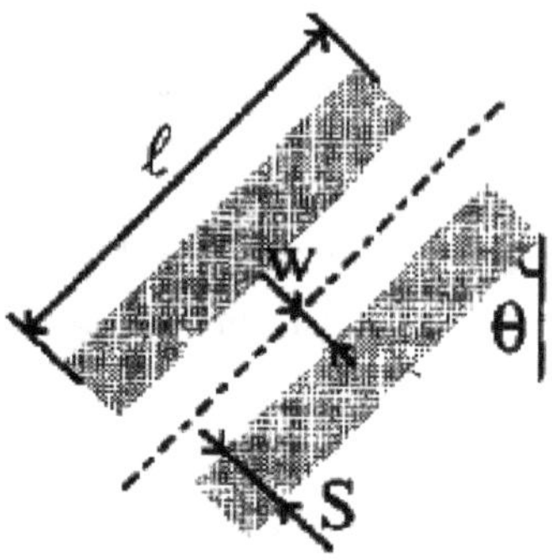

Figure 1: Geometrical configuration of the slit model

Figure 2 shows the proposed pellet scale models with square and triangular configurations.

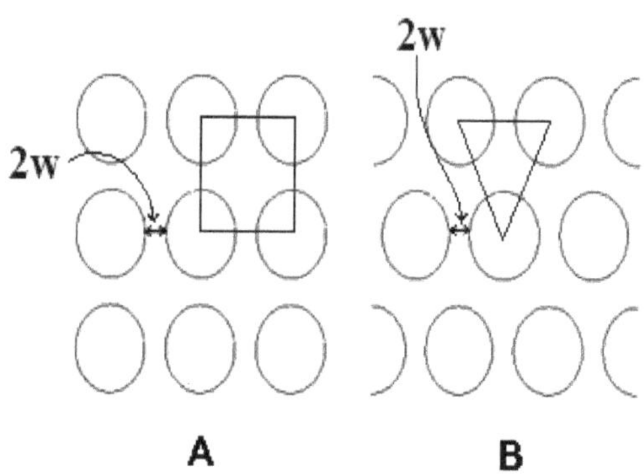

Figure 2: Proposed pellet scale model for two different cylindrical particle arrangements: A) Square; B) Triangular

Since cylindrical particles are symmetric along the axis, the proposed model is considered in a 2D space, in which the effects of cylinder ends are assumed to be negligible. Although this assumption, which is in agreement with other studies of flow over circular cylinders [17, 24], is more applicable to flow regimes with low Reynolds numbers, the 2D and 3D simulations are preformed and the obtained results are then compared. It is concluded that the errors of measurement in a 2D model compared to a 3D one are less than 4% (Figure3).Therefore, due to the reduction in the number of calculations, the 2D model is selected and cylindrical particle arrangement is considered in a horizontal state due to higher mechanical stability.

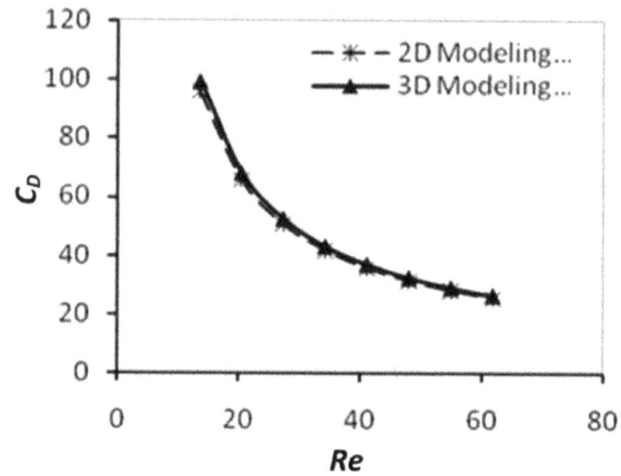

Figure 3: Comparison of pellet scale model drag coefficient for 2D and 3D CFD simulations

The particle arrangement in the pellet scale model should exhibit the tortuosity of the bed. Thus, two particle arrangements (square and triangular) are considered. A simplified symmetrical model for both square and triangular configurations is also presented in Figure 4.

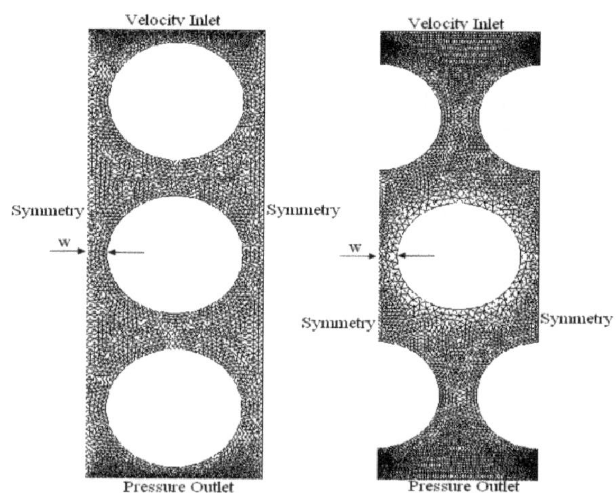

Figure 4: Comparison of pellet scale model drag coefficient for 2D and 3D CFD simulations

As it is shown in Figure 4, triangular mesh was used for the flow domain. Primarily, calculations were carried out to achieve grid-independent results, which were done with different mesh sizes. The continuity and momentum equations (Equations 1 and 2) were discretized by finite volume formulation [22]. UPWIND method was used for the discretization of convective terms and a segregated solver with implicit linearization was applied to solve the momentum equations. Also, the SIMPLE algorithm was used for pressure velocity coupling [22].

The results show that the number of catalyst particles does not affect the drag force values. Therefore, the pellet scale model containing three particles with the drag force imposed on the middle one was considered. The proposed method can also be used to study the wetting effect of catalyst particle on the drag coefficient values. According to the slit model, when catalyst particles are wetted, it is surrounded by a layer of liquid with a thickness of $\delta_s$which is calculated as follows:

$$\delta_s = \frac{h_{ls}}{\eta_s a_s} \quad (10)$$

where, η_s is the static wetting efficiency that equals to 1 for full wetted particles. In this case, the static liquid holdup h_{ls} is calculated as follows:

$$h_{ls} = \left(20 + 0.9E\ddot{O}\right)^{-1}; \; E\ddot{O} = \frac{\rho_l g d^2 \varepsilon^3}{\sigma_l \left(1-\varepsilon\right)^2} \quad (11)$$

Figure 5 shows three catalyst particles surrounded by a layer of static liquid.

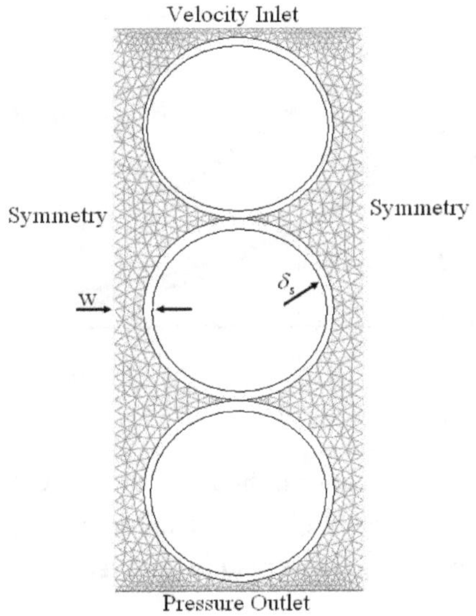

Figure 5: Square arrangements of wetted particles

Evaluation of Slip Condition on Wetted Particles

Studying the boundary conditions on the surface of the wetted catalyst particles is of great importance. Accordingly, the boundary condition of the particle which is surrounded by a layer of liquid has been investigated in the present work using Volume of Fluid (VOF) model. This model was used to evaluate the slip boundary condition at gas-liquid interface on the wetted particles. The VOF model enables the computation of multiphase flows in which gas-liquid interfaces are clearly identified. In the VOF model, the variables such as pressure and velocity are shared by both phases and correspond to volume-averaged values. The volume-averaged conservation equations for mass and momentum describing the two immiscible incompressible fluid hydrodynamics are respectively given by Equations 12 and 13:

$$\frac{\partial \rho}{\partial t} + \nabla.\left(\rho u\right) = 0 \quad (12)$$

$$\frac{\partial}{\partial t}\left(\rho u_j\right) + \frac{\partial}{\partial x_i}\left(\rho u_i u_j\right) = -\frac{\partial P}{\partial x_j} + \frac{\partial}{\partial x_i}\mu\left(\frac{\partial u_i}{\partial x_j} + \frac{\partial u_j}{\partial x_i}\right) + \rho g_i + f_\sigma \quad (13)$$

Where, mixture density and dynamic viscosity are determined by volume fraction averaging equation 14.

$$\begin{aligned} \rho &= \alpha\rho_g + \left(1-\alpha\right)\rho_l \\ \mu &= \alpha\mu_g + (1-\alpha)\mu_l \end{aligned} \quad (14)$$

In VOF model, tracking the interface between gas and liquid phases is accomplished by the solution of a continuity equation for the gas volume fraction (α) as follows (Equation 15):

$$\frac{\partial \alpha}{\partial t} + u.\nabla\alpha = 0 \quad (15)$$

The liquid volume fraction is equal to 1-α. When computational cell is completely filled with the gas phase, α is equal to one and in reverse it is zero; consequently, the interface can be found in the cells with $0<\alpha<1$.

RESULTS AND DISCUSSION

Validation of Pellet Scale Model Predictions

At first, in order to study the predictability of the model developed by CFD techniques, the drag coefficient for a single cylindrical particle under different air flow rates conditions were calculated and the results were compared with obtained values from Haider and Levenspiel equation [25].

As shown in Figure 6 the average error of the model based on CFD techniques is less than 6%. This good agreement indicates that our CFD model can accurately predict the drag coefficients for cylindrical particles.

Dry Particles Drag Coefficient

According to the pellet scale model, a single-phase flow of air (constant density) was passed through a bed loaded with the cylindrical particles (d=1.5 mm and L=3.11 mm).

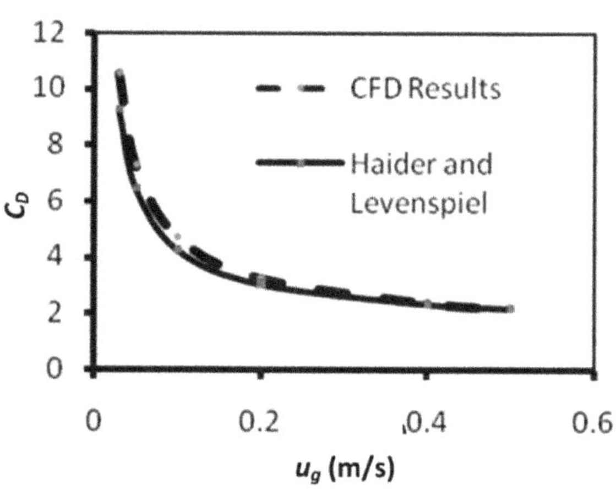

Figure 6: Drag coefficients of cylindrical particle obtained from CFD model and Haider and Levenspiel equation [25]

The drag coefficient was calculated using the related equations and CFD techniques, and the results were then compared to those of the K-C equation. Figure 7 shows the influence of gas phase velocity on drag coefficient values in a medium with constant porosity (i.e. 0.4) for both triangular and square configurations alongside K-C equation results.

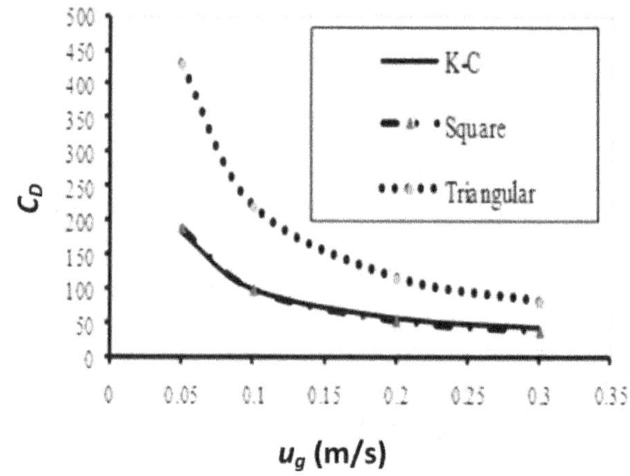

Figure 7: Influence of gas velocity on CFD drag coefficient of triangular and square configurations alongside K-C equation (Porosity= 0.4)

As it can be seen in Figure 7, good agreement between the square configuration and K-C equation (Equations 5 and 6) results is observed, while the predicted triangular values are higher. It must be noted that the two configurations, namely square and triangular, have the same porosity but different tortuosity. Figure 8 shows drag coefficients versus porosity for both triangular and square configurations alongside K-C equation results (gas velocity= 0.1 m.s^{-1}).

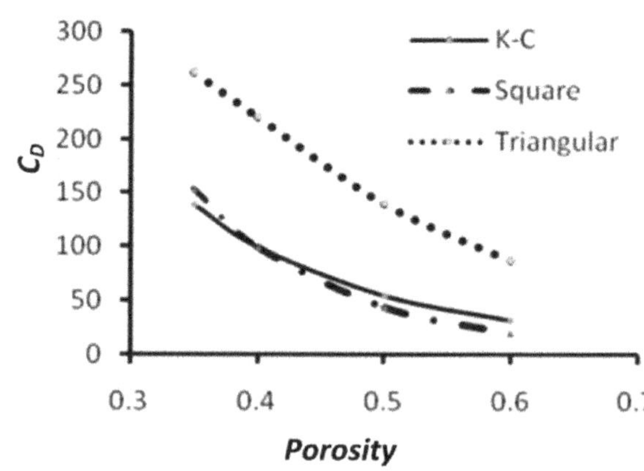

Figure 8: Influence of bed porosity on CFD drag coefficients for triangular and square configurations alongside K-C equation results (gas velocity=0.1 m.s^{-1})

Porosity variation in a trickle bed reactor is usually between 0.37 and 0.55. In this region, the square configuration exhibits very good agreement with K-C equation; however, the triangular configuration results are over estimated. The major drag force in the square configuration is imposed on the walls around cylindrical particles, while in the triangular configuration it is mainly imposed on the upper zone of cylindrical particles. Since drag force is defined as the imposed force on the particle along the direction of the flow, the triangular results are higher than the square configuration. The square and triangular arrangements are considered as the upper and lower limits of tortuosity and represent the lowest and highest drag force applied to respectively. As mentioned earlier, K-C equation and the square configure-tion are in very good agreement.

The Effects of No-slip Boundary Condition on the Simulation of Wetted Particles

In order to study the validity of the no-slip boundary condition for wetted particles, a thin liquid film with 0.02 mm in thickness is patched around a particle with a diameter of 1 mm and a length of 4 mm. Afterwards, the simulations were done by use of VOF governing equations (Equations 12-15) at different gas velocities. The velocity contours for these conditions are shown in Figure 9. As it can be seen, the gas velocity at gas-liquid interface is zero for the different velocities of inlet gas. Based on these results, it is confirmed that the no-slip condition for a wetted particle is an appropriate assumption for the CFD simulation.

Drag Coefficient for Fully-wetted Particles

A single-phase flow of air (constant density) was passed through wetted cylindrical particles (d=1.5 mm and L=3.11 mm) in a porous medium (Figure 4) and drag force values were calculated. Figures 10 and 11 show the drag force values for the square configuration and K-C results at different velocity and porosity values. It can be seen similar to dry particles, good agreement is observed between K-C equation results and the square configuration for wetted particles. Figures 12 and 13 show the drag coefficient of dry and wetted particles with the square configuration. As it can be seen, the difference between dry and wet curves decreases at higher gas velocity and bed porosity values. It can be seen that in pre-wetted beds drag coefficient has a higher value and pressure drop relative to dry beds.

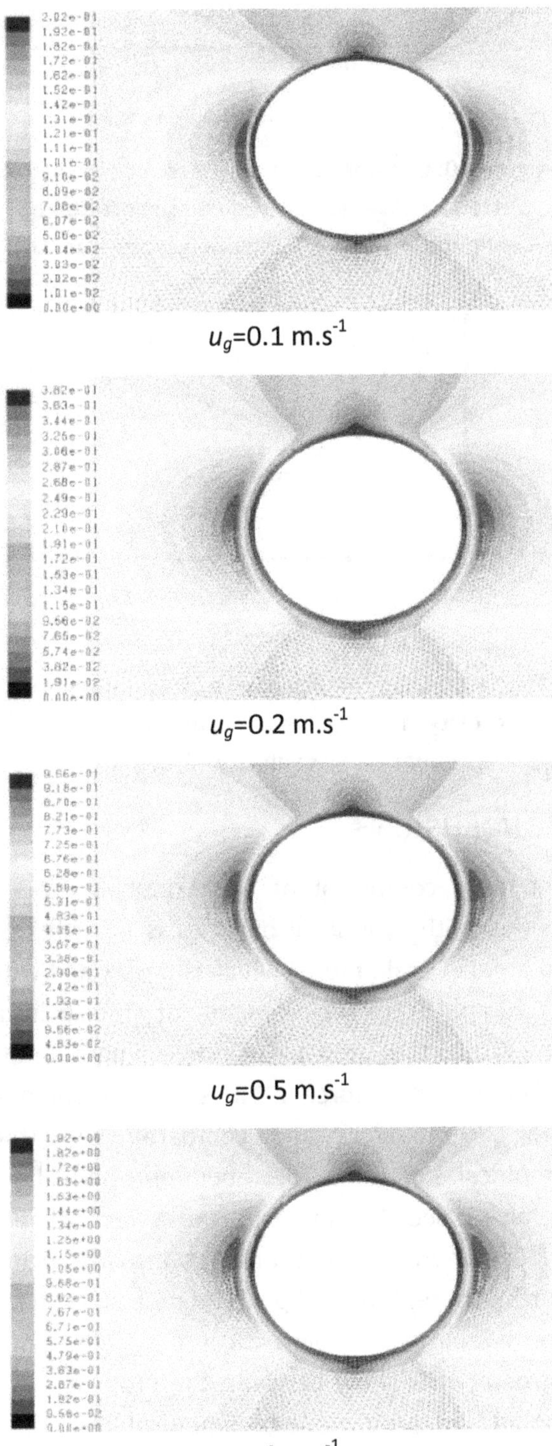

Figure 9: Velocity changes around a wet catalyst particle at different velocities

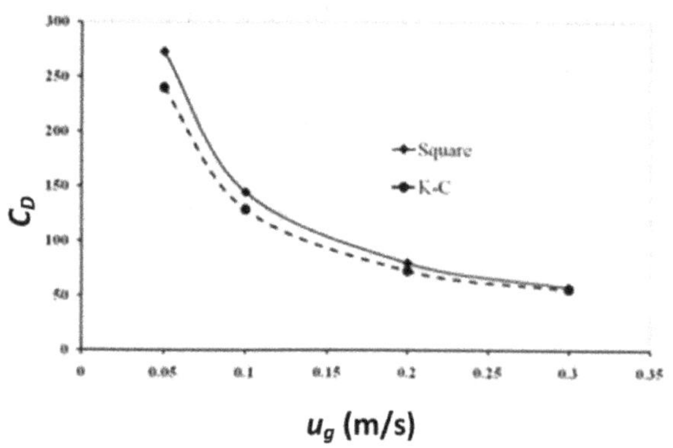

Figure 10: Comparison of drag force coefficient of wetted particles vs. gas velocities in the square configuration with K-C equation (porosity= 0.4)

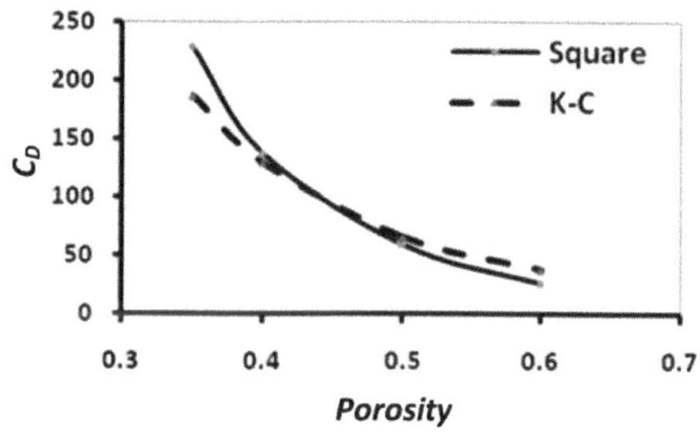

Figure 11: Drag coefficient of wetted particles vs. porosity in the square configuration and K-C equation (gas velocity= 0.1 m.s^{-1})

CONCLUSIONS

The drag coefficient of a trickle bed reactor loaded with cylindrical particles is investigated by a pellet scale model which is developed with a CFD approach. The arrangement of Particles is one of the most effective parameters influencing drag force in TBR's this is studied in the current work and compared with the empirical equations. The model presented here is also used to find out the effect of the triangular and square configurations of particles. Furthermore, the no-slip boundary condition for wetted particles is investigated herein. Good agreement is show between the proposed CFD model and predictions of empirical equations. The model presented here is also used to find out the effect of the triangular and square equations. It is concluded that the pellet scale model with the square configuration makes a better prediction of drag force in the bed compared to the triangular one.

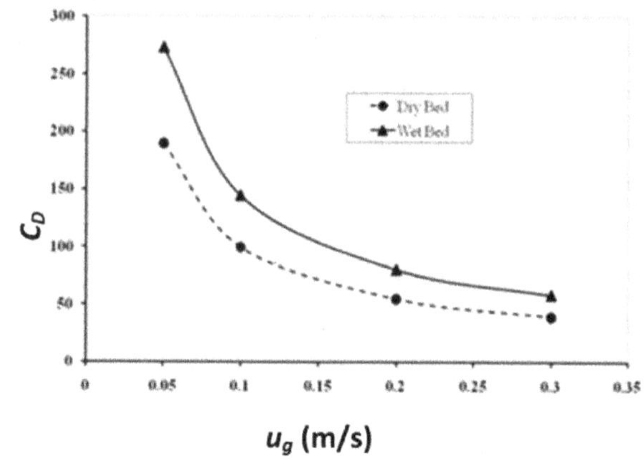

Figure 12: Drag coefficient of dry and wet particles vs. gas velocities in the square configuration (porosity= 0.4)

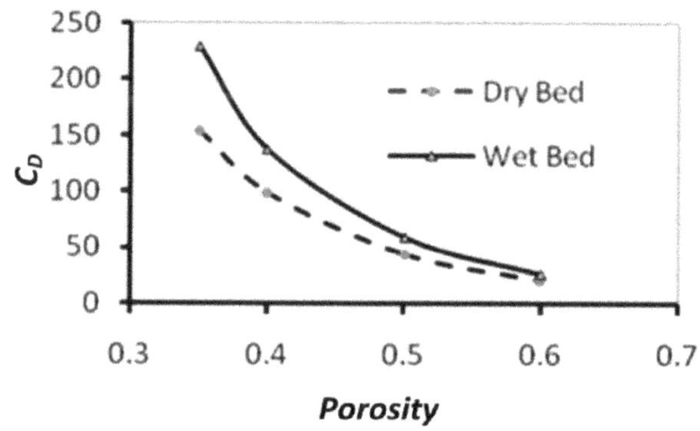

Figure 13: Drag force coefficient of dry and wet particles vs. porosity in the square configuration (gas velocity= 0.1 m.s^{-1})

NOMENCLATURE

A	: Particle surface area (m^2)
C_D	: Drag coefficient (-)
d_P	: Equivalent Diameter (mm)
E_1, E_2	: Ergun Constants (-)
F_D	: Drag force between gas and solid phases
G	: Gravity acceleration (m.s^{-2})
h_e	: External liquid holdup
P	: Pressure (Pa)
S	: Slit half-wall thickness (mm)
t	: Time(s)
u	: Velocity (m.s^{-1})
u_k	: Velocity of k phase (m.s^{-1})

w : Slit half-void thickness (mm)

Subscripts

G : Gas

L : Liquid

S : Solid

Greek Symbol

α : Mean fraction of gas phase

δ_S : Wet thickness (m)

ρ : Density ($kg.m^{-3}$)

μ : Viscosity ($kg.m^{-1}s^{-1}$)

ε : Porosity

τ : Stress tensor ($N.m^{-2}$)

REFERENCES

[1] Attou A., Boyer C., Ferschneider G., "Modelling of the Hydrodynamics of the Cocurrent Gas-liquid Trickle Flow through a Trickle-bed Reactor", *Chem. Eng. Science*. **1999**, 54, 785-802.

[2] Boyer C., Volpi C., Ferschneider G., "Hydrodynamics of Trickle Bed Reactors at High Pressure: Two-phase Flow Model for Pressure Drop and Liquid Holdup, Formulation and Experimental Validation", *Chem. Eng. Science*. **2007**, 62, 7026-7032.

[3] Lopes R. J. G., Quinta-Ferreira R. M., "Three-Dimensional Numerical Simulation of Pressure Drop and Liquid Holdup for High-pressure Trickle-bed Reactor", *Chem. Eng. Journal*. **2008**, 145, 112-120.

[4] Bazmi, M., Hashemabadi, S. H., Bayat M., "Modification of Ergun Equation for Application in Trickle Bed Reactors Randomly Packed with Trilobe Particles Using Computational Fluid Dynamics Technique", *Korean J. Chem. Eng.* **2011**, 28(6), 1340-1346.

[5] Bazmi, M., Hashemabadi, S. H., Bayat, "CFD Simulation and Experimental Study for two-Phase Flow through the Trickle Bed Re-actors, Sock and Dense Loaded by Trilobe Catalysts", *International Communications in Heat and Mass Transfer*, **2011**. 38, 391-397.

[6] Carbonell R. G., "Multiphase Flow Models in Packed Beds", *Oil & Gas Science and Technology* **2000**, 55(4) 417-425.

[7] Lakota A., Levec J., Carbonell R. G., "Hydrodynamics of Trickling Flow in Packed Beds: Relative Permeability Concept", *AIChE Journal*. **2002**, 48(4), 731-738.

[8] Nemec D., Levec J., "Flow through Packed Bed Reactors: 2. Two-phase Concurrent Downflow", *Chem. Eng. Science,* **2005**, 60, 6958-6970.

[9] Iliutal., Larachi F., "Modeling the Hydrodynamics of Gas-liquid Packed Beds Via Slit Models: Review" .**2005**, *The Brekeley Electronic Press*, 3.

[10] Iliutal., Larachi F., Al-Dahhan M. H., "Multiple-zone Model for Partially Wetted Trickle Flow Hydrodynamics", *Chem. Eng. Res. Des. (Trans IChemE part A)* **2000**, 78, 982-990.

[11] Attou A., Ferschneider G., "A Two-fluid Hydrodynamic Model for the Transition between Trickle and Pulse Flow in a Cocurrent Gas-liquid Packed-bed Reactor", *Chem. Eng. Science*. **2000**, 55, 491-511.

[12] Attou, A., Ferschneider, G., "A. Two-fluid Model for Flow Regime Transition in Gas-Liquid Trickle-bed Reactors". *Chemical Engineering Science*. **1999**, 54, 5031-5037.

[13] Gunjal P. R., Kashid M. N., Ranade V. V., Chaudhari R. V., "Hydrodynamics of Trickle-Bed Reactors: Experiments and CFD Modeling", *Ind. Eng. Chem. Res*. **2005**, 44, 6278-6294.

[14] Lopes R. J. G., Quinta-Ferreira R. M., "Evaluation of Multiphase CFD Models in Gas-liquid Packed-bed Reactors for Water Pollution Abatement", *Chem. Eng. Science*, **2010**, 65, 291-297.

[15] Souadnia A., Latifi M. A., "Analysis of Two-phase Flow Distribution in Trickle-bed Reactors", *Chem. Eng. Science*. **2001,** 56, 5977-5985.

[16] Nemec D.,Levec J., "Flowthrough packed bed reactors: 1. Single-phase flow", *Chem. Eng. Science.* **2005**, (60), 6947–6957.

[17] Trivizadakis M. E., Giakoumakis D., Karabelas A. J., "A Study of Particle Shape and Size Effects on Hydrodynamic Parameters Of Trickle Beds", *Chem. Eng. Science*. **2006**, 61, 5534–5543.

[18] Lanfrey P. Y., Kuzeljevic Z. V., Dudukovic M.P., "Tortuosity Model for Fixed Beds Randomly Packed with Identical Particles", *Chem. Eng. Science*. **2010**, 65, 1891-1896.

[19] Hellstrom, J. G., Lundstorm T. S., "Flow through Porous Media at Moderate Reynolds Number", *International Science Colloquium, Modeling for Material Processing*. **2006**, Riga, June 8-9.

[20] Lopes R. J. G., Quinta-Ferreira R. M., "Numerical Simulation of Trickle-Bed Reactor Hydrodynamics with RANS-Based Models Using a Volume of Fluid Technique", *Ind. Eng. Chem. Res.* **2009,** 48, 1740-1748,.

[21] Gunjal P. R., Ranade V. V., Chaudhari R. V., "Computational Study of a Single-phase Flow in Packed Beds of Spheres", *AIChE Journal*. **2005,** 51 (2), 365-378.

[22] Patankar S. V., "Numerical Heat Transfer and Fluid Flow", *Taylor and Francis* **1980**.

[23] Macdonald I. F., El-Sayed M. S., Mow K., Dulllen F. A. L., "Flow through Porous Media: the Ergun Equation Revisited", *Ind. and Eng. Chem. Fundamentals*. **1979**, 18, 199-208.

[24] M. Rahman, Karim, Md. M., Alim, Md. A., "Numerical investigation of unsteady flow past a circular cylinder using 2D finite volume method" *Journal of Naval Architecture and Marine Engineering*. **2007**, 4, 27-42.

[25] Haider A., Levenspiel O., "Drag Coefficients and Terminal Velocity of Spherical and Nonspherical Particles", *Powder Technology*. **1989**, 58, 63-70.

12

Evaluating Different Approaches to Permeability Prediction in a Carbonate Reservoir

Farhad Khoshbakht

Department of Reservoir Study and Field Development, Research Institute of Petroleum Industry (RIPI), Tehran, Iran

ABSTRACT

Permeability can be directly measured using cores taken from the reservoir in the laboratory. Due to high cost associated with coring, cores are available in a limited number of wells in a field. Many empirical models, statistical methods, and intelligent techniques were suggested to predict permeability in un-cored wells from easy-to-obtain and frequent data such as wireline logs. The main objective of this study is to assess different approaches to the prediction of the estimation of permeability in a heterogeneous carbonate reservoir, i.e. Fahliyan formation in the southwest of Iran. The considered methods may be categorized in four groups, namely a) empirical models (Timur and Dual-Water), b) regression analysis (simple and multiple), c) clustering methods like MRGC (multi-resolution graph-based clustering), SOM (self organizing map), DC (dynamic clustering) and AHC (ascending hierarchical clustering), and d) artificial intelligence techniques such as ANN (artificial neural network), fuzzy logic, and neuro-fuzzy.

This study shows that clustering techniques predict permeability in a heterogeneous carbonate better than other examined approaches. Among four assessed clustering methods, SOM performed better and correctly predicted local variations. Artificial intelligence techniques are average in modeling permeability. However, empirical equations and regression methods are not capable of predicting permeability in the studied reservoir. The constructed and validated SOM model with 6×9 clusters was selected to predict permeability in the blind test well of the studied field. In this well, the predicted permeability was in good agreement with MDT and core derived permeability.

Keywords: Permeability, Carbonate Reservoir, Clustering, Intelligent, Experimental Correlation

INTRODUCTION

In addition to being porous, a reservoir rock must be able to allow fluids to flow. The ability of rock to conduct fluids is defined as permeability. Core is the most reliable source of matrix permeability. Usually in a field, core is available in a limited number of wells because of costs and technical limitations. Permeability as one of the most important petrophysical properties of a reservoir is the main input to static and dynamic models. In order to construct a realistic model of a reservoir, accurate permeability data with an appropriate distribution ought to be available. Thus it is essential to predict permeability from easy-to-obtain and

***Corresponding author**

Farhad Khoshbakht

Email: khoshbakhtf@ripi.ir

frequent data such as logs. The prediction of permeability is not straightforward. The difficulty of predicting permeability is due to the fact that carbonate rocks are very heterogeneous and factors such as diagenesis (e.g. cementation), texture, grain size variation and fabric, cementation, and clay content control permeability [10]. Unlike core, wireline logs are available in almost all wells in a field. By establishing a robust model which relates permeability to wireline log responses permeability in un-cored wells can be predicted.

More than 40 percent of the world oil reservoirs are placed in carbonate rocks. Because of the economic importance of carbonates, their accurate characterization is vital for exploration and production. The ultimate goal of reservoir characterization is to improve production efficiency and oil recovery by understanding and modeling the reservoir. To correctly model the flow behavior in a carbonate reservoir, it is essential to understand the permeability profile [20].

Carbonate rocks are unstable and diagenetic processes chemically and physically alter a somehow homogeneous formation into a completely heterogeneous one. The permeability of a carbonate reservoir often is severely affected by tectonic and diagenetic phenomena like fracturing, dissolution, and leaching, which makes the prediction of permeability very difficult.

In the last decade, many researchers have tried to estimate permeability from wireline logs. Because of availability and frequency of wireline logs, they have become a popular source for estimating permeability. Archie (1942) introduced relationships which estimated permeability via core analysis data such as porosity and formation resistivity factor [3]. Porosity, water saturation, capillary pressure, formation resistivity factor, and NMR T1 & T2 parameters, derived from wireline logs, are used to estimate permeability. Leverett (1941), Tixier (1949), Wyllie and Rose (1950), Timur (1968), and Coates and Dumanoir (1974) developed correlations based on well log measurements to determine permeability [5, 21-23].

In recent years, artificial intelligence techniques have widely been used to estimate permeability from wireline logs. Balan et al. (1995), Mohaghegh et al. (1997), Lin et al. (1994), Zhang et al. (1996), and Huang et al. (1996) predicted permeability by means of neural networks. Cuddy (1998) and Finol et al. (2001) used the fuzzy approach to estimate permeability [4,7-9,13,16,25].

In this study, different approaches to permeability prediction are examined in order to establish a robust model to predict permeability in un-cored wells. The method with the best performance, i.e. SOM, was selected to predict permeability in un-cored wells. The predicted permeability in the blind test wells was compared with MDT-(*Modular Formation Dynamics Tester*) derived permeability and core permeability to ensure that the model worked properly.

MATERIALS AND METHODS

The studied field is located in Zagros fold-thrust belt, the southwest of Iran (Figure 1). Zagros is a part of Tethys Ocean and is one of the most important petroleum basins in the world [1]. This basin is located in the south west of Iran and north of Iraq. The geological history of this basin includes long time subsidence and deposition interrupted by short time uplift. Folding process of this basin occurred in Miocene and Pliocene and continued until now, which has formed long anticlines [17]. These anticlines constitute most of oil traps in this basin.

Fahliyan (the uppermost Jurassic-Cretaceous) is one of the major reservoir formations of Iranian oil fields. Fahliyan is a clean limestone comprised of massive oolitic or pelletal limestone [2]. In the studied field, Fahliyan formation is divided to eight reservoir zones.

Figure 1: Location of the studied field

In this study, the wireline and core data of four wells obtained from Fahliyan formation were used. The datasets of three wells (A, B, and C) were used for building models and the forth well (D) left out for the test. A suit of logs including neutron porosity (NPHI), sonic log (DT), resistivity log (LLD), photoelectric (PEF), and natural gamma ray (GR) was selected as the input to all the models (Figures 2 and 3). These logs record the properties of the reservoir, which control permeability such as shale content, effective porosity, water saturation, and lithology.

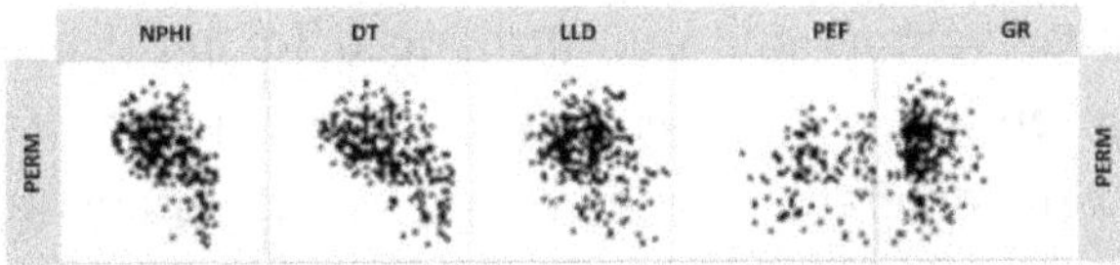

Figure 2: Cross plot of core permeability versus selected wireline logs (From left to right: NHPI, DT, LLD, PEF, and GR)

Permeability Modeling

In order to predict permeability in un-cored wells, models should be constructed. The inputs to these models are well logs and the output is permeability. In wells with core data (A, B, and C), the models were constructed and tuned, and then they were used to predict permeability in un-cored wells. In the following, all the models are briefly described and the predicted permeability by means of each one is presented.

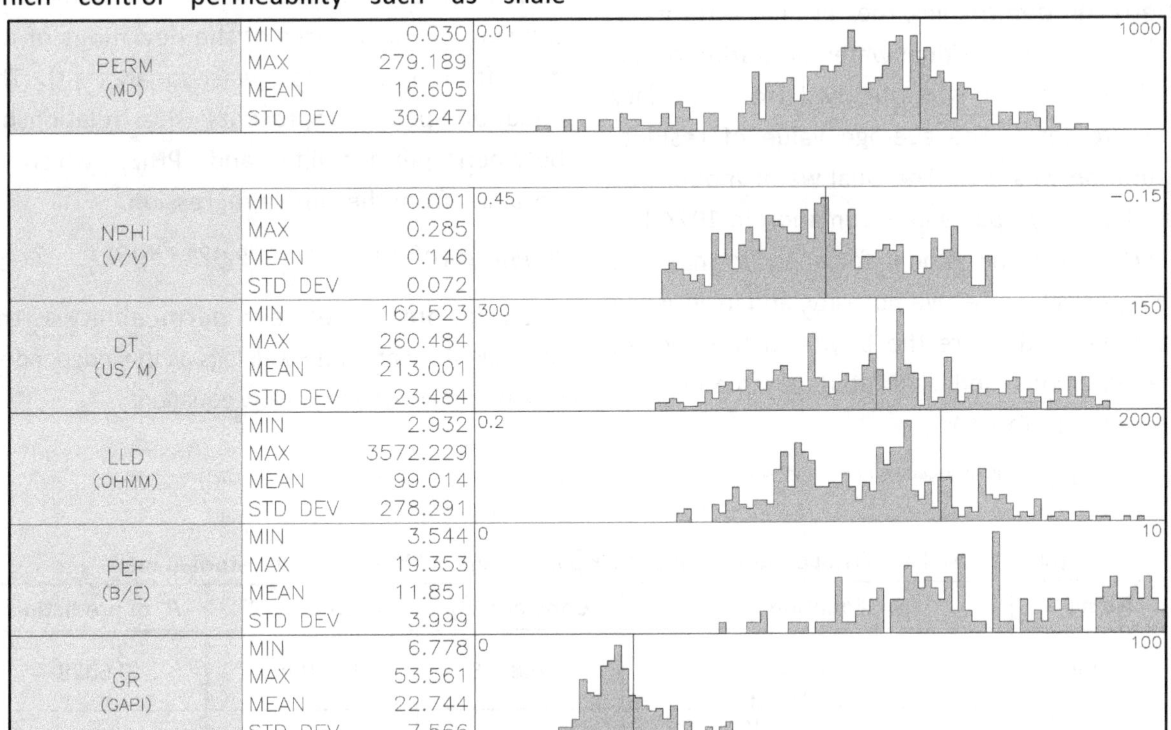

Figure 3: Statistics and histogram of permeability as the output of the models and selected wireline logs (NHPI, DT, LLD, PEF, and GR) as the inputs to the prediction models

Empirical Models

Empirical models are based on experiential observations and are introduced for particular reservoirs. For applying them to other reservoirs, with different properties, it is necessary to calibrate their parameters. Generally empirical equations relate permeability to porosity and water saturation. First, we calibrated the models by using the core data of train wells. In train wells, the inputs (porosity and irreducible water saturation) and the output of the model (permeability) were available and the only unknown of the equations (*A* and *B*) can be calculated. Then, in blind-test well the permeability was predicted using the calibrated models. In this study, we used two common empirical equations for predicting permeability: Timur and Dual-Water.

Timur (1968) found a relationship for estimating the permeability of sandstones from the in-situ measurements of porosity and residual saturation [21]. His model is applicable where the condition of residual saturation exists. The main source of uncertainty in Timur model is the high amount of error in determining the residual saturation. In our case, the residual water saturation of the studied reservoir was obtained from capillary pressure tests. The average value of residual saturation was 0.18. The Dual water model was developed by Coates and Dumanoir in 1974 [5]. They suggest this model for shaly formations. Total porosity, effective porosity, and irreducible water saturation are the inputs to this model. The equations and calibrated parameters are presented in Table 1.

Regression Analysis

Regression analysis is a statistical technique that identifies the relationship between two or more quantitative variables: a dependent variable whose value is to be predicted, and independent variable(s). Regression analysis is used to understand the statistical dependence of one variable on other variables.

We utilized simple and multiple regressions in order to estimate permeability from wireline logs. In the simple regression, the effective porosity (PHIE) was the input (independent variable). In the multiple regression, DT, LLD, NPHI, PEF, and GR were selected as the inputs. In the multiple regression, the data is constituted using the below equation:

$$Log\ (perm) = a_0 + a_1 \times (DT) + a_2 \times (LLD) + a_3 \times (NPHI) + a_4 \times (PEF) + a_5 \times (GR)$$

The values between brackets are known and the coefficients (a_0, a_1, a_2, a_3, a_4, and a_5) are unknown. A system of matrix (*Y=aX*) representing the above linear equations is constructed and solved for unknown coefficients by minimizing the sum of the squares of the deviations of the data from the model (least-squares fit). The equation below represents the relationship between permeability and PHIE, which is obtained from the simple regression.

$$Perm = 10^{(-0.0357117+6.340498(PHIE))}$$

The relationship between permeability as the dependent variable and well logs as the independent variables is given by the below equation:

$$Perm = 10^{(-1.748+0.00884115\times(DT)-0.132608\times Log(LLD)+1.64347\times(NPHI)+385849\times(PEF)+0.00575193\times(GR))}$$

Table 1: Used empirical equations and their parameters calibrated for the studied wells

Name	Equation	Constant (*A*)	Exponent (*B*)	R^2 of prediction
Timur	$K = \left[A \dfrac{\varphi_t^B}{S_{wirr}}\right]^2$	6.758595	1.230943	0.6028
Dual-Water	$K = \left[A.\varphi_e^B.\dfrac{\varphi_t - V_{wirr}}{V_{wirr}}\right]^2$	8.001080	1.176190	0.5787

Clustering Methods

The aim of cluster analysis is to classify a dataset into groups that are internally homogeneous and externally isolated on the basis of a measure of similarity or dissimilarity among groups [12]. Clustering methods are widely used for electro-facies analysis and the prediction of petrophysical properties. By integrating clustering methods with intelligence techniques, some new methods such as SOM and MRGC were born [11,24]. In this study, we utilized two classic clustering methods, namely DC and AHC, and two new clustering methods, i.e. SOM and MRGC, to predict permeability from wireline logs.

Predicting permeability by means of clustering methods is carried out in three steps:

1) Partitioning log data of the train wells into somehow similar groups and determining the properties of each groups;

2) Identifying clusters of test data using cluster properties, determined in the previous step;

3) Predicting permeability in each cluster of the blind-test well by means of KNN method.

Each clustering method tries to partition data into stable and homogeneous groups on the basis of its particular algorithm. Selecting the optimal number of clusters is important, because variation in the number of clusters strongly affects the results. There are no universal criteria for choosing the optimal number of clusters in the examined methods. Therefore, we tried different values for the number of clusters and chose the optimal values via trial and error. NPHI, DT, LLD, PEF, and GR were the inputs to all the clustering methods.

MRGC is based on non-parametric K-nearest neighbor and graph data representation. MRGC is a tool which analyzes the structure of the complex data and partitions natural data groups into different shapes, sizes, and densities [24]. MRGC automatically determines the optimal number of clusters. We assigned 4 and 35 respectively as the minimum and maximum of the desired clusters for MRGC. After executing MRGC, the log data of train wells were partitioned into six orders of clusters (5, 7, 10, 14, 16, and 18 clusters). The highest R^2 (0.5978) was obtained when MRGC model with 16 clusters was used for permeability prediction (Figure 4-a).

The SOM is a computational method for the visualization and analysis of high-dimensional data, especially experimentally acquired information [11]. We tested SOM networks with different dimensions to partition the datasets of train wells. Then, the permeability of the blind-test well was predicted using each model (Figure 4-b). As shown in Figure 4-b, the optimal dimension of SOM network is 6×9, which gives the highest R^2 (0.6936) (Table 2 and Figure 5-f).

Dynamic Clustering or *k*-means clustering is a non-hierarchical method to classify a dataset on the basis of a pre-defined number of clusters. The algorithm assigns each object into a cluster by iteration. It minimizes the sum of distance from each object to its cluster centroid, over all clusters. This algorithm moves objects between clusters until the sum cannot be decreased further [15]. This clustering method is suitable for datasets with a large amount of data. We examined different values for the number of clusters and predicted permeability in the blind-test well. Permeability prediction on the basis of 12 clusters (Figure 4-c) has the highest R^2 (0.6631) among DC models with different numbers of clusters (Figure 5-g).

Hierarchical clustering method partitions data over a variety of scales by creating a cluster tree or dendrogram [15]. Ascending hierarchical cluster (AHC) analysis is a statistical method for finding relatively homogeneous clusters of cases based on measured characteristics [14]. AHC produces different levels of clusters and user can choose the most appropriate level for the dataset. Generally, clustering of a dataset can be executed by means of AHC in three steps: 1) measuring the distance between every pairs of objects in the dataset, 2) linking pairs of objects which are very close together, and 3) cutting the hierarchical tree into clusters [15]. Hierarchical trees were cut at different levels and the results showed that AHC model with 8 clusters has the highest R^2 (Figure 4-d) among levels of clusters. AHC models with less or more than 6 clusters decrease the R^2 value (Figure 5-h). Increasing the number of clusters does not make sharp changes in the results, because AHC combines relatively similar high order clusters and makes low order clusters.

Artificial Intelligence Techniques

ANN, fuzzy logic, and neuro-fuzzy are three artificial intelligence techniques which were used in this study to predict permeability from wireline logs.

Neural network is a modeling technique which models systems in a brain-like way. The main feature of neural network is that it can learn the internal characteristics of a system by analyzing datasets.

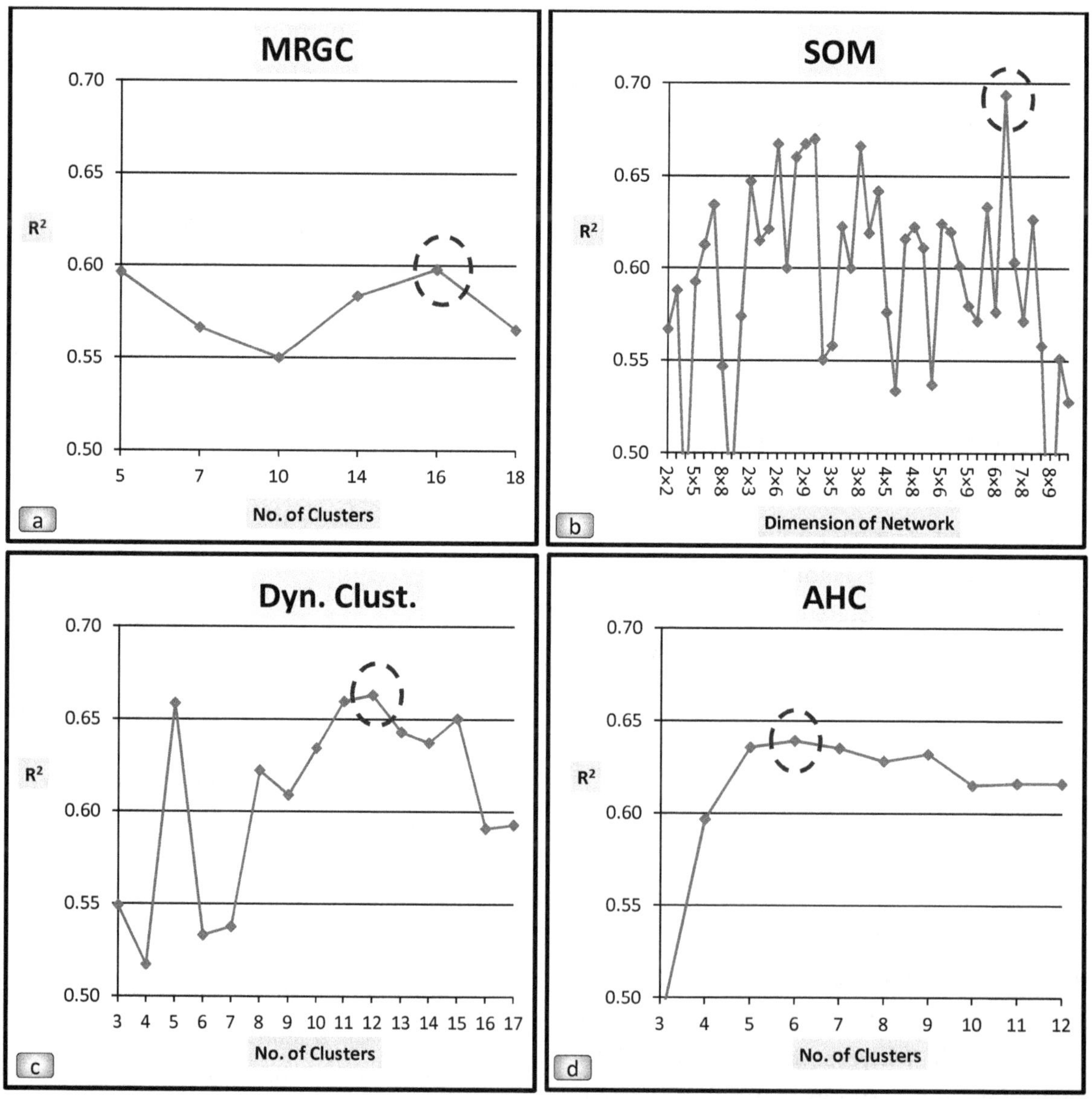

Figure 4: Variation of R^2 by changing the number of clusters for a) MRGC, b) SOM, c) DC, and d) AHC

In order to predict permeability using ANN, a back propagation neural network with two hidden layers have been designed. A trained network was used to predict the permeability of the test well. The R^2 of the predicted permeability by means of neural network versus core permeability is 0.6279 (Table 2 and Figure 5-i).

In this paper, a fuzzy approach proposed by Cuddy (2000) was utilized for permeability prediction [7]. This approach works by assigning a probability to the quality of the prediction from each log, then combines the probabilities, and predicts the most likely permeability [7]. We examined different numbers of classes and predicted the permeability for each case. Fuzzy model with 10 classes predicted permeability with the highest R^2 (0.6275) among models with different numbers of classes (Figure 5-j).

We used a locally linear neuro-fuzzy (LLNF) model which utilized local linear model tree learning algorithm (LOLIMOT) to predict the permeability. The fundamental approach to a locally linear neuro-fuzzy model is dividing the input space into small linear subspaces with fuzzy validity functions. Any produced linear part with its validity function is described as a fuzzy neuron. Thus the total model is a neuro-fuzzy network with one hidden layer, and a linear neuron in the output layer, which simply calculates the weighted sum of the outputs of locally linear models [18]. By plotting the mean square error for the train and test data, we found that the optimal number of hidden layer neurons is 29, and constructed our neuro-fuzzy model with 29 neurons. The R^2 of permeability prediction by means of neuro-fuzzy model was 0.5900 (Figure 5-k).

RESULTS AND DISCUSSION

The prediction of permeability in un-cored wells is one of the difficult tasks in reservoir characterization, especially in heterogeneous and complex carbonates. Wireline logs are frequent and easy-to-obtain data, which are usually available in almost all the wells. We launched this study to examine four common approaches to predicting permeability, namely empirical methods, statistical analysis, clustering methods, and artificial intelligence techniques. The method with the best performance was selected to predict permeability in un-cored wells. The predicted permeability in one of the un-cored wells of the studied field was compared with MDT- (*Modular Formation Dynamics Tester*) derived permeability to ensure that the model worked properly.

The first examined approach was empirical equation: Timur and Dual-Water models. The R^2 of the measured permeability versus predicted permeability using Timur and Dual-Water models are 0.6028 and 0.5787 respectively (Table 2). Even though the predicted permeability by means of Timur model is closer to the measured permeability than the one measured by Dual-Water model, both models performed weakly (Figures 5-a and 5-b). In spite of our efforts to properly calibrate the parameters of these models, the permeability prediction was not satisfactory. Because of the heterogeneity of the reservoir, empirical equations failed to predict permeability properly. Usually the use of empirical equations is restricted to the region(s) where the models are developed for, and in different reservoirs, imperial models should be employed with caution.

Regression analysis is simpler and easier to practice than other methods. The R^2 values of permeability prediction by means of simple and multiple regressions are 0.4920 and 0.5279 respectively (Table 2). The distribution shape of the predicted data via simple and multiple regressions are similar and both overestimate low permeabilities and underestimates high permeabilities.

Table 2: The R^2 of the predicted versus measured permeability for all the used methods

Type	Empirical		Regression		Clustering				Artificial Intelligence		
Model	Timur	Dual-Water	Simple	Multiple	MRGC 16	SOM 6×9	DC 12	AHC 6	ANN	Fuzzy 10	Neuro-Fuzzy
R^2	0.6028	0.5787	0.4920	0.5279	0.5978	0.6936	0.6631	0.6391	0.6279	0.6275	0.5900

The cross plots of permeability versus logs (Figure 2) show that there is not a simple and linear relationship between them. Consequently, we cannot expect methods like empirical equations and regression analysis to explore complex relationships. Empirical methods simplify problems by reducing the number of variables, while different factors influence the permeability of reservoirs. In addition, regression methods tend to average data and ignore local variations, which are important in thin layered reservoirs. Figures (5-c) and (5-d) present the plots of the predicted permeability versus the measured permeability by regression analysis.

The aim of cluster analysis is to classify a dataset into groups that are internally homogeneous and externally isolated on the basis of a measure of similarity or dissimilarity between groups [12]. For proper permeability prediction in complex reservoirs, different criteria have been proposed for data classification such as reservoir layers, litho-facies, rock types, and hydraulic flow units. But in order to predict permeability from logs, methods should be employed which can handle the dimensionality and complexity of log data. Because of the variety of wireline logs and the heterogeneity of carbonates, it is usually difficult to easily recognize natural groups of data.

Generally, clustering methods predict permeability better than the other examined methods. Even though they are not perfect, their results are acceptable. Selecting the optimal number of clusters is one of the important tasks in using clustering techniques, because variation in the number of clusters strongly affects the result of permeability prediction. In addition, there are no general criteria for choosing the optimal number of clusters in the examined methods. We examined different values for the number of clusters and predicted permeability in each case. Each method has its own value for the optimal number of clusters, e.g. the optimal number of clusters for MRGC, SOM, DC, and AHC are 16, 54, 12, and 6 respectively. Predicting permeability on the basis of SOM has the highest R^2 among the examined clustering methods, even though SOM is not much more robust than the other clustering methods; but, the results of the prediction are slightly comparable. However, SOM predicts low and high permeabilities better than the other methods. The performance of ANN and fuzzy logic for predicting permeabilities between 1 to 10 mD is acceptable, but both methods fail to predict permeabilities higher than 10 mD correctly. Contrary to our expectation neuro-fuzzy performance was weaker than the performance of ANN and fuzzy logic. Neuro-fuzzy only predicts very high values (~100 mD) properly, but for the range 1 to 10 mD, at which most methods do well, neuro-fuzzy was disappointing.

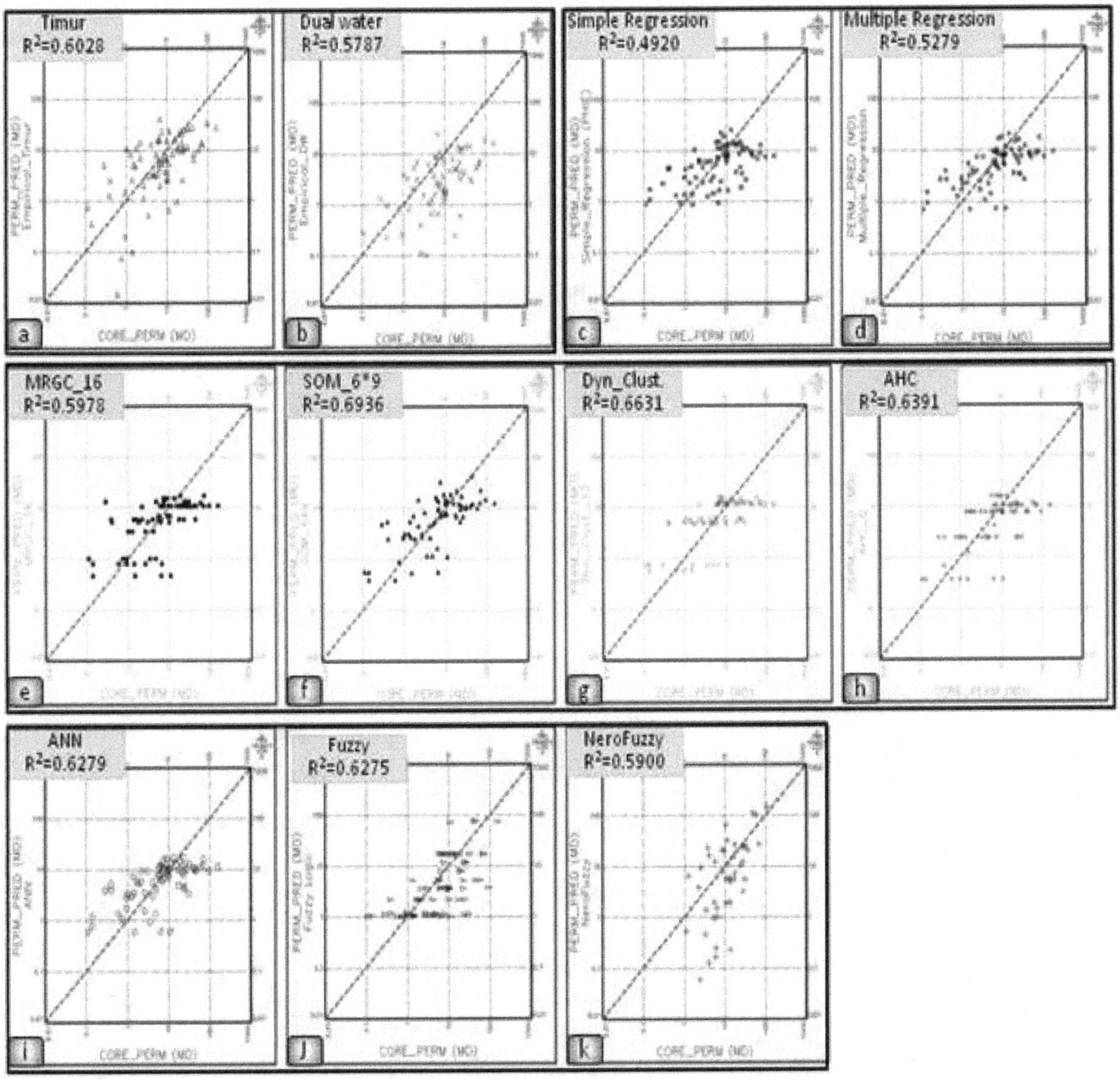

Figure 5: Cross-plots of the measured permeability versus the predicted permeability by means of different methods, the methods and the R^2 of the prediction are presented in the up-left box of each cross plot.

The main preference of clustering methods is their ability to predict local variations. Most examined methods, especially regression and empirical methods, ignore local variations. These methods predict medium permeabilities well, but the prediction of extremes is associated with high errors. Statistically the number of samples in the extremes is low and most techniques cannot learn the pattern of these parts of data in training. Consequently, they tend to average the data and ignore local variations.

The best performed method (SOM with 6×9 clusters) was used to predict permeability in one of un-cored wells of the studied field. In this well, the permeability derived from MDT test was available. As it is clear in the right column of Figure 6, the predicted permeability matches well with MDT permeability. This interval comprises limestone with a limited amount of shale and its porosity is almost constant. Utilizing empirical methods which relate permeability to porosity leads to somehow constant predicted permeability, assuming that irreducible water saturation is also constant in this interval.

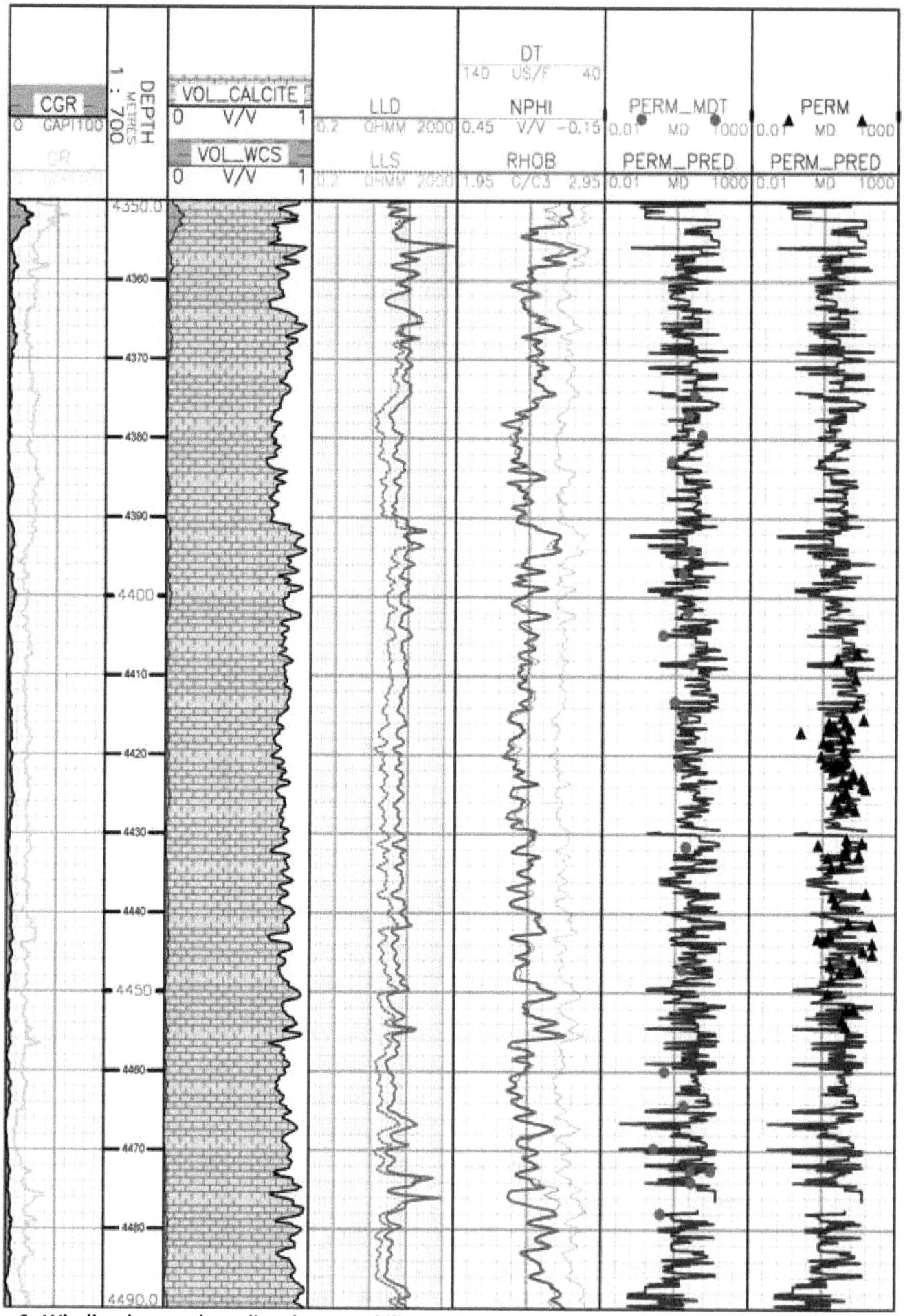

Figure 6: Wireline logs and predicted permeability in the blind test well; From left: Track 1) CGR and GR, Track 2) depth scale, Track 3) lithology, Track 4) resistivity logs, Track 5) NPHI, DT & RHOB, Track 6) predicted permeability by means of SOM 6×9 (solid line) and MDT permeability (dots), Track 7) predicted permeability by means of SOM 6×9 (solid line) and measured permeability (square).

CONCLUSIONS

Numerous methods have been proposed by researchers to predict permeability in un-cored wells. Common approaches to predicting permeability are empirical models, regression analysis, clustering methods, and artificial intelligence techniques. We examined these methods to predict permeability in a carbonate reservoir. The study showed that among the examined methods, regression analysis was the worst and clustering techniques were the best in predicting permeability. Accordingly, among the four assessed clustering methods, SOM performed better and could properly predict local variations. Since the clustering methods classify the datasets into homogeneous subclasses and then predict the permeability in each sub-class, it is expected that the clustering methods perform better compared with the other methods which treat the dataset as a whole; in fact, in each subclass the relationship between permeability and the log responses is simpler.

Artificial intelligence techniques were average in modeling permeability. Empirical equations and regression techniques were not capable of predicting permeability in the studied reservoir. The successful prediction of permeability in an un-cored well, confirmed by MDT, justified the capability of SOM as a clustering technique.

REFERENCES

[1] Alavi M., "Tectonics of the Zagros Orogenic Belt of Iran: New Data and Interpretations," *Tectonophysics*, **1994**, *229*, 211-238.

[2] Alavi M., "Regional Stratigraphy of the Zagross Fold-thrust Belt of Iran and its Proforeland Evolution," *American Journal of Science*, **2004**, *304*, 1-20.

[3] Archie G. E., "The Electrical Resistivity Log as an Aid in Determining Some Reservoir Characteristics," *Petroleum Transactions*, **1941**.

[4] Balan B., Mohagheg S., and Ameri S., "State of the Art in Permeability Determination from Well Log Data: A Comparative Study, Model Development," paper SPE 30978 presented at *SPE Eastern Regional Conference & Exhibition held in Morgantown*, West Virginia, U.S.A, **1995**.

[5] Coates G. R. and Dumanoir J. L., "A New Approach to Improved Log Derived Permeability," *The Log Analyst*, **1974**.

[6] Cuddy S. J., "The Application of the Mathematics of Fuzzy Logic to the Geosciences," *Paper SPE 49470*, **1998**.

[7] Cuddy S. J., "Litho-Facies and Permeability Prediction from Electrical Logs Using Fuzzy Logic," *paper SPE 65411*, **2000**.

[8] Finol J., Guo Y. K., and Jing X. D., "A Rule Based Fuzzy Model for the Prediction of Petrophysical Rock Parameters," *Journal of Petroleum Science and Engineering*, **2001**, *29*, 97-113.

[9] Huang Z., Shimeld J., Williamson M., and Katsube J., "Permeability Prediction With Artificial Neural Network Modeling in the Venture Gas Field," Society of Exploration Geophysics, **1996**, *61*(2), 422-436.

[10] Jennings J. W. and Lucia F. J., "Predicting Permeability from Well Logs in Carbonates with a Link to Geology for Interwell Permeability Mapping," *Paper SPE 84942-PA*, **2003**.

[11] Kohonen T., *"Self-Organizing Maps,"* 3th ed, Springer, **2001**, 501.

[12] Lee S. H. and Datta-Gupta A., "Electro-facies Characterization and Permeability Predictions in Carbonate Reservoirs: Role of Multivariate Analysis and Nonparametric Regression," *Paper SPE 56658*, Houston, Texas, **1999**.

[13] Lin J. L. and Salisch H. A., "Determination from Well Logs of Porosity and Permeability in a Heterogeneous Formation," paper SPE 28792, presented at *the SPE Asia Pacific Oil & Gas conference*, Melbourne, Australia, **1994**.

[14] Lukasova A., "Hierarchical Agglomerative Clustering Procedure," *Pattern Recognition*, **1979**, *1*, 365-381.

[15] Matlab, Statistics Toolbox user's Guide, Mathworks, Inc, **2007**, 1561.

[16] Mohaghegh S., Balan B. and Ameri S., "Permeability Determination from Well Log Data," *SPE International Journal*,

1995, 17-21.

[17] Mottie H., "Geology of Iran; Petroleum Geology of Zagros," *Geological Society of Iran Publication*, **1995**, *37* (2), 1009.

[18] Nelles O., "Nonlinear System Identification," Springer-Verlag, New York, **2002**, *13*(4), 785.

[19] Rabiller P. and Ye S. J., "The Iterative use of Clustering and Modeling to Improve Permeability Prediction," *SPWLA 42th Annual Logging Symposium*, Houston, Texas, **2001**.

[20] Akbar M., Vissapragada B., Dhabi A., Alghamdi A. H., et al., "A Snapshot of Carbonate Reservoir Evaluation," *Schlumberger-oil Field Review*, **2000**, *12*(4), 20-41.

[21] Timur A., "An Investigation of Permeability, Porosity, and Residual Water Saturation Relationships for Sandstone Reservoirs," *The Log Analysts*, **1968**, 9(4).

[22] Tixier M. P., "Evaluation of Permeability from Electric-Log Resistivity gradients," *Oil and Gas Journal*, **1949**, *17*, 68-73.

[23] Wyllie M. R. J. and Rose W. D., "Some Theoretical Considerations Related to the Quantitative Evaluation of the Physical Characteristics of Reservoir Rock from Electrical Log Data," *Petroleum Transactions, AIME*, **1950**, *2*(4), 105-118.

[24] Ye S. J. and Rabiller P., "A New Tool for Electro-Facies Analysis: Multi-Resolution Graph-Based Clustering," *SPWLA 41st Annual Logging Symposium*, **2000**.

[25] Zhang Y. J. and Lollback P. A., "A Methodology for Estimating Permeability from Well Logs in a Formation of Complex Lithology," paper SPE 37025, presented at *the SPE Asia Pacific Oil & Gas Conference*, Adelaide, Australia, **1996**, 28-31.

13

Utilizing Conceptual Modeling in the Study of One of The Iranian Fractured Carbonate Reservoirs

Seyed Majid Hashemi* and Gholamreza Bashiri

Institute of Enhanced Oil Recovery, NIOC, Tehran, Iran

ABSTRACT

A typical Iranian carbonate matrix block surrounded by an open fracture was modeled in order to understand the fracture-matrix interaction and realize how to model the interaction best. The modeling was carried out by using a fine-scaled Eclipse model in the single porosity mode (the fractures were explicitly modeled). The model was extended to a stack of 6 matrix blocks to understand block-to-block interaction under both water and gas injection scenarios. The results and conclusions obtained from the single porosity single block and 6-block model were used in order to optimize the full field model. The simulation results showed that gas injection worked as a major recovery mechanism for medium and good rock types at all block heights and for poor rock type with block heights of more than 2 meters. The oil recovery results were in the same range for the single and six-stack block model, but the delay of oil recovery by gas gravity drainage is clearly seen in the six-stack block results indicating a period of at least 8-10 years required for reaching ultimate oil recovery for typical block heights. The water injection simulations showed that water imbibes drained the matrix block in a short time dependent on water front advancement in the fracture system surrounding the block. From the simulation results, it was concluded that the water imbibition process was a fast recovery mechanism for all block heights dependent on injection rate and fracture volume which had to be flooded to achieve full recovery of the block. The expected recovery factor of drained oil to initial oil in place after water injection was in the range of 15-35 % for poor to good rock types and different block sizes in the range of 1-5 meters. The results were highly influenced by wettability conditions in the reservoir.

Keywords: Iranian Fractured Carbonate Reservoir, Conceptual Modeling, Water Injection, Gas Injection, Drainage and Imbibitions.

INTRODUCTION

Current practices in the numerical simulation of fractured reservoirs rely on the construction of both static and dynamic conceptual models from which one may create an integrated reservoir model to be simulated. Commercial numerical models capable of handling flow in fractured reservoirs have been present in the industry for quite some time. However, the correct application of those simulators for representative reservoir models is not easy. Early dual porosity models include those of Kazemi et al. [1] and Saidi [2]. Saidi modeled a fracture reservoir by dividing it into sectors in

*Corresponding author

Seyed Majid Hashemi

Email: smh2006@gmail.com

which the fracture was assumed to have infinite transmissibility. Understanding the main physical processes, on which the performance and IOR potential of one of Iranian carbonate reservoirs are based, is one of the main challenges in this evaluation and in general for carbonate fields. The use of conceptual models is a widely accepted method, which can help to improve the estimation of potential IOR attempts. Normally, the physical understanding is based on the single-porosity simulation of the problem on small scale and the results are matched in a dual porosity model on the same scale. Then, pseudo-capillary and relative permeability curves together with recommendations for dual porosity modeling options can be specified for full field reservoir simulation studies.

This paper will discuss parametric ways to improve the construction of representative models by using a fine grid simulation and utilizing it in full field modeling.

Literature Review

In order to investigate the physics of fluid flow in reservoir, small scale simulation is necessary. The models need a high degree of discrepancy and input data on core scale.

Yamamoto was one of the first researchers who made a conceptual model for compositional simulation in one of Iranian hydrocarbon south reservoirs [3]. He concluded that for a given rock type, recovery from a block was a function of block height, environment in the fissures, and the rate of pressure decline in the fissures.

Fung described the effect of block to block processes in naturally fractured reservoirs by utilizing conceptual modeling [4]. He showed that most current dual porosity models use a single matrix block approach to calculate matrix/fracture transfer within a grid block. Kossack described realistic numerical models for fracture reservoirs [5]. In this study a fracture reservoir undergoing a water flood is simulated with a single porosity formulation where the matrix blocks are subdivided into core plug size grid blocks and the fractures are subdivided into even smaller blocks. He concluded that the single porosity simulation of the displacement of oil from a matrix block is sensitive to the size of the numerical grid blocks.

Saidi et al. discussed the gas /oil gravity drainage process in fractured reservoirs in Iran [6]. Early dual porosity models include those of Kazemi et al. and Saidi. Saidi modeled a fracture reservoir by dividing it into sectors in which the fracture was assumed to have infinite transmissibility.

Vidal et al. discussed oil reimbibition between stacked matrix blocks in naturally fractured reservoirs [7]. They concluded that the capillary continuity between blocks tended to minimize the reimbibition of the oil from the fracture to the matrix and consequently could enhance the produced oil flowing in fractures network to the wellbore.

Field Specification

This field is an onshore oil field located in the foothills of the Zagros Mountains in Iran. The Asmari, Pabdeh, and Bangestan reservoirs consist of fractured carbonates with a minor amount of clastic sediments. The production is mainly from the fractured carbonates. The original oil column was some 1000 meters thick, above the original WOC at around 1900 meters (sub-sea). An original gas cap in the Asmari formation overlaid with the oil column and the crest of the structure is around 250 meters (sub-sea). Lithologically, the two main reservoirs, namely Asmari and Bangestan, are composed of carbonates with a well developed fracture system. The oil has a gravity of 30 API.

Procedure

Grid

The simulations have been performed with two

main grid types, namely 1-matrix block grid and 6-matrix block grid. The matrix blocks are in each model subdivided into 6 numerical grid blocks in x-, y-, and z-direction. No dip is assumed in the simulations; but the general understanding is that a high dip could improve the recovery of oil from the matrix blocks, especially for the poorer rock types.

One layer of grid blocks on each side, representing the fracture system, surrounds all matrix blocks. Figures 1 to 3 show the outline of the single matrix block and the 6-stacked matrix block models and saturation distribution describeing the modeling of matrix, fracture, and dummy layers in more detail.

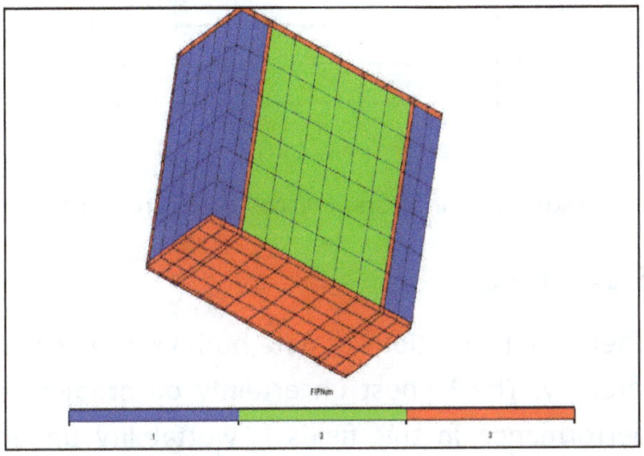

Figure 1: Single matrix block model grid showing the matrix block (green), the fracture system (red), and the dummy layers (blue) at the top and the bottom of the grid with an x-z cross section at y = 6

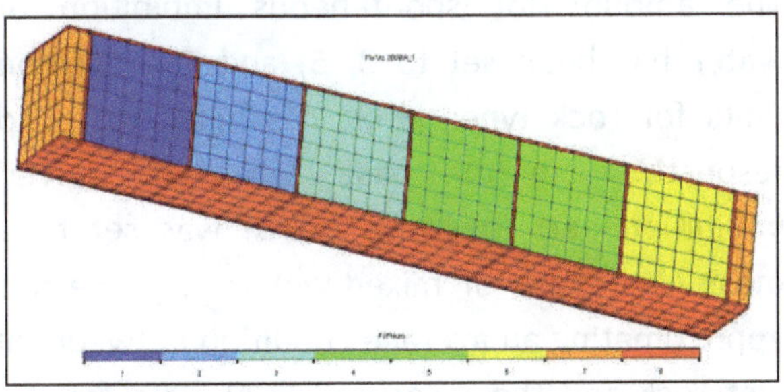

Figure 2: 6-stacked matrix block model showing the numerical grid block division with an x-z cross section at y=6 (dummy blocks, orange; fractures system, red; matrix blocks, remaining colors)

The grid block dimensions and porosity/permeability properties of the matrix block(s) investigated can be found in Table 1 and Table 2, respectively. The effect of grid block dimensions of different block sizes is assumed insignificant in these simulations. The fractures are assigned an effective permeability according to a fracture aperture of 0.3 mm giving a permeability value of 7560 mD. According to parallel plate theory, a fracture porosity of 0.01% and a total porosity of approximately 8% give an effective fracture permeability of 604.9 mD.

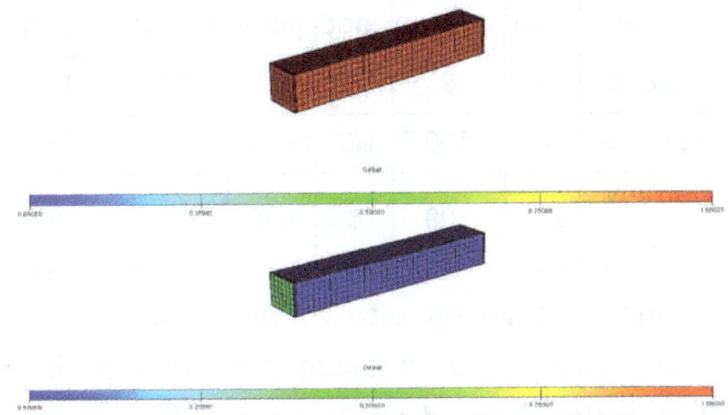

Figure 3: 6-stacked matrix block model showing the 3D saturation in the initial state (a) and after 20 years of production (b)

Table 1: Numerical grid block dimensions for one matrix block surrounded by fractures for all sensitivities (single matrix block and stack of matrix blocks).

Blocksize	DX (meter)	DY (meter)	DZ (meter)
1 meter	0.17	0.17	0.17
2 meter	0.33	0.33	0.33
3 meter	0.5	0.5	0.5
5 meter	0.83	0.83	0.83

The fractures were modeled using fracture grid cells of 3 cm to avoid numerical problems due to high throughput in very small grid cells, and the results were assumed not to be influenced by this modification since the rates of the advance of the WOC (2 meter/year) and GOC (3 meter/year) were modeled approximately equal to the field performance in the fracture system by manipulation of the injection and production rates in the small scale simulations.

PVT Properties

The single block and 6-block models are

assumed to be located in the highly fractured sectors of the reservoir at a depth of 1000 m. Thus the oil is undersaturated under initial reservoir conditions.

Table 2: Matrix and fracture property specifications (porosity and permeability) in single-porosity simulations

Property	Porosity (%)			Horizontal permeability (md) (permz = 0.5 x permx)		
Rock type	poor	medium	good	poor	medium	good
Matrix	6	9	12	0.1	0.5	1.0
Fracture	100	100	100	604.8	604.8	604.8
Dummy block	100	100	100	600	600	600

The model is set up with water, oil, and gas as active phases, and dissolved gas and vaporized oil as options turned on. Therefore, the amount of vaporization of oil in the presence of under-saturated gas is set by default parameters in VAPPARS, and the resolution of gas is set to infinity. The main PVT properties of the oil used in this study are displayed in Table 3 and Figures 4 to 6.

Table 3: Properties of oil used at 160 °F

Bubble point Pressure (psia)	2231
Density of total gas evolved (g/litre)	1.1046
Density of stock tank oil (g/litre)	0.8807
Density of reservoir fluid at Pb	0.7528
Average compressibility factor(5021-pb)	556.5

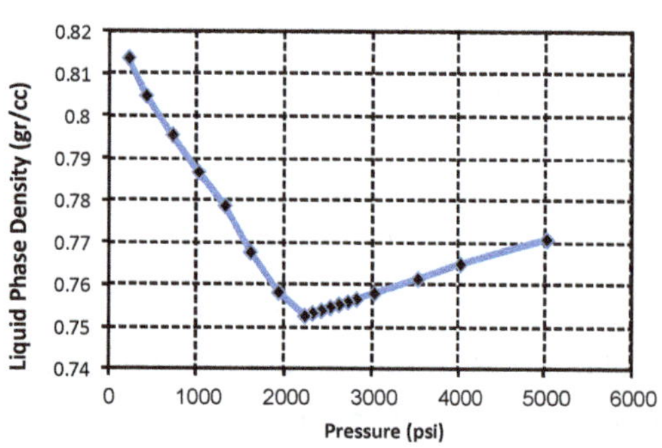

Figure 4: Variation of oil density vs. pressure

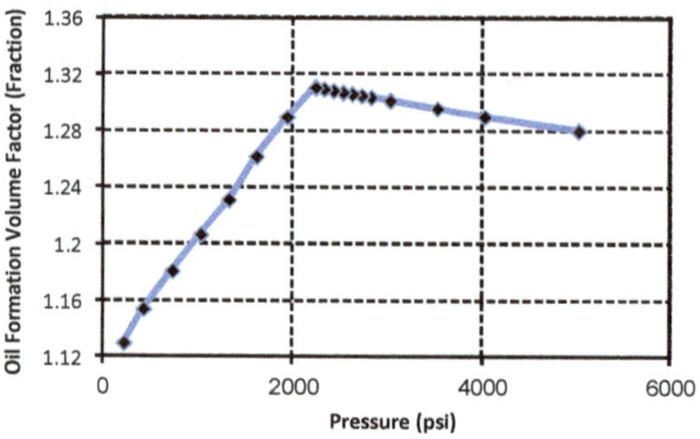

Figure 5: Variation of oil formation volume factor vs. pressure

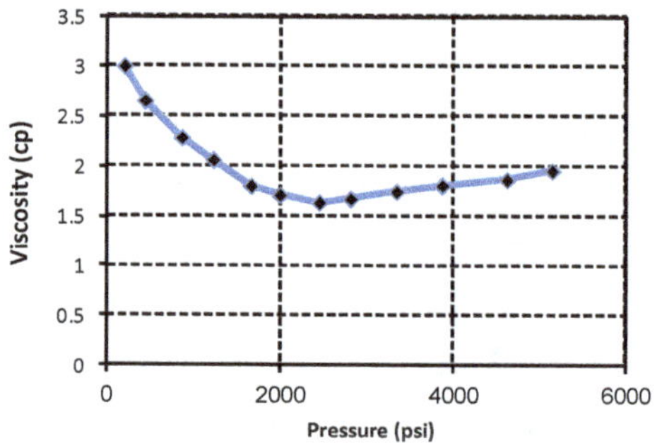

Figure 6: Variation of viscosity vs. pressure

SCAL Data

There are restrictions of data both vertically and laterally. The highest uncertainty on predicting performance in this fields is wettability under reservoir conditions (no wettability measurements of the reservoir rock is available in this field).

The amount for spontaneous imbibition of water has been set to 3, 5, and 7 saturation units for rock types poor, medium and, good respectively in the base case. The forced imbibition part (negative Pcow) was set to a most likely case of mixed-wet rock properties approximating an average condition between oil wet rock and the measured SCAL data for the field (assuming core material was water-wet after cleaning). The water-oil drainage and imbibition and the gas-oil drainage curves are given in Figures 7-12 (medium rock type).

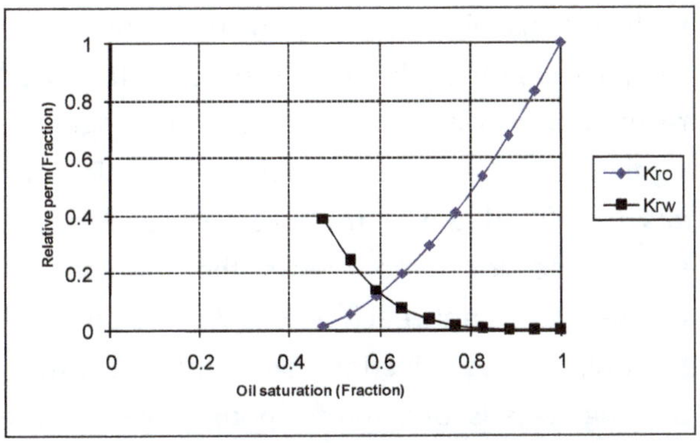

Figure 7: Water-oil drainage relative permeability curve (medium rock type)

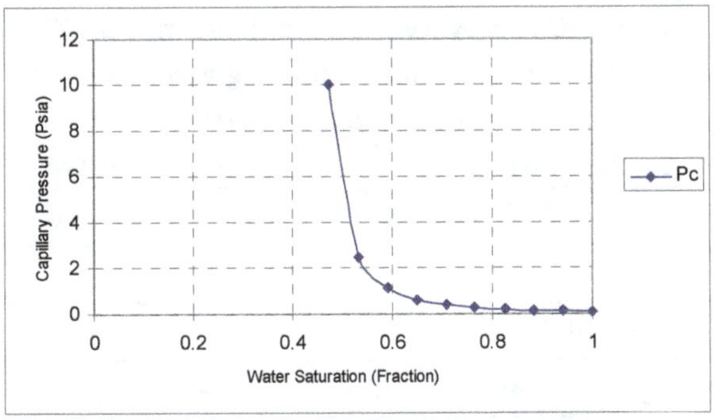

Figure 8: Water-oil drainage capillary pressure curves for medium rock type (threshold pressure= 0.1 psia)

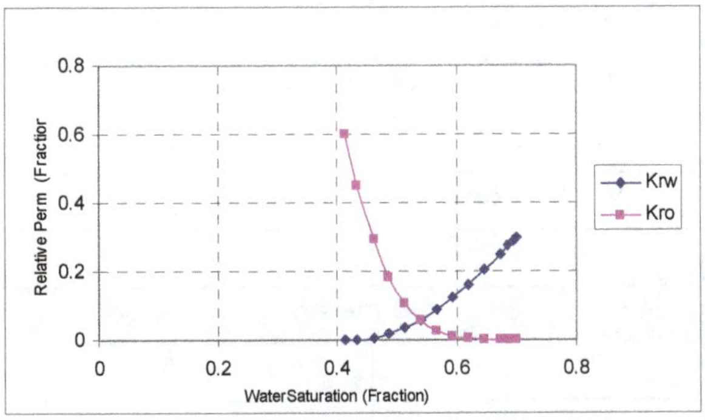

Figure 9: Water-oil imbibition relative permeability curves (medium rock type)

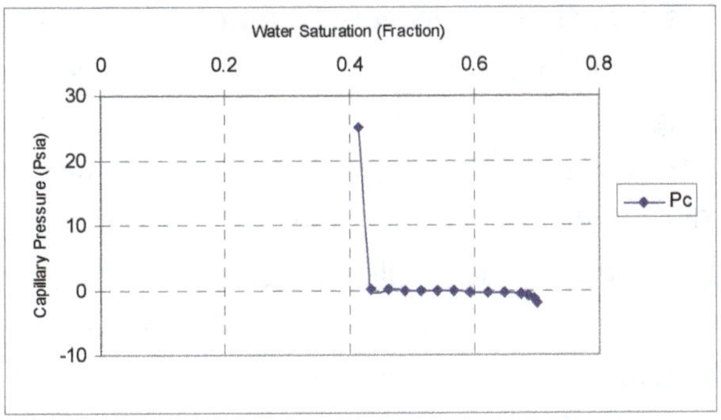

Figure 10: Water-oil imbibition capillary pressure curves (medium rock type)

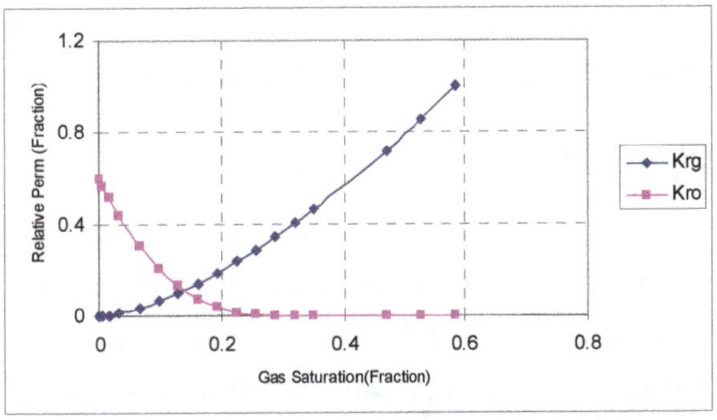

Figure 11: Gas-oil drainage relative permeability curves (medium rock type)

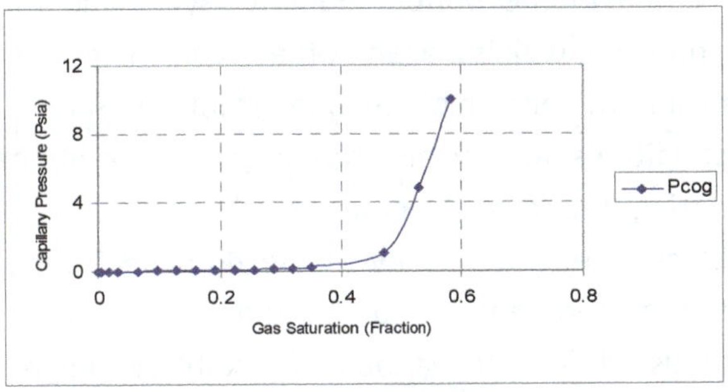

Figure 12: Gas-oil drainage capillary pressure curves for medium rock type (threshold pressure= 0.03 psia)

Water injection simulations have been performed with the hysteresis option switched on. The default hysteresis options have been used in Eclipse. Gas-oil hysteresis has been investigated assuming no imbibition gas-oil capillary pressure curve; however, since no significant differences with the non-hysteresis simulation were observed, this option was turned off for gas injection sensitivities.

Less favorable conditions for water imbibition have also been studied in these models, and the water displacement efficiency is as expected highly dependent on the amount of spontaneous and forced imbibition specified in the input SCAL curves. Since these parameters are poorly defined in this carbonate rocks due to limited and uncertain data in the SCAL database, water displacement efficiencies in the range of 10-40% for the three rock types in the base case could be considered as normal values for this field. Water displacement efficiency could be very low assuming more oil-wet reservoir conditions; values below 10% are normal in other carbonate reservoirs in the Middle East.

Well Specifications

In all the simulations, 2 wells have been specified: one producer and one injector. When performing gas injection, the injector is placed in the top dummy layer and the producing well is placed in the bottom dummy layer. For water injection sensitivities, the water injector is located in the bottom dummy layers and the producing well is placed at the top of the grid. In order to be able to find stable numerical conditions and mimic the physical conditions during the displacement of oil in the matrix blocks (i.e. the velocity of the flooding fronts), the injection and production rates needed to be adjusted for the various matrix block heights (see Table 4). These injection and production rates imply small pressure depletion during flooding, and the effect of pressure on the ultimate recovery will be checked in the column model and full field model.

Table 4: Specified injection and production rates in the Eclipse runs

Block size	Injection rate (m^3/day)	Production rate (m^3/day)
1 meter	0.0005	0.0005
2 meters	0.001	0.001
3 meters	0.003	0.003
5 meters	0.008	0.008

For all the simulations, the BHP target for the injector is defined as 250 bars. No BHP limit is set for the producer. The simulations were conducted using no lift curves for well performance.

RESULTS AND DISCUSSION

Single Matrix Block Model Results after 7 Years

Table 5 shows the final oil recovery after 7 years of gas injection in the single matrix block model with various block sizes. Saidi et al. discussed the gas/oil gravity drainage process in fractured reservoirs in Iran [6]. Gas gravity drainage is a significantly slow process going on long after the gas has flooded the fracture surrounding the block, and for large blocks the ultimate oil recovery does not reach before 7 years after the start-up of gas injection. Thus the gas-gravity drainage rate is low, and for poor rock types the gas has problems entering the block for all matrix blocks less than 2 meters.

Table 5: Oil recovery factor (drained matrix oil to initial matrix oil in place) for single matrix block after 7 years of gas injection

Single matrix block	Recovery factors (%)
1 meter	
Poor Rock	2.1
Medium Rock	9.4
Good Rock	20.2
2 meters	
Poor Rock	7.0
Medium Rock	22.1
Good Rock	33.2
3 meters	
Poor Rock	7.0
Medium Rock	26.2
Good Rock	35.4
5 meters	
Poor Rock	5.4
Medium Rock	25.8
Good Rock	34.8

The expected oil recovery factor of drained oil to initial oil in place after gas injection is in the range of 10-35% for poor to good rock types and different block sizes in the range of 2-5 meters. The effect of gas injection on oil recovery is governed by the capillary pressure curve given for each rock type and block height. The maximum recoverable oil is to some extent influenced by the residual oil saturation S_{org} given in the relative permeability curves. From the simulation results, it is concluded that the water imbibition process is a fast recovery mechanism for all block heights dependent on injection rate and fracture volume that has to be flooded to achieve full recovery of the block. A typical cubic matrix block is drained in approximately 0.5-2 years (dependent on block size) to residual oil saturation under a typical field water front, advances 1-2 meters per year. The expected recovery factor of drained oil to initial oil in place after water injection is in the range of 15-35% for poor to good rock types, and different block sizes in the range of 1-5 meters. The results will be highly influenced by the wettability conditions in the reservoir.

The 6-stacked Matrix Block Model Results after 20 Years

Table 6 shows the final oil recovery in the 6-stacked matrix block model after 20 years of gas injection. The recovery is reported in the first block below the dummy layers with the gas injector.

Table 6: Oil recovery factor (drained matrix oil to initial matrix oil in place) for 6-stack matrix block model after 20 years of gas injection

6- Stack matrix block	Recovery factors (%)
1 meter	
Poor Rock	2.5
Medium Rock	10.0
Good Rock	21.9
2 meters	
Poor Rock	10.9
Medium Rock	24.4
Good Rock	36.9
3 meters	
Poor Rock	16.0
Medium Rock	31.5
Good Rock	41.9
5 meters	
Poor Rock	17.1
Medium Rock	33.8
Good Rock	44.30

Gas injection works as a major recovery mechanism for medium and good rock types at all block heights; it is also suitable for poor rock type with block heights of more than 2 meters. The oil recovery results are in the same range as for the single block model but the delay of oil recovery by gas gravity drainage is clearly seen in the results indicating at least 8-10 years to reach an ultimate oil recovery for typical block heights.

The re-imbibition process and its consequence of delaying oil recovery in the stack compared with single matrix block recovery have not yet been fully quantified for this field. The simulation results showed that the delay in oil recovery (the time required to reach a specific oil recovery value in a stack) by assuming no dip and no capillary continuity is linear with the number of block assumed contributing to the oil production at the bottom of the stack. This is due to the full re-imbibition of oil from a block into its neighboring block beneath it. Thus the oil from a top block has to drain through all the blocks below it to reach the producer.

Sensitivity Simulation Results in the Gas Injection Case

Grid Sensitivity

It is of theoretical interest to see how the oil recovery from the matrix changes as the discretization of the matrix block becomes coarser. Three additional discretizations were simulated as shown in Table 7. To determine the oil recovery from the regions of the reservoir, a comparison plot of the regional oil efficiency (ROE) from the matrix block vs. time for the four cases is given in Figure 13. As the results show, the single porosity model is sensitive to numerical block size. Grid sensitivity shows that the small grid blocks cause variation in oil recovery. Since a high capillary contrast, the initial flow from matrix to fracture is high.

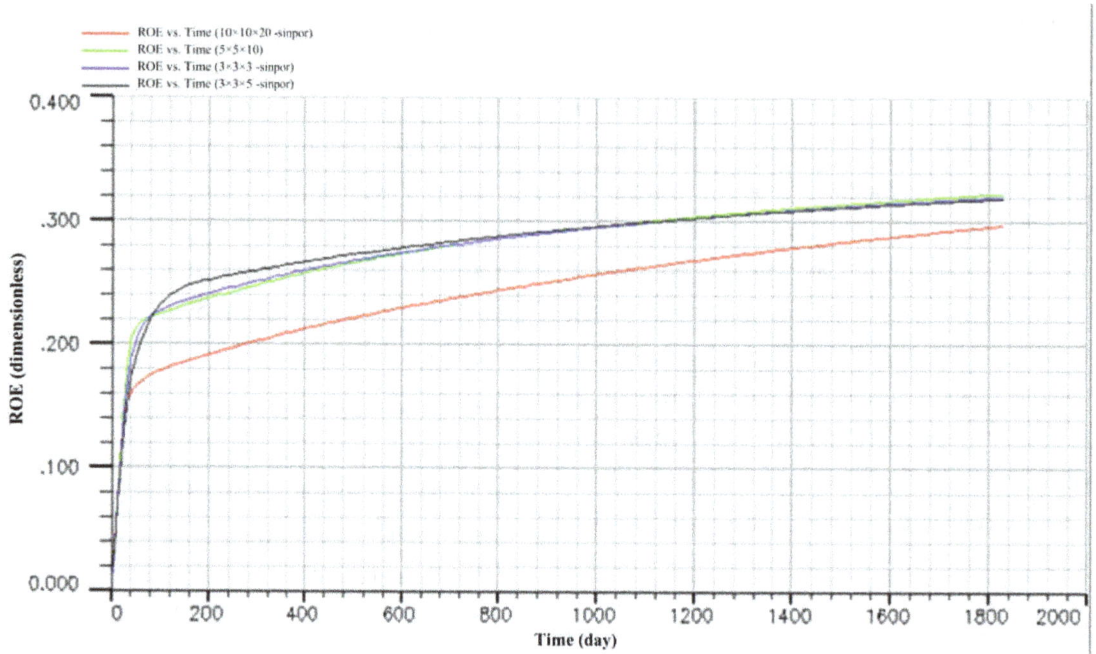

Figure 13: Comparison of oil recovery for various grid refinements in single porosity simulation

Block-to-Block Flow

Oil produced from one matrix block may often flow into, and enter, the under laying matrix block. This oil re-infiltration (often referred to as block-to-block flow) may take place due to actual physical contact between the blocks (permeable contact points), by oil droplets produced from one block falling on top of the next block and entering due to gravity and/or capillary forces, or through liquid bridges and film flow

Table 7: Grids for single porosity simulation

Number of grid cells in matrix	Total number of grid cells in matrix and fracture	Oil recovery at 1860 days
8x8x18	10x10x20	29.8 %
3x3x8	5x5x10	32.2 %
1x1x3	3x3x5	32%
1x1x1	3x3x3	31.2%

Block-to-block flow can easily be modeled in the simulation model by introducing a stack of matrix blocks either completely separated by horizontal fractures or with some degree of contact between them. In the model, with a stack of matrix blocks separated by horizontal fractures, all oil produced from one matrix block will re-infiltrate into the next matrix block (100% block-to-block flow). This results in a significant delay in the oil production, because a whole stack of blocks never has a production rate that is significantly higher than that of a single block. This delay in the recovery is illustrated in Figure 14, which shows the recovery vs. time for various matrix block positions in a stack that consists of 50 matrix blocks completely separated by horizontal fractures.

The time to reach 50% recovery for the top block is 60 days, similar to the single-block base case, since this block is unaffected by the production from the underlying blocks. However, all oil produced from this block flows into the underlying blocks causing a delay in the recovery in the blocks beneath; block number 5 in the stack reaches 50% recovery in about 500 days, whereas it takes more than 5500 days for the bottom block to reach a recovery of 50%. Clearly, the block-to-block flow results in a significant delay in the oil production; however, it should be noted that the ultimate recoveries are not directly affected. This means that in this field, where the estimated oil recoveries from the gas invaded zones are relatively modest (around 30% of IIOP), a delay effect could merely be caused by the re-infiltration; this means there is a significant potential for additional oil recovery.

The cases in Figure 14 represents 100% re-infiltration. It is likely that the degree of re-infiltration is less in the field. Factors that might significantly reduce the block-to-block flow are interbeded layers of non-fractured rock, impermeable fractures (shales and mineral deposits), sloped fractures, and viscous forces (pressure gradients in the fractures).

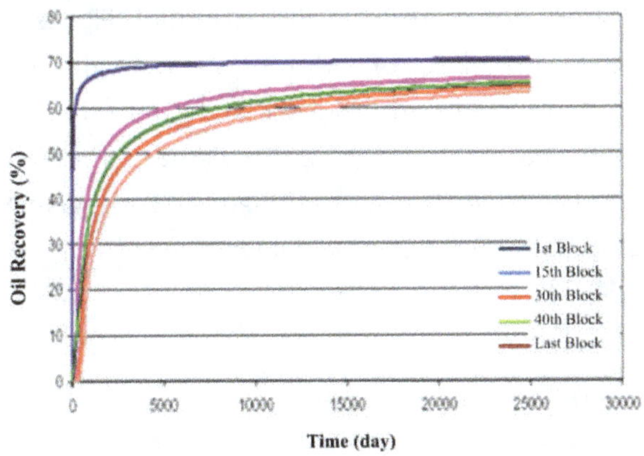

Figure 14: Stacks of 50 matrix block- Oil recovery vs. time for different matrix block position

Matching Single Porosity Results with Dual Porosity/Dual Permeability Models

The single matrix block model is matched in dual porosity mode with a dual porosity/dual permeability grid block model of 1x1x3 blocks, one for the matrix block in the middle and 2 extra blocks for the dummy injector and producer located in the fracture cells above and below. The 6-stacked matrix block model is matched with a dual porosity model with a dimension of 1x1x8 having the same requirement of 6 matrix blocks with fracture connections and 2 dummy fracture blocks above and below for the location of producer and injector.

The water displacement rate and final recovery of a single matrix block under water flooding can be easily matched by tuning the spontaneous imbibition part of the water-oil capillary pressure curve. The same conclusion can be drawn from matching the 6-stacked model. No further pseudoization to obtain correct oil recovery rate and ultimate final oil recovery are observed at this stage; however, additional work has to be done to confirm this conclusion for all matrix block geometries, fracture apertures, and fluid systems.

During gas injection, the single porosity solution of oil rate versus time and ultimate oil recovery in the block cannot be matched with the normal gravity drainage mode set in Eclipse. The invading gas saturation during gas gravity drainage in low permeability matrix blocks does not follow the assumption of vertical displacement. Some components of gas front movements in the horizontal directions at the side faces of the block are observed, and thus the match of the ultimate recovery during gas injection is easily achieved by using the alternative gravity drainage mode in Eclipse. The initial oil rate when the fractures surrounding the block are gradually filled by advancing gas and the gas gravity drainage rate when the matrix block is fully surrounded by gas is not easily matched in the simulations at this stage and further work is needed to find the final solution of this pseudoization problem. The forces and conditions affecting the recovery factor are summarized in Table 8.

The rate of production depends on the transmissibility between matrix blocks and fracture. High stack heights, low interfacial tension, and low capillarity tend to reduce the rate. In the case of gas injection, the residual oil swells, which increases the oil zone thickness and reduces the oil viscosity and hence improves the overall oil production.

Psuedoization in a Single-porosity Model

According to scale up theory, one can modify

the rock curves so that the simulation with a coarse grid will match the fine grid solution. To test this theory, a history matching process is made with the 1x1x1 grid case. Here the matrix block is modeled with one numerical grid block 4.9x4.9x4.9 m.

Table 8: The forces and conditions affecting the recovery factor

Mechanism	Gas Flooding	Water Flooding
Capillary Drainage	High Drainage Pc	-
Capillary Pressure Imbibition	-	Low Imbibition Pc Oil Wet Condition
Gravity Forced Drainage	Small Stack high Small Fluid Density Difference	

Adjustments to both the relative permeabilities (Krow and Krw) and the capillary pressure (Pcow) were made. Only a change in the capillary pressure provided a change in the flow, which allowed the match of the oil recovery results. A comparison of the solution oil recovery from the matrix block and the matched result is given in Figure 15.

Psuedoization in a Dual-porosity Model

In this section, attempts were made to match the calculated dual porosity oil recovery from the matrix block with the fine grid single porosity results which were the solution. Attempts were made to match the solution with changes in stack height, relative permeability, sigma, and capillary pressure. In order to study the effects of stack height, 6 different stack height sizes were used in the model. Figure 16 shows the effect of different stack height sizes on oil recovery. In order to study the effects of sigma parameter, 4 different sigma parameters were used in the model. Figure 17 shows the effect of different sigma values.

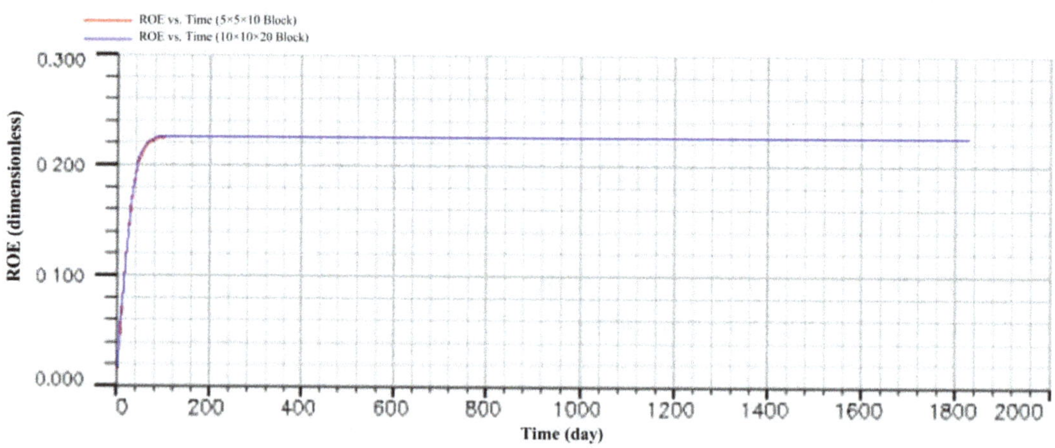

Figure 15: Comparison of oil recovery for various grid refinements in dual porosity simulation

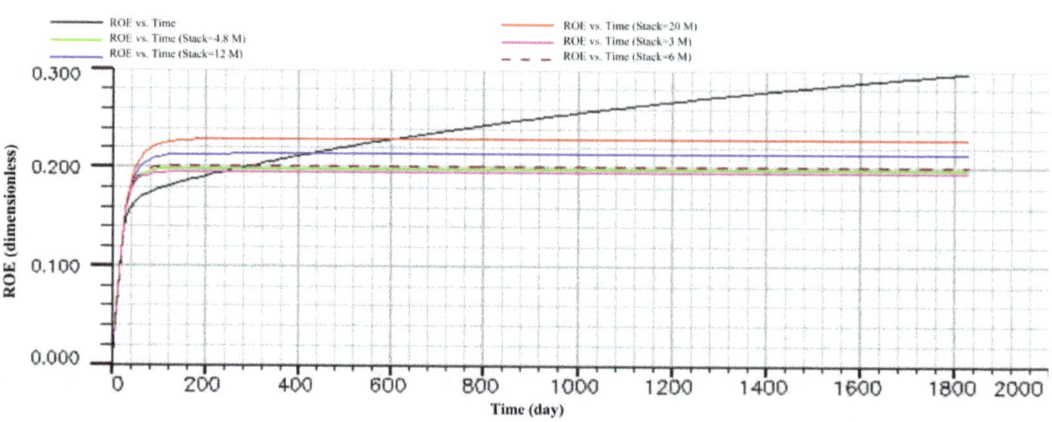

Figure 16: Effect of stack height block on oil recovery

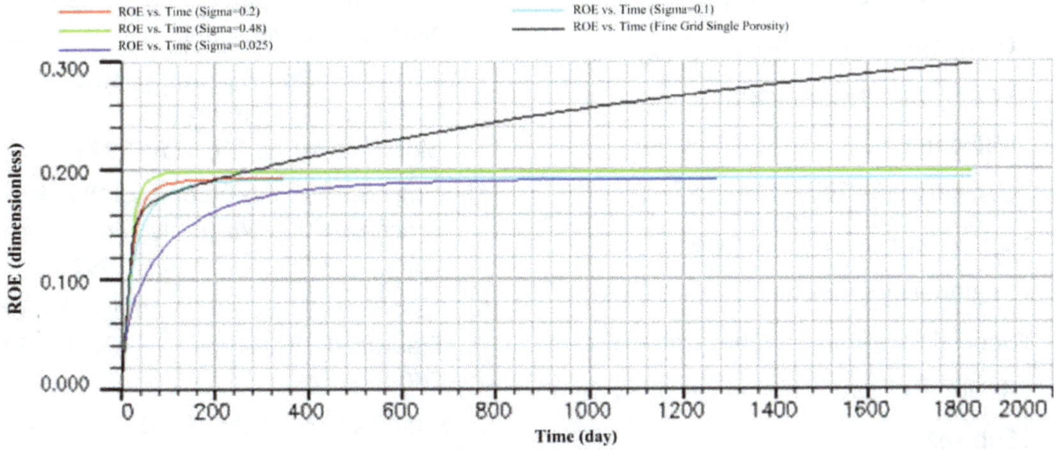

Figure 17: Effect of sigma parameter on oil recovery

CONCLUSIONS

Based on conceptual studies, the following conclusions and recommendations can be drawn for field oil recovery mechanisms.

Block-to-block flow results in a significant delay in the oil production; however, it should be noted that the ultimate recoveries are not directly affected. This means that in this field, where the estimated oil recoveries from the gas invaded zones are relatively modest (around 30% of OIIP), a delay effect could merely be caused by the re-infiltration; this means there is a significant potential for additional oil recovery.

The field performance of matrix oil recovery around 35-38% behind the advancing gas front implies no capillary continuity between individual matrix blocks with block sizes in the range of 2-4 meters. Small matrix block sizes and efficient stack heights of around 2-4 meters is also a possible scenario that will obey the field performance.

Carbonates of poor rock type quality are not drained by gas for matrix block sizes of 1 meter; a minimum block size of approximately 2 meters is necessary to force gas into the block. Expected oil recovery factors in the gas flooded zone is 11% for 2-meter block sizes and 17-18% for 5-meter block sizes for this rock type.

Expected oil recovery factors behind the gas front for medium and good rock types are respectively in the range of 10% and 22% for matrix block sizes of 1 meter and respectively 34% and 45% for block sizes of 5 meters.

The effect of pressure on gas-oil interfacial tension, gas and oil phase densities, and thus the resulting gas-oil capillary pressures in the matrix block during gas displacement decreases the above stated recovery estimates for gas gravity drainage (because of a decline in pressure).

Water imbibition ultimate recovery is expected to be in the range of 15-35% for poor to good carbonate rock with the additional effect of block size. These results are highly uncertain, with a large range of possible outcomes (2-50%) based on available SCAL data and experience from other carbonate fields in the region.

Grid sensitivity shows that the small grid blocks cause variation in oil recovery. Since a high capillary contrast, the initial flow from matrix to fracture is high.

NOMENCLATURE

BHP : Bottom hole pressure

Dx : Block length in x direction

Dy : Block length in y-direction

Dz : Block length in z direction

GOC : Gas oil contact

Krg : Gas relative permeability

Kro : Oil relative permeability

Krw : Water relative permeability

MD : Millidarcy

Pc : Capillary pressure

Pcow : Oil-water capillary pressure

ROE : Regional oil efficiency

SCAL : Special core analysis laboratory

S_{org} : Residual oil saturation in gas oil system

SS : Sub sea

WHS : Wellhead Separator

WOC : Water oil contact

REFERENCES

[1] Kazemi, H. et al., "Numerical Simulation of Water-oil Flow in Naturally Fractured Reservoirs," *SPEJ*, USA, December, **1976**.

[2] Saidi, A. M., "Mathematical Simulation Model Describing Iranian Fractured Reservoirs and Its Application to Haft Kel Field," *Ninth World Pet. Cong.*, Tokyo, **1975**.

[3] Yamamoto, Roy H., "Compositional Reservoir Simulator for Fissured System: The Single Block Model," *SPEJ*, May, **1972**.

[4] Fung, S. K., "Simulation of Block to Block Processes in Naturally Fractured Reservoirs," *SPERE*, November, **1991**.

[5] Kossak, H., "Realistic Numerical Models for Fractured Reservoirs," *SPERE*, February, **2000**.

[6] Saidi, A. M., Tehrani, D. H, and Wit, K., "Mathematical Simulation of Fractured Reservoir Performance Based on Physical Model Experiments," *Tenth World Pet. Cong.*, Bucharest, **1979**.

[7] Vidal, J, Aquitaine, E, "Gas/Oil Gravity Drainage: Laboratory Study of Oil Reimbibition and Streaming between Stacked Matrix Blocks," *SPE 25074, European Petroleum Conference*, France, November, **1992**.

[8] Eclipse Technical and User Manuals, **2010**.

14

Alkyl Imidazolines and their Ethoxylated Derivatives as Antioxidants for Hydrocarbon Products

Dariush Farkhani*, Habibolah Rezaei, Nader Gholami, Mitra Sina, Ali Asghar Khalili, and Ali Mehdizadeh

Chemical Science and technology Research Division, Research Institute of Petroleum Industry, West Blvd., Azadi Sports Complex, Tehran, 1485733111, Iran

ABSTRACT

Alkyl imidazolines have been reportedly used in a wide range of industrial formulations with different applications. Ethoxylated alkyl imidazolines with appropriate ethoxylation degrees can be used as antioxidants and retarders in the formation of peroxides resulting from oxidation in hydrocarbon media. In this work, ethoxylated imidazolines were shown to be more effective in hydrocarbon media in comparison with reference antioxidants. According to the experimental results, ethoxylated alkyl imidazolines (12 moles EO), as an antioxidant, were twice as efficient as zinc dialkyldithiophosphoric acid (ZDDP).

Keywords: Antioxidants, Corrosion Inhibitors, Ethoxylated Alkyl Imidazolines, Peroxides

INTRODUCTION

Alkyl imidazolines are prepared by reacting a polyethylenepolyamine structure (ethylenediamine, diethylenetriamine, triethylenetetraamine, and tetraethylenheptentamine) and a carboxylic acid such as tall oil fatty acid or acid cuts [1-7].

In general, in addition to their applications in pharmaceutical industry as anti-bacterial agents, bitumen and asphalt as emulsifiers, detergents, and demulsifiers, alkyl imidazolines have also been reported to function as additives in oil products with different applications.

Some of these functions include dispersing agents [8-10], metal inactivators [11], corrosion inhibitors [12-18], antirust agents [19], detergents [20-21], anti-gum formation agents [22], and antiabrasive agents [20]. Ethoxylated and propoxylated imidazolines have been used as multipurpose fuel additives [23]. These compounds prevent the decomposition of organic compounds accelerated in the presence of a metal. Antioxidants are one of fuel additives. An antioxidant is an organic compound used in rubbers, natural fats, food products, and fuels such as gasoline and diesel [24-25] to prevent and retard the oxidation and decomposition of molecules as well as gum formation. For example, an alkyl imidazoline has been used to prevent explosion. Furthermore, alkyl imidazolines have been applied in lubricants for corrosion and oxidation inhibition [26].

The most pronounced outcome of not using an appropriate additive in petroleum products such as lubricants and different organic-based fuels is their conversion into peroxides and hydroperoxides, which is also known as auto-oxidation.

***Corresponding author**

Dariush Farkhani

Email: farkhanid@ripi.ir

The acidification of the medium and corrosion and knocking in engines are some of the adverse effects of hydrocarbon degradation and conversion to peroxide, the mechanism of which is shown in Figure 1.

The oxidation of hydrocarbon substrates initiated by AIBN occurs according to a free radical mechanism shown in Figure 1.

Chain initiation:

$$\text{initiator} \longrightarrow r^{\bullet} \longrightarrow R^{\bullet}$$

Chain propagation:

$$R^{\bullet} + O_2 \longrightarrow RO_2^{\bullet}$$

$$RO_2^{\bullet} + RH \longrightarrow ROOH + R^{\bullet}$$

Chain termination:

$$RO_2^{\bullet} + RO_2^{\bullet} \longrightarrow \text{Products}$$

$$In + RO_2^{\bullet} \longrightarrow \text{Products}$$

$$In^{\bullet} + RO_2^{\bullet} \longrightarrow \text{Products}$$

Figure 1: Hydrocarbon degradation

where, $r^{\bullet}$ is a radical formed from the initiator; RH is a hydrocarbon; $R^{\bullet}$ and $RO_2^{\bullet}$ are free alkyl and peroxide radicals respectively; In stands for an inhibitor and $In^{\bullet}$ represents a radical resulted from the inhibitor [27].

A small amount of an antioxidant is added to prevent or retard the oxidation and formation of explosive peroxide intermediates. Depending on the effectiveness of the antioxidant, the amount of absorbed oxygen increases. Thus, antioxidants, which are soluble in hydrocarbon media are favorable.

Ethoxylated alkyl imidazolines have been synthesized and used as fuel antioxidants and their performance is evaluated in this work.

EXPERIMENTAL

Instrumentation and Chemical

Cumene was distilled over Na in a nitrogen atmosphere for 2 hours to obtain a fraction with a boiling point of 152 °C. Azobisisobutyronitrile (AIBN) was purchased from Fluka. Zinc dialkydithiophosphoric acid (ZDDP) was prepared from Lubrizol Company. The imidazoline used herein was supplied by a local chemical factory.

Oxidation Rate Measurement

The apparatus used was previously described elsewhere [28]. In order to evaluate the oxidation rate, the following general method was applied (the proper amount of the chemicals needed for each experiment will be described later in each case). 25 ml of the testing sample was dissolved in purified cumene and 0.35 g of AIBN was added to a double-wall borosilicate reactor. Then the reactor was immediately purged with pure oxygen gas under 1 atmospheric pressure to empty out the any gases in the reactor. In this case, the oxidation reaction starts as soon as the temperature of the flask reaches 60 °C and since then the internal pressure of the reactor is measured as the indication of the amount of consumed oxygen in the reaction. The initial reactor pressure made as a result of the released heat of the reaction should be reduced by adjusting the mercuric manometer; another manometer filled with a chemical having a lower density such as dibutyl phthalate is applied. The measured pressures are corrected by the application of a conversion factor against the relative densities of dibutyl phthalate to mercury.

Synthesis

Ethoxylated alkyl imidazolines are prepared by reacting alkyl imidazolines and ethylene oxide using a base catalyst at a known temperature and pressure (Figure 2).

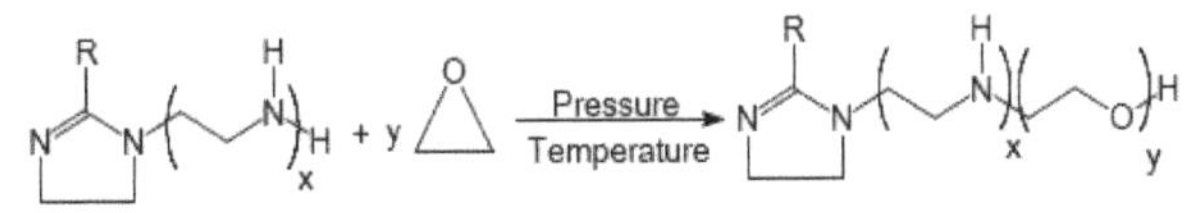

R=C_{12}-C_{18}

x=1-3

y=4, 8, 12, 18

Figure 2: Ethoxylated alkyl imidazolines

Evaluation

To evaluate the antioxidant performance, the amount of oxygen consumption and oxidation retardation time in isopropyl benzene (cumene) media in the presence of Azobisisobutyronitrile (AIBN) as a radical initiator are measured. Cumene is converted into cumene hydroperoxide in this process (Figure 3). The influence of ZDDP and ethoxylated imidazolines on the oxidation retardation and peroxide formation is subsequently studied.

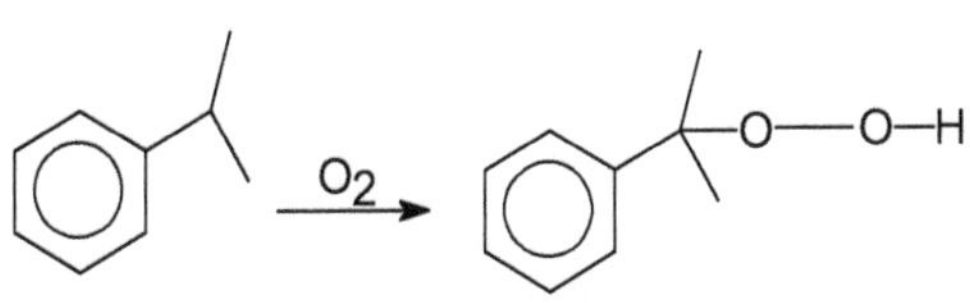

Figure 3: Isopropyl benzene(cumene) oxidizied to cumen hydroperoxide

A typical experiment of evaluation procedure is described as follows. The antioxidant agent and AIBN (0.35 g) were dissolved in pure cumene in a two-wall borosilicate reactor [28]. After purging the reactor with oxygen, the oxygen consumption rate was measured at atmospheric pressure at 60 °C by the difference of initial and final pressures of O_2. The effectiveness of ethoxylated imidazoline as antioxidant was expressed in terms of oxygen consumption as shown in Table 1 and Figures 4-9.

Table 1: Effectiveness of ethoxylated imidazoline as antioxidants in terms of oxygen consumption

Antioxidant type	Retardation time (min)
Lubrizol ZDDP	28
Alkyl imidazoline	30
Alkyl imidazoline 4 EO	25
Alkyl imidazoline 8 EO	38
Alkyl imidazoline 12 EO	52
Alkyl imidazoline 18 EO	24

As observed in Figures 4-9, oxygen is completely used at a retardation time of zero in the absence of any antioxidant and its consumption rate varies depending on antioxidant structure. For example, Lubrizol ZDDP, as a standard antioxidant, gives the same retardation time as alkyl imidazoline, while ethoxylated alkyl imidazolines of up to 12 EO increases this to 52 minutes, which is an appreciable enhancement.

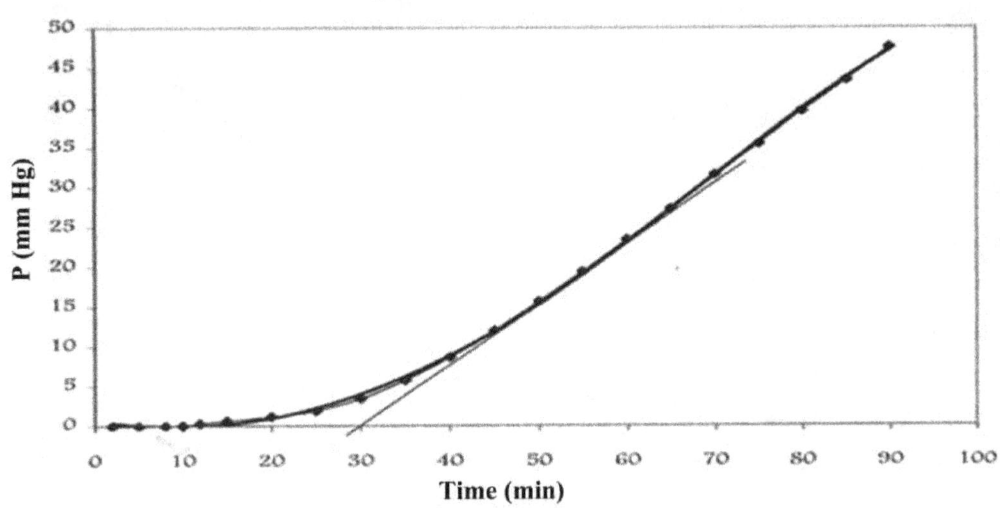

Figure 4: Oxygen consumption as a function of time using 0.001 moles alkyl imidazoline in cumene oxidation (25 ml) in the presence of 0.35 g of AIBN at 60 °C

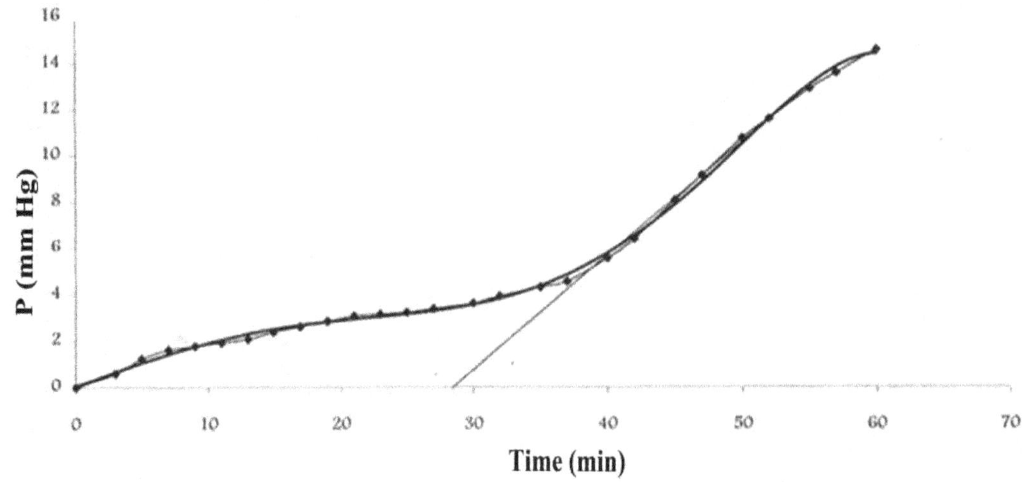

Figure 5: Oxygen consumption as a function of time using 0.001 moles Lubrizole ZDDP in cumene oxidation (25 ml) in the presence of 0.35 g of AIBN at 60 °C

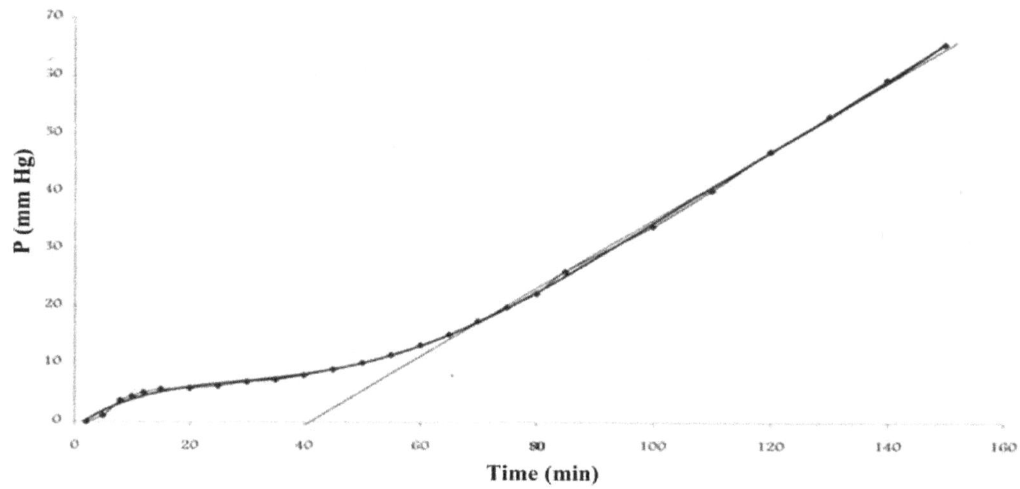

Figure 6: Oxygen consumption as a function of time using 0.001 moles ethoxylated (4 ethylene oxide units) alkyl imidazoline in cumene oxidation (25 ml) in the presence of 0.35 g of AIBN at 60 °C

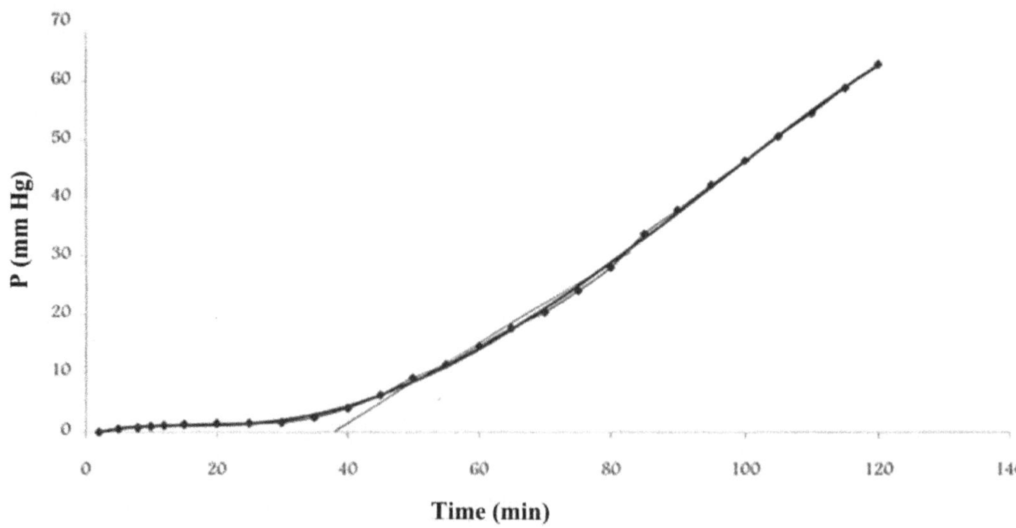

Figure 7: Oxygen consumption as a function of time using 0.001 moles ethoxylated (8 ethylene oxide units) alkyl imidazoline in cumene oxidation (25 ml) in the presence of 0.35 g of AIBN at 60 °C

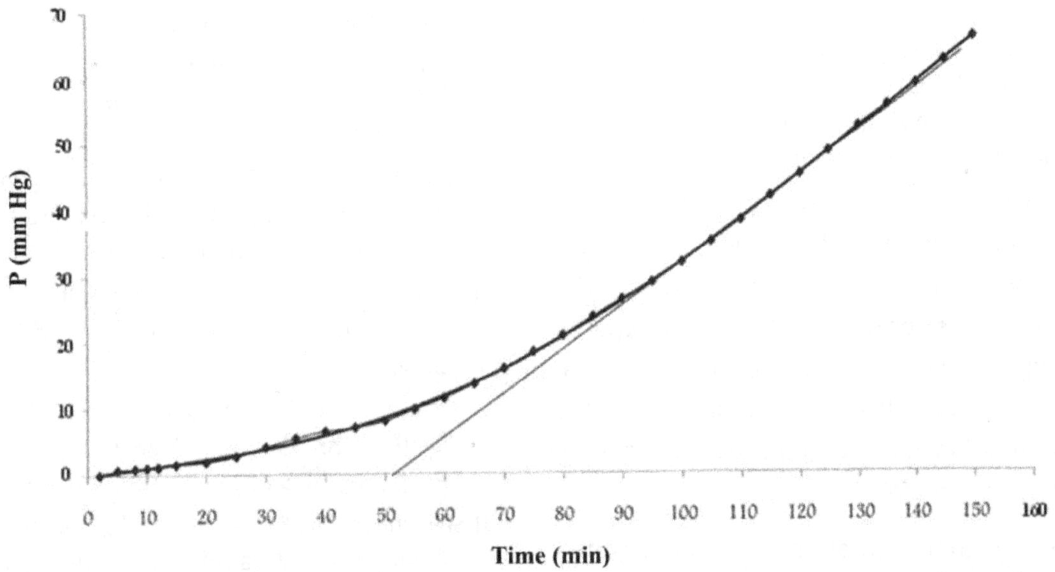

Figure 8: Oxygen consumption as a function of time using 0.001 moles ethoxylated (12 ethylene oxide units) alkyl imidazoline in cumene oxidation (25 ml) in the presence of 0.35 g of AIBN at 60 °C

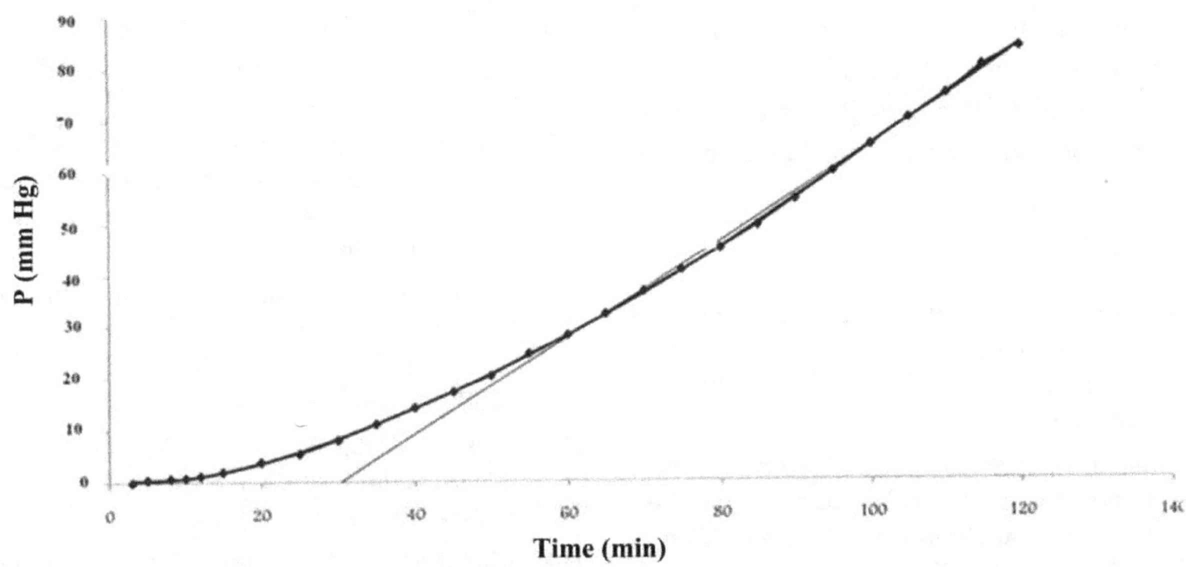

Figure 9: Oxygen consumption as a function of time using 0.001 moles ethoxylated (18 ethylene oxide units) alkyl imidazoline in cumene oxidation (25 ml) in the presence of 0.35 g of AIBN at 60 °C

RESULTS AND DISCUSSION

As previously mentioned, Alkyl imidazolines can be used as corrosion inhibitors, antirust, anti gum formation, and dispersing agents for various media in storage tanks and pipelines in hydrocarbon products such as fuels, gasoline, gasoil, etc. In the current work, ethoxylated alkyl imidazolines are used as antioxidants for the first time. The results obtained show that ethoxylated alkyl imidazolines containing 12 EO units are almost twice as effective as the commercial Lubrizol ZDDP.

Increasing ethoxylation chain length results in a higher HLB value which decreases oil solubility; this seems to inversely affect oxidation retardation. Based on the initial results and considering the relationship between the solubility of ethoxylated alkyl imidazolines and alkyl chain length, ethoxylation degree, and the specifications of hydrocarbon media, the design of a high performance antioxidant is plausible depending on hydrocarbon type.

CONCLUSIONS

A commercial alkyl imidazoline was used to prepare the ethoxylated derivatives with different ethoxylation degrees and the products thus obtained were investigated as antioxidants in hydrocarbon media. Alkyl imidazolines was highly reactive in the ethoxylation reaction. The antioxidancy performance of the ethoxylated alkyl imidazolines with an appropriate ethoxylation degree was satisfactory compared to ZDDP and its oxidation inhibition time was about twice as long as that of ZDDP.

A limitation in the application of ethoxylated alkyl imidazolines is the degree of ethoxylation. The higher the degree of ethoxylation is, the lower the oil solubility and antioxidancy performance of the product become. Considering the simplicity of ethoxylation and the availability of ethylene oxide, it is therefore recommended that this compound should be field tested as an antioxidant.

REFERENCES

[1] Sridhar, S. & Carter, R.G., "Diamine and Higher Amines, Aliphatic", pp. 485-518, Vol. 8, Kirk-Othmer Encyclopedia of Chemical Technology, **2001**.

[2] Bajpai, D., Tyagi, V. K., "Fatty Imidazolines: Chemistry, Synthesis, Properties and their Industrial Applications," *J. Oleo Science*, **2006,** *55*, 7, 319-329.

[3] Bistline, R. G., Hamps, J. W., LinField, W. M., "Synthesis and Properties of Fatty Imidazolines and their N-(2-aminoethyl) Derivatives," *JAOCS*, **1983**, *60*, 4, 823-828.

[4] Bajpai, D., Tyagi, V. K., "Synthesis and Characterization of Imidazolinium Surfactants Derived from Tallow Fatty Acids and Diethylenetriamine," *European Journal of Lipid Science and Technology*, **2008**, *110*, 19, 935-945.

[5] Chheda, "Asphaltene Dispersants for Petroleum Products," US Patent 2007/0124990, **2007**.

[6] Dow Chemical Co., *Ethylenediamine*, [online], (**2001, August**), 48, Available: http://www.dow.com/amines/pdfs/108-01347.pdf.

Author(s), *Book Title* (edition), [Type of medium], (**year, Month)**, *Vol.* (issue), pp., site/path/file [date accessed]

[7] Hopkins, D. L., "Asphalt Additive Compositions", WO Patent 8,807,066, **1988**.

[8] Makoto T. S., Tatsuya K. K., "Induction System for Vehicle Powered by an Air Propeller," US Patent 4,726,328, **1988**.

[9] Sakamoto, N., Kawashimo, H., Wakayama, K., Ikenaga, T., Wataanbe, Y., Shigemi, W., and Wakayama, H., "Stabilized Fuel Oil Containing a Dispersant," US Patent 4,746,328, **1988**.

[10] Guido, G., Henri, S., Groebke, Z. K., and Roger, N., "4-Imidazoline," US Patent 200901-2138(A1), **2009**.

[11] Buitrago, "Lubricating Oil Composition which Decreases Copper Corrosion and Method of Making Same," US Patent 7,056,871, **2006**.

[12] Oppenlaender et al., "Imidazoline-based Corrosion Inhibitors which Inhibit Corrosion Caused by CO_2 and H_2S," US Patent 4,388,214, **1983**.

[13] Ohlsen, J. R., Brown, J. M., "Corrosion Inhibitor Composition and Method of Use", US Patent 5,459,125 (A), **1995**.

[14] Mc Gregor, W., Libassi G., "Corrosion Inhibitors for the Petroleum Industry," EP Patent 1333108 (A3), **2003**.

[15] Shakirovich, R. U., Rais, Z., Ralovich, R. D., Evna, D. G., Evna, A. E., Karamovna, A., "Procedure for Production of Corrosion Inhibitors," RU Patent 2394941(C1), **2008**.

[16] Braga, T. G., Martin, R. L., McMahon, J. A., Arnold, Bernardus, A. Benjie, T., "Combinations of Imidazolines and Wetting Agents as Environmentally Acceptable Corrosion Inhibitors," US Patent 6,338,819B1, **2002**.

[17] Meyer, G. R., "Imidazoline Corrosion Inhibitors," US Patent 7,057,050, **2006**.

[18] Martin E. C. and Titaworth, H. R., "Rust Inhibitor for Gasoline," US Patent 2,919,979, **1960.**

[19] Davis M. E., and Dille, K. L, "Antirust Fuel Composition," US Patent 4,263,014, **1981**.

[20] Piotrowski, A.B., "Detergent compositions", US Patent 4,292,046, **1981**.

[21] Owiti, C. A., Khan, A., "Methods and Compositions for Stabilizing Fatty Acid Imidazoline Solutions," US Patent 5,530,137, **1996.**

[22] DeBenneville, P. L., Exner, L. J., Cheltenham, P., "Specific N-vinyl-2-imidazolines," US Patent 3,360,506, **1967**.

[23] Thayer Oakmont, H. I., "Ethoxylated or Propxylated 1,2,4- or 1,2,5-trisubstituted Imidazolines and Lecithin Mixtures and Reaction Products thereof," US Patent 3,615,291, **1971**.

[24] Russel, G. A., "Fundamental Processes of Autoxidation," *J. Chem. Educ.*, **1959**, *36*. 111-118.

[25] Karavalakis, G. S. and Stournas, S., "Impact of Antioxidant Additives on the Oxidation Stability of Diesel/Biodiesel Blends," *Energy Fuels*, **2010,** *24*, 6, 3682-3686.

[26] Lissant, K. J., Del, V., "Thixoprotic Oil-in-Water Emulsion Fuels," US Patent 3,490,237, **1970**.

[27] Oleynik, A. S., Trubinikova, Y. N., Kandalinsteva, N. V., and Grigoriev, "Studies on Antioxidative Properties of Nitrones from 3-Imidazoline-3-oxide, Dihydro-pyrazine-1,4-dioxide and 2H-imidazole-1-oxide Series in the Reactions with Peroxy radicals," *Chemistry for Sustaiable Development*, **2007**, *15*, 555-559.

[28] Bateman, L. and Cunneen, J. I., "Oxidation of Organic Sulfides: Part III. A Survey of Autoxidizability of Monosulfides," *Rubber Chemistry and Technology*, **1956**, *29*, 1, 71-82.

15

An Investigation into Lithofacies of the Cretaceous Chishan Group in the Wubao Fault Zone of Gaoyou Sag, Subei Basin, China

Liu Xuan[1,2], Zhong Jianhua[2,3], Yang Yuping*[2,4], Li Yong[2], and Ni Liangtian[2]

[1] SINOPEC Exploration and Production Research Institute, Beijing 100083, China

[2] School of Geo-Sciences, China University of Petroleum, Qingdao 266580, Shandong Province, China

[3] Guangzhou Institute of Geochemistry, Chinese Academy of Sciences, Guangzhou 510640, Guangdong, China

[4] SINOPEC Zhongyuan Oilfield Exploration and Development Institute, Zhengzhou 450003, Henan, China

ABSTRACT

This study aims to identify favorable oil–gas reservoir facies in the Chishan group of the Wubao fault zone (Gaoyou sag, Subei basin, China) using the methods of outcrops and cores observation, granularity analysis, scanning electron microscope, log data, etc. The results suggest that the Chishan group in the studied area mainly develops desert sedimentary system, and contains five kinds of facies from bottom to top, namely dry salt lake, aeolian sand, intermittent river, fan delta, and salt lake facies; the Chishan group was divided two members according to the lithology and a sedimentary cycle of base-level. The lower member of the Chishan group in the Wubao fault zone contained dry salt lake, intermittent river, and aeolian sand facies. The upper member of the Chishan group in the Chenbao area consisted mainly of aeolian sand. The southwestern Zhousong area contained (from east to west) a succession of aeolian sand, intermittent river, fan delta, and salt lake facies. Among these facies, aeolian sand was divided into aeolian sand dune, interdune, and aeolian sand sheet three sub-lithofacies. In the aeolian sand dune, the sand dune was a typical micro-lithofacies and had a scattered distribution. The fan delta had a bead-like distribution along the main fault, in which mainly underwater distributary channel and sand bank were developed. The sand bodies of the sand dune, underwater distributary channel, and sand bank were all well developed, and had average porosity values of >20%, meaning that they were favorable oil–gas reservoirs. The interdune sediments were fine-grained and had low porosity and permeability. Hence the reservoir properties of the interdune sediments were poor, and they could represent either fluid interlayers of reservoirs or source beds.

Keywords: Typical Micro-lithofacies, Facies, Chishan Group, Wubao Fault Zone, Gaoyou Sag

INTRODUCTION

The Cretaceous Chishan group is an extensively developed desert sedimentary system in the Subei basin, China (Figure 1). The discovery of aeolian sandstone reservoirs of the Chishan group in the Chenbao oilfield, located in the Wubao fault zone of Gaoyou sag, has focused research attention on the sedimentary environment of the Chishan group in adjacent areas. Various authors have studied the sedimentary characteristics of the

***Corresponding author**

Yang Yuping

Email: 379707644@qq.com

desert sedimentary system of the Chishan group, especially the aeolian sand in the Jiangsu and adjacent regions [1-4]. However, no detailed studies have been done on the sedimentary environment of the transition zone from the aeolian sand to the salt lake. The Wubao fault zone, located in the transitional zone, has an area of about 650 km^2, but has been drilled for hydrocarbon exploration over an area of only about 150 km^2. The Chishan group is found in the depression of Gaoyou sag, and is distributed mainly in the Wubao low uplift, situated in reversed stair-step fault blocks and in the fault-nose zone. In addition, the low uplift is located on the right side of a steep slope zone. Because the sedimentary environment is poorly understood and the division of sedimentary facies is not sufficiently detailed, effective guidance has yet to be provided regarding the exploitation of oil and gas in the Wubao fault zone. Combined with information from outcrops in the city of Jurong and from cores in the Wubao fault zone, this study analyzes the sedimentary types and characteristics of the Chishan group in the Wubao fault zone, and aims to inform the exploration and development of reservoirs of the Chishan group in the studied area.

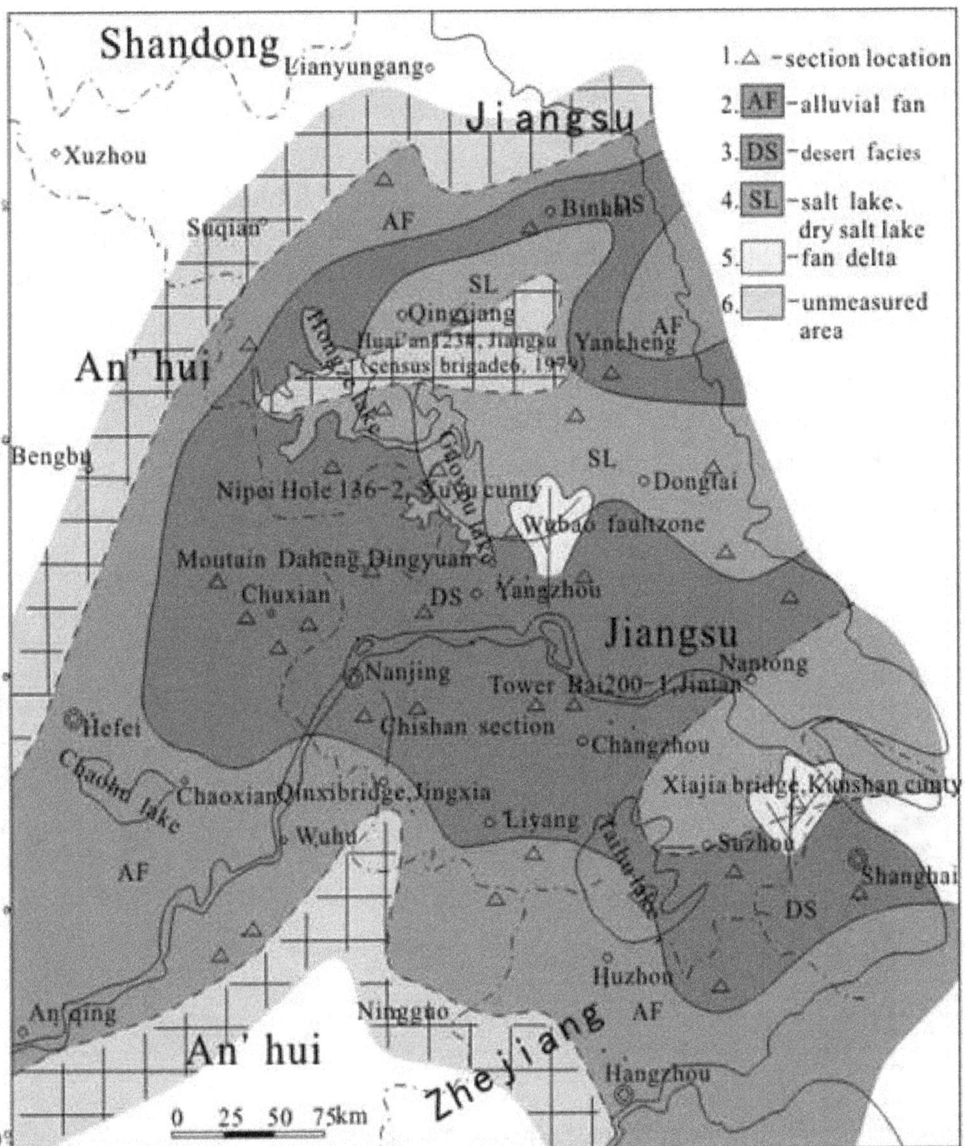

Figure 1: Sketch map showing the sedimentary facies of the upper Cretaceous Chishan group in the Subei basin and neighboring areas (modified from Yue et al., 1999 [5])

Lithologic Characteristics

The Chishan group outcrops as a set of brick-red sandstones in Jurong. The lower part of the group comprises interbedded or rhythmically bedded brick-red or purple, fine-grained sandstone, siltstone, pebbly sandstone, calcareous clay rock and clay rock, rich in calcium. The upper part contains red-brown or purple lithic quartz sandstone, graywacke and siltstone, and locally contains elongate lenticular beds that differ from the surrounding lithology; the bottom of this part commonly contains fine gravels. The outcrops of the Chishan group are scattered, and the thickness of the group varies between 300 and 700 m [6-7]. Drilling of the Chishan group in the Wubao fault zone, in the area of the Wubao low uplift, indicates that the thickness of the group varies greatly from 151 m in Well zhou 14 to 811 m in Well zhou 38-2, with the thickness on average being about 400 m. The lithologic characteristics of the group in some wells are similar to those of outcrop sections, but in other wells, the top of the group shows gray or dark-gray, massive mudstone, silty mudstone intercalated with thin-layered siltstone, or fine-grained sandstone; the middle to lower part is composed of gray or dark-gray, massive, fine-grained sandstone, and medium-grained sandstone intercalated with gray or dark-gray argillaceous clumps or strips. The Chishan group in the Wubao fault zone comprises a sedimentary cycle of base-level shallowing and then deepening in higher stratigraphic levels (Figure 2), and is separated by a large Stokes surface in the aeolian sand or a marked erosion surface in the salt lake. Based on all the above, the Chishan group is divided into two lower and upper members.

Division and Characteristics of Facies

The Chishan group in the Subei basin consists of three sedimentary environments: alluvial fan, desert, and lake [2,5]. The Wubao fault zone is located in a zone of desert–lake transition (Figure 1). On the basis of the results of the previous investigations of the paleodesert deposits [8-11], and through a comparative study with the sections of the Chishan group in Jurong city, the Chishan group in the Wubao fault zone is subdivided into five facies from bottom to top, namely dry salt lake, aeolian sand, intermittent river, fan delta and salt lake facies. The aeolian sand dune, interdune, and fan delta are further subdivided into several micro-lithofacies. Their characteristics are as follows:

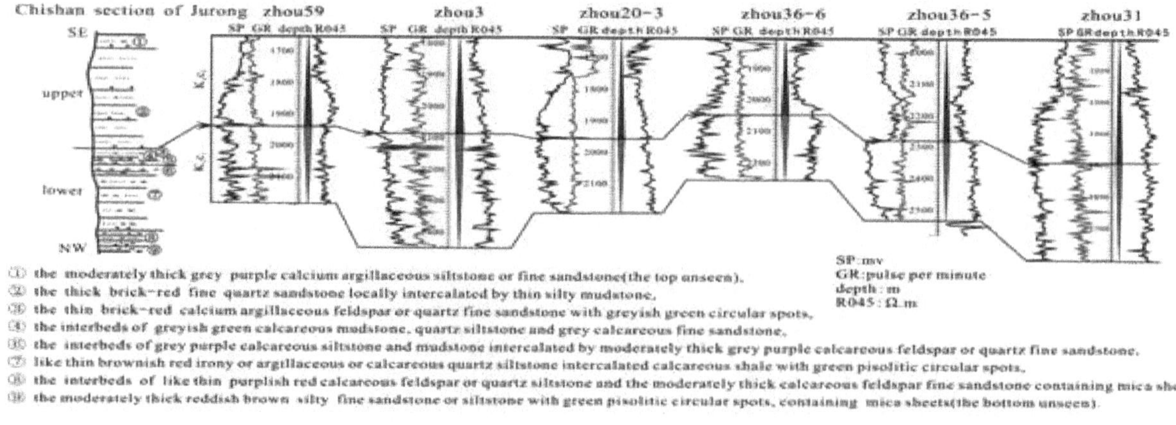

Figure 2: Correlation of the Chishan group in the Wubao fault zone and Jurong city

Dry Salt Lake

The dry salt lake is developed at the base of the Chishan group and has a thickness of about 10 m. The rock layer is dip-dyed by iron oxide and is mainly composed of gypsum-salt rock containing calcite, dolomite, magnetite, hematite, and limonite, with local zircon and other heavy minerals. The gypsum is mostly medium to coarse crystals, presenting microlaminations or stripes, which make the rock display a layered structure. The calcite is present as fine grains and the dolomite as powder crystal. All of the sediments described above are typical sediments and characteristics of a salt lake that formed during evaporation.

Aeolian Sand

Aeolian sand could be divided aeolian sand sheet, aeolian sand dune, and interdune in the current study.

Aeolian Sand Sheet

The aeolian sand sheet reaches a thickness of about 20 m in cores, and contains mainly siltstone and fine-grained sandstone. Parallel bedding is developed and horizontal laminae are observed in both outcrops and cores (Figure 3). The laminae become thicker upwards, and the interlamination is intercalated with thin layers of argillaceous siltstone and mudstone, which are laterally inconsistent and vary in thickness from 0.5 to 20 cm. The lower part of well log shows a finger, gradually changing upwards into a trend with only minor oscillations (Figure 4).

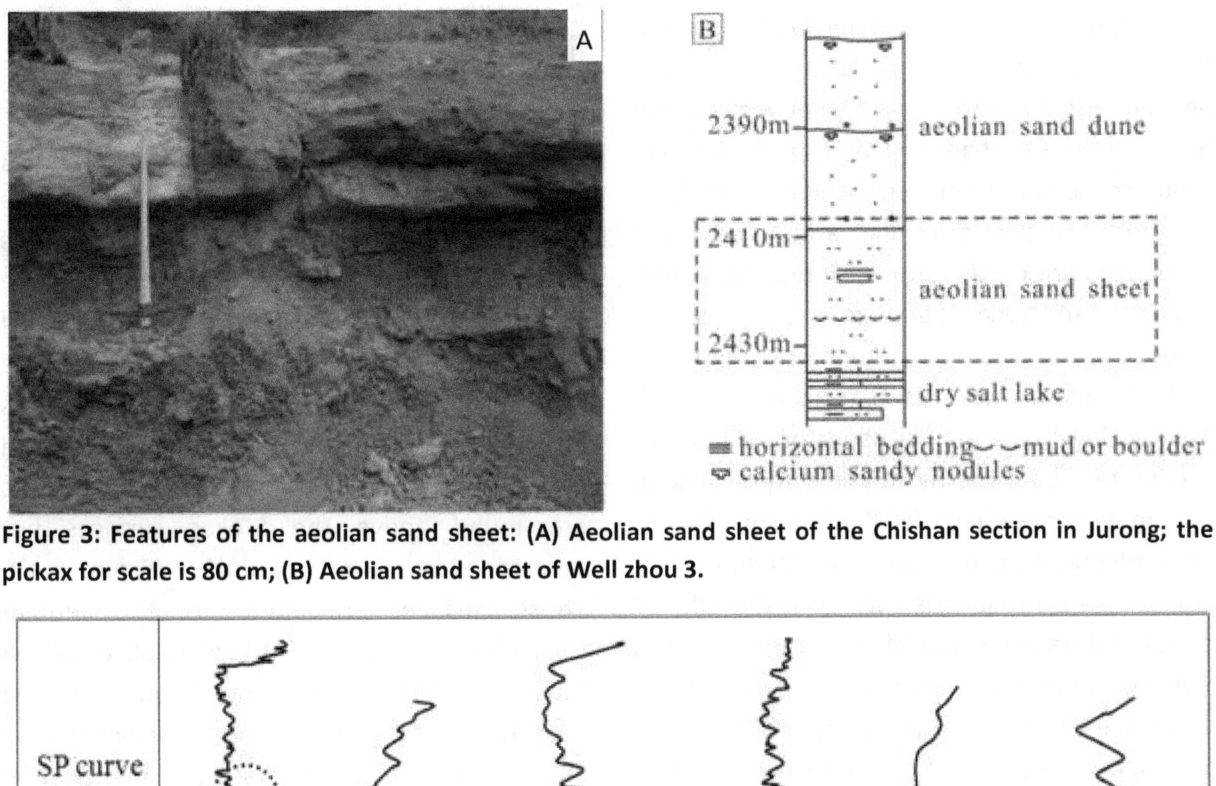

Figure 3: Features of the aeolian sand sheet: (A) Aeolian sand sheet of the Chishan section in Jurong; the pickax for scale is 80 cm; (B) Aeolian sand sheet of Well zhou 3.

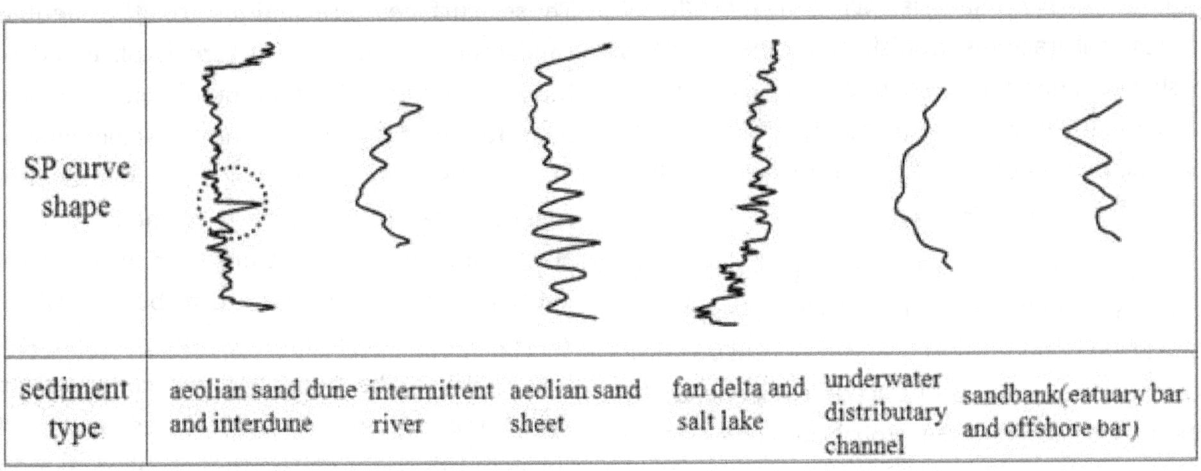

Figure 4: Features of electrofacies

The sub-lithofacies is located on the flat part of the transition between the desert edge and lake, where wind strength could have been weak, humidity has been relatively high, and the landscape could periodically have been first drowned and then exposed. The argillaceous siltstone and mudstone represent the sediments deposited during drowned, and the mudstone represents the dried clay exposed, which was later covered by thin sheets of sandy sediment.

Aeolian Sand Dune

The aeolian sand dune can be divided into the sand drift, sand cover, and sand dune. Of these, the sand dune is the main sedimentary type of this sub-lithofacies. The sand dune is the most typical and recognizable of the desert facies, and is widely distributed in the Subei basin, observed in both outcrops and cores. The thickness of the aeolian sand dune varies widely between 15 and 50 m in the Chenbao area and in the Chishan group sections in Jurong city, and reaches 50 m or more in the Zhousong area. The main lithology is brick-red or brownish-red, thick, fine-grained sandstone. Sand-grain surfaces are covered by an iron-rich membrane (hematite and limonite), known as "desert varnish" [1]. The packing of fine-grained sandstone is loose to moderate, but locally tight. The mineral components are quartz (60%-70%), subordinate lithic fragments, and feldspar, with minor dark-colored heavy minerals. Wind has produced conchoidal fractures, dish-shaped impact craters, v-shaped impact craters, direct impact ditches, and frost-like surfaces on quartz grains (Figure 5). The degree of mineral maturity is lower than that of modern desert sand.

The grain-size probability curves generally show three segments (Figure 6), which represent the saltation component, a small suspension component, and a small traction component. The inclinations of grain-size probability curves are in the range of 65 °-75°, which accounts for the tight distribution of sizes and good sorting. The roundness of sandstone particles is high in the Chishan group sections in Jurong city, but medium in cores. The rock is particle-supported and contacts with porous cemented. Mudstone matrix and cements are uncommon, both making up less than 7% of the rock volume, and there are many intergranular pores. Large-scale cross-bedding is developed in the aeolian sand dune, and biological remains and organism drill-holes are also found (Figure 7A–C). The dips of large-scale cross-bedding range between 26° and 35°, and the inclinations are bimodally distributed, between 80° and 145° and between 0° and 40°, in the Chishan sections in Jurong city. Combining the cross-bedding data with desert wind-direction data [12-13], the direction of the dominant wind is inferred to have been from the southwest.

Thin layers or stripes of well-rounded, coarse sand, and granular gravel are found on bedding surfaces (Figure 7D-E), and have the characteristics of ventifacts. In some sections, boundary surfaces of various dimensions cut through the beds; these are Stokes surfaces, which are the surfaces of aeolian sandstone representing erosion to the level of underground water. Gray calcareous spots and strips are found on these surfaces, and extend into the mudstone of the interdune. Sandstones are found above the Stokes surfaces but mudstones are placed below them (Figure 7F-G). These surfaces are unique to the aeolian sandstone. A layered structure is observed on the lee side of aeolian sand dunes, and may have been formed from the deferential sedimentation of a debris flow [11]. Our present investigation suggests that this feature is formed from the alternating effects of strong and weak wind, because its particle size distribution shows values that change abruptly, but not gradually (Figure 7H). All the aforementioned characteristics are typical of aeolian sand dunes, and such characteristics differentiate these sediments from those of other environments.

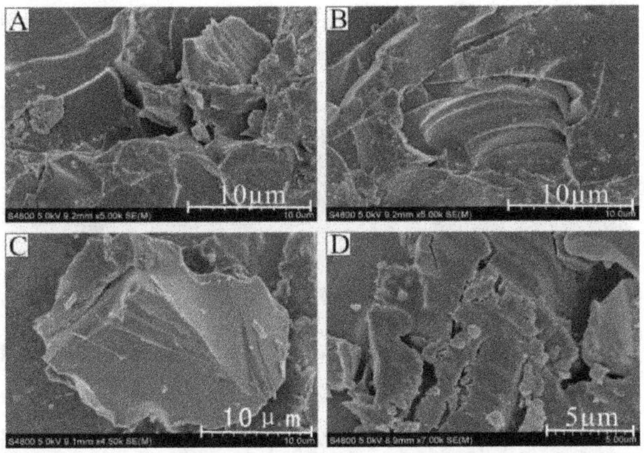

Figure 5: Scanning electron micrograph images showing the surface features of quartz grains; (A) Conchoidal fractures; (B) Dish-impact craters and frost-like surface; (C) V-shaped impact craters; (D) Direct-impact ditches.

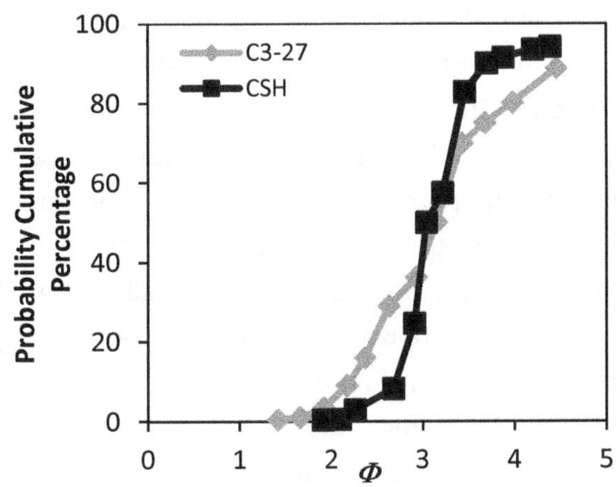

Figure 6: Probability accumulative curve of aeolian sand dune sediments. Sample CSH is from a member of the Chishan group in the Chishan sections (taken from Yue et al., 1999); sample C3-27 is from a member of the Chishan group in the Chenbao oilfield.

Interdune

The interdune can be divided into the wet interdune, dry interdune, and evaporative interdune. The wet interdune is developed mostly in the upper Cretaceous sediments of the Subei basin and represents the interdune depression that formed from ephemeral surface water derived either from fluctuations in the water table or from intermittent rain. The beds are purple mudstone to siltstone and are mainly thin (Figure 8A) or are lenses intercalated with thick sets of coarse-grained sandstone.

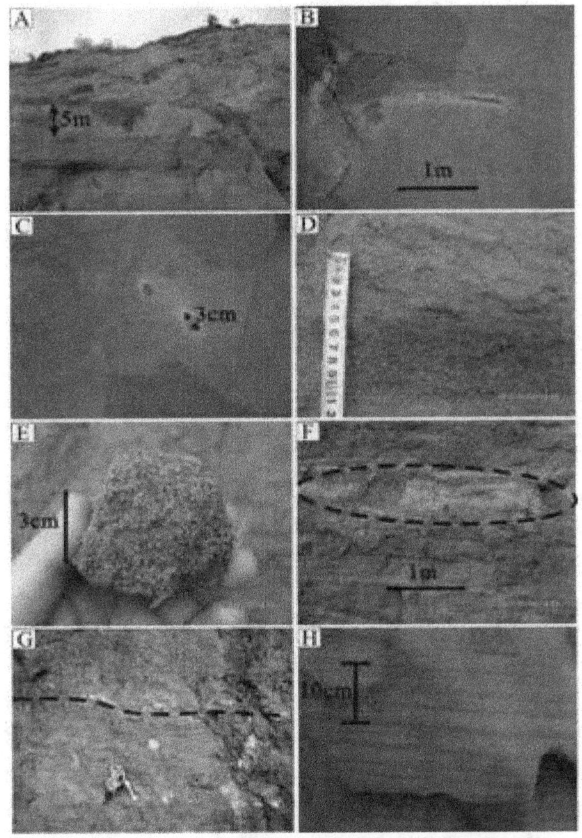

Figure 7: Features of aeolian sand dune sediments in the Chishan group in Jurong city; (A) Large-scale cross-bedding; (B) Biological remains on the lee side of sand dune; (C) Organism drill-holes on the lee side of sand dune; (D–E) Ventifacts; (F) A Stokes surface characterized by gray calcareous spots and strips; (G) Mudstone of the interdune into which the Stokes surface extends; (H) The layered structure on the lee side of an aeolian sand dune.

Ripple bedding and microscale horizontal bedding are observed in the Chishan sections in Jurong city (Figure 8B), and fractured fragments of clay and mud indicate intense evaporation thorough desiccation (Figure 8C). The mud fractures are mostly filled by aeolian sand, reflecting the typical sedimentary tectonics of a sand column or sand vein in a wet interdune (Figure 8D). The thickness of the wet interdune deposits ranges from several centimeters to more than 10 m, and they are irregularly distributed and limited in extent. The wet interdune is commonly overlain by sediments of the sand dune and is characterized by abnormal-amplitude finger logging, which contrasts strongly with the low-amplitude straight logging of sand dune sediments (Figure 4).

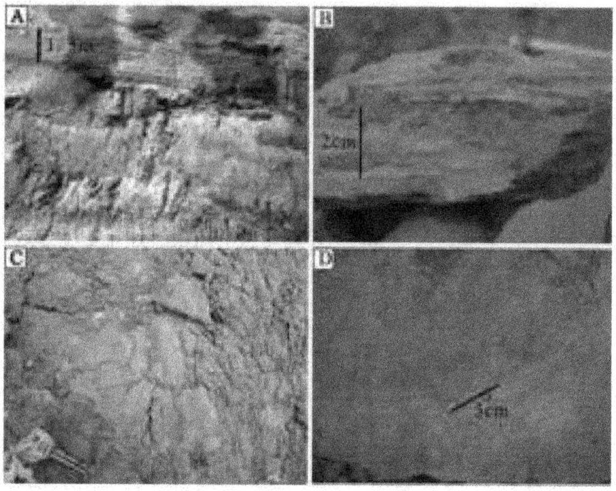

Figure 8: Sedimentary features of the interdune in the Chishan group in Jurong city; (A) Horizontal bedding of the interdune; (B) Ripple bedding of the wet interdune; (C) Shrunken mud fractures and dried clay fragments; The key for scale is 5 cm long; (D) Sand column or sand vein formed by younger aeolian sand filling fractures developed in mud.

Intermittent River

The intermittent river is found in both outcrops and cores, and the average thickness is about 10 m. The main lithologies are siltstone and fine-grained sandstone intercalated with mudstone, and the sediments are moderately to poorly sorted. The mineral components are predominantly quartz, with subordinate lithic fragments and feldspar. The grains are angular to subrounded. Both scour-fill and bioturbation structures are developed (Figure 9A, B), and both wave-bedding and small scale cross-bedding occur locally. Crumbly muddy gravels, pelitic strips, and calcium nodules are observed in cores (Figure 9C). All the aforementioned sediments are characteristic of intermittent river channel deposits.

Fan Delta

The fan delta is a large-scale facies found in both outcrops and cores, and attains a thickness of 100 m or more. The lithologies are gray, fine-grained sandstone, siltstone, and argillaceous siltstone.

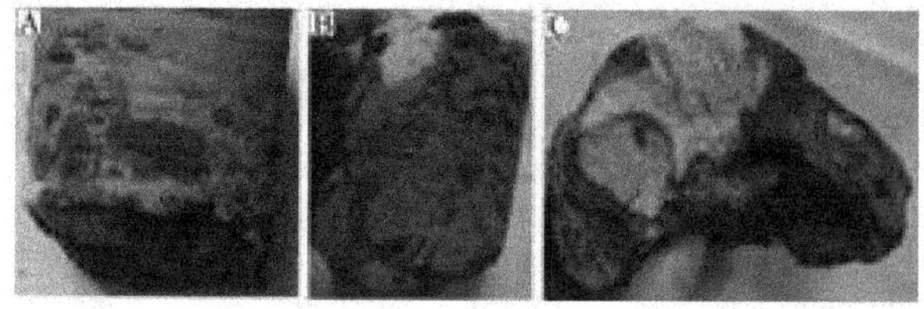

Figure 9: Sedimentary features of the intermittent river from Well zhou 3, 2231.75 m; (A) Scour in the riverbed; (B) Bioturbation; (C) Mud gravels and calcium nodules.

The mineral components are dominated by quartz (50-65%) with subordinate rock fragments (20-30%) and feldspar (15-25%), which reflects the higher rate of sedimentation compared with other facies and shorter transport distances of the sediments. The textural maturity of the sediments is low. The sediments are grain supported and contact with porous cemented, and they show moderate to tight packing. The cements are mainly dolomite and contain iron locally. Parallel bedding, ripple bedding, and convolute bedding are all widely developed (Figure 10A, C, and D), and graded bedding and reverse-graded bedding are observed locally (Figure 10E and D). Grayish-green, muddy gravels with particle sizes up to 3 cm containing irregularly shaped pyrite accretions are also found (Figure 10B).

Given the aforementioned characteristics, together with the small size of the drilled area (150 km^2) in the Wubao fault zone, the steep slope of the terrain (1.3-7.4%, calculated as the height difference between adjacent topographic contours, divided by the distance between the contours), and data from other studies [14-17], we suggest that these sediments belong to a fan delta adjacent to the aeolian sandstone, which was formed by an intermittent river running into a lake. In fact, these deposits are similar to the landward sedimentary facies (river and aeolian sand dune facies) of the Yallahs fan delta in the coastal area of southeast Jamaica. In accordance with the sedimentary characteristics of the Wubao fault zone, three micro-lithofacies can be distinguished within the fan delta: the underwater distributary channel, estuary dam, and far bar. The cores of the underwater distributary channel sediments are gray in color, show parallel bedding (locally graded bedding), and their logging curves are bell shaped. Both the estuary dam and offshore bar sediments consist of gray, fine-grained sandstone and siltstone, and their combined logging curves are funnel shaped and continuous. The amplitude of oscillations in the logging curves decreases downwards from the estuary dam to the offshore bar sediments (Figure 4).

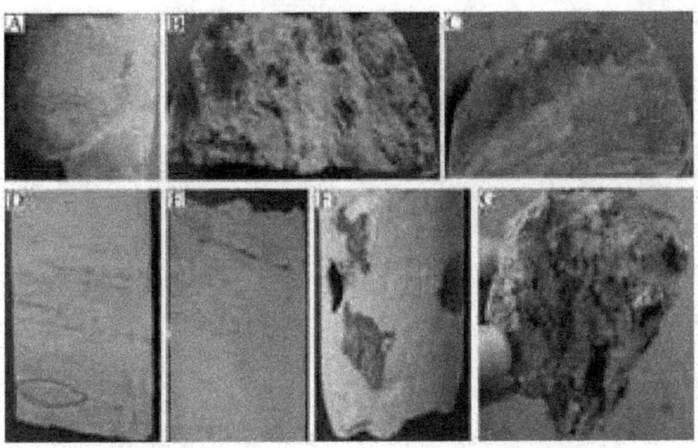

Figure 10: Sediments of the underwater distributary channel and estuary bar micro-lithofacies in the fan delta facies; (A) Parallel bedding in Well zhou 44-6, 1820.84 m; (B) Grayish-green, muddy gravels intercalated with sandstone, Well zhou44-6, 1822.3 m; (C) Ripple marks on the bedding surface, Well zhou 44-6, 1874.4 m; (D) Convolute bedding, Well zhou 44-6, 1822.1 m; (E) Graded bedding, Well zhou 44-6, 1822.3 m; (F) Pyrite accretions, Well zhou 44-6, 1826 m; (G) sandy mudstone, Well zhouxie 44-7, 1877.66 m.

Salt Lake

Salt lake sediments are gray or dark gray in color and are composed of gypsum, halite, dolomite, gypsiferous mudstone, and argillaceous glauberitic siltstone; glauconite is observed in siltstone. The cements are dolomitic and contain iron locally. High magnesium contents in some of the sediments indicate an affinity with sea water. Based on the above characteristics, we infer that the sediments belong to a terrigenous offshore salt lake environment.

Distribution of Facies and Typical Micro-lithofacies

Five facies, namely the dry salt lake, aeolian sand, intermittent river, fan delta, and salt lake, are found in the Chishan group in the Wubao fault zone of Gaoyou sag (Figure 11). The succession reflects a sedimentary cycle of base-level shallowing and then deepening in higher levels of the sedimentary column. On the basis of the sedimentary features observed in outcrops and cores, logging data, and relationships between lithology and electrical properties, the distributions of facies and typical micro-lithofacies are shown in Figure 12. The lower member of the Chishan group in the Wubao fault zone of Gaoyou sag comprises aeolian sand sheet, intermittent river, aeolian sand dune, interdune, and dry salt lake only around Well zhouxin 38-2. The thickness of the sand dune reaches 50 m or more, and the thickness of the sand sheet ranges from about 15 to 30 m. The upper member in the Chenbao area consists mainly of the aeolian sand dune, the thickness of which is 15 to 50 m, whereas the Zhousong area in the southwest of the Chenbao area contains the aeolian sand dune, interdune, aeolian sand sheet, intermittent river, fan delta, and salt lake from the east to the west. The distribution of sediments of the sand dune is relatively scattered, and it is developed only around wells zhou 12, zhou 3, and zhou 41-9, with a thickness reaching 50 m or more. The distribution of the fan delta is controlled by a reverse stair-step fault and shows a bead-like pattern along the trend of the main fault. In the fan delta front, the thickness of sandstone is 10 to 30 m, and the shale content increases upwards through the facies.

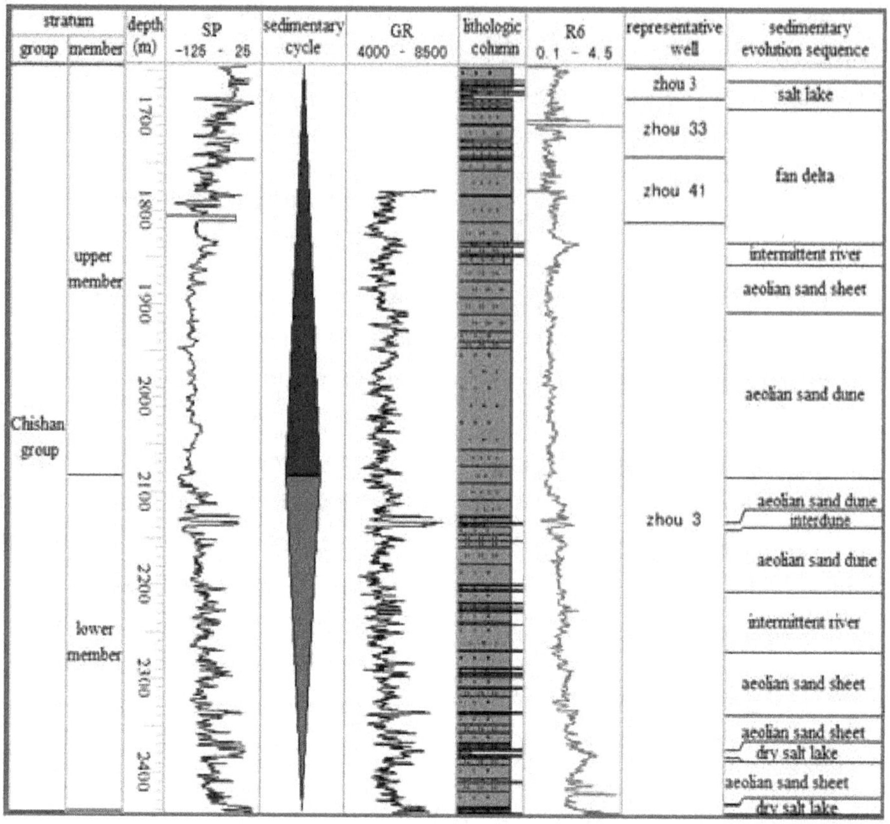

Figure 11: Sedimentary characteristics and inferred sedimentary environments of the Chishan group

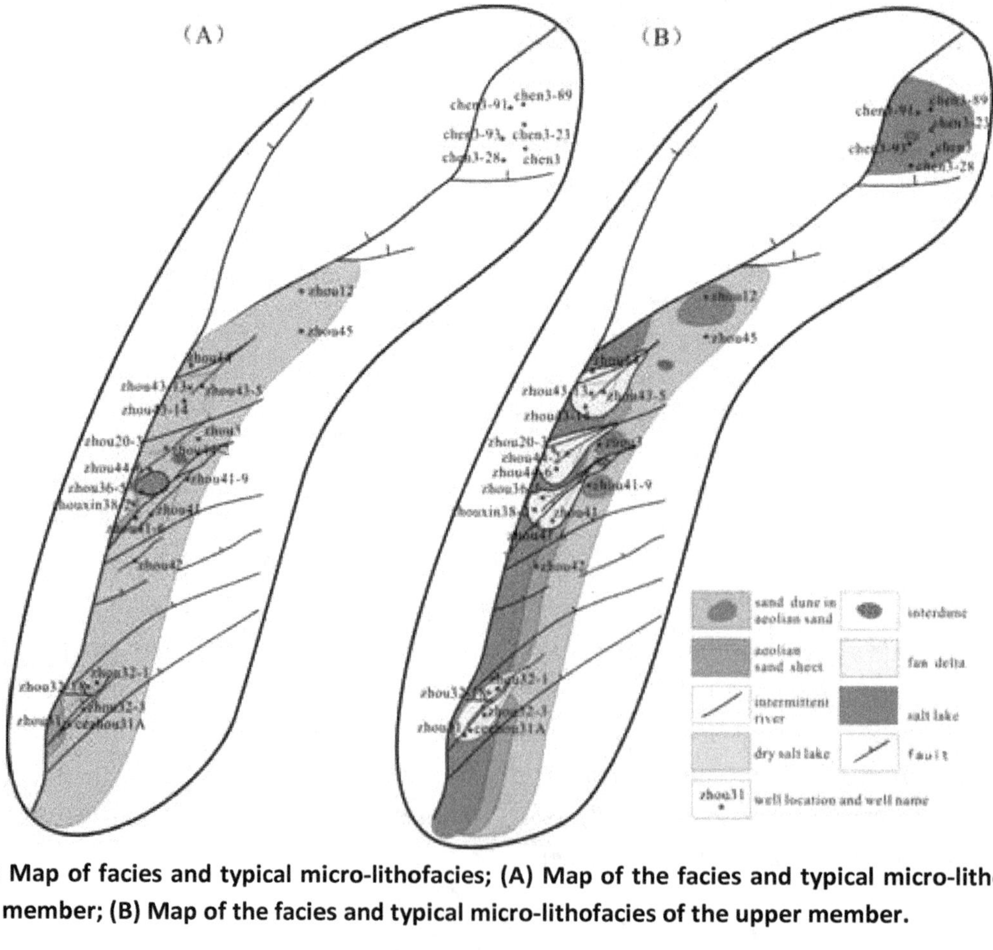

Figure 12: Map of facies and typical micro-lithofacies; (A) Map of the facies and typical micro-lithofacies of the lower member; (B) Map of the facies and typical micro-lithofacies of the upper member.

Geological Significance

Over the past two decades, oil and gas reservoirs related to paleodesert environments have been discovered in various locations both in China and elsewhere. Such reservoirs include large gas fields in Groningen in the Netherlands, Lehmann in England, and Unyielding in the North Sea [18]. Cretaceous desert sediments are widely distributed across China and are found in the Tarim, Ordos, Sichuan, Jianghan, and Subei basins. In 1997, the Chenbao oilfield was discovered in the Subei basin [19].

The key controls on hydrocarbon accumulation are the thickness of sandstone and the properties of reservoirs. The sediments of the interdune are fine grained, and the porosity and permeability are low; thus the reservoir properties of this sub-lithofacies are poor (Table 1). The sand dune, the underwater distributary channel, and the estuary bar are thick sedimentary bodies. In addition, the porosity of sandstone in the sand dune is about 20% on average, and in the underwater distributary channel and estuary bar about 30% on average. Furthermore, the permeability of these three sedimentary bodies is high (Table 1). Therefore, the sand dune, underwater distributary channel, and estuary bar are good reservoirs for hydrocarbon production, as confirmed by the oil production of wells Zhou44-6 and Zhou 3.

Table 1: Physical properties of samples of the Chishan group in the Wubao fault zone and Jurong city

Well	Sample depth (m)	Porosity (%)	Horizontal permeability (md)	Sedimentary environment	Outcrop sample	Porosity (%)	Horizontal permeability (md)	Rock density (g/cm^3)	Sedimentary environment
zhou 3	2031.65	19.6		Sand dune	1	11.86	9.52	2.34	Interdune
	2232.8	20.4							
zhou 44-6	1824.14	29	60.4	Underwater distributary	2	10.28	0.818	2.37	
	1825.29	31.6	72.5						
zhouxie 44-7	1874.3	26.8		Sand bank	3	26.97	155	1.88	Sand dune
	1876.96	37	587						
	1880.31	32.3			4	25.65	81	1.93	
	1881.31	32.4	311						
	1881.54	32.8	172		5	24.39	243	1.96	

CONCLUSIONS

The Chishan group in the Wubao fault zone contains five facies, namely dry salt lake, aeolian sand, intermittent river, fan delta, and salt lake, representing a sedimentary cycle of shallowing base level followed by deepening upwards through the sedimentary column. The sand dune is the predominant micro-lithofacies in the aeolian sand. The sandstone in the aeolian sand dune deposits shows typical aeolian features in terms of petrology, structure, quartz-particle surfaces, Stokes surfaces, large-scale cross-bedding, and layered structure. The fan delta is a new facies proposed in the studied area and is distinguished on the basis of facies characteristics, including the steep slope of the paleo-terrain, near-source deposition, rapid changes in depositional phase, the limited extent of the deposit, and the high gravel content of the sediment.

The lower member of the Chishan group in the Wubao fault zone contains dry salt lake, aeolian sand sheet, intermittent river, aeolian sand dune, and interdune. The upper member in the Chenbao area is mainly comprised of aeolian sand dune deposits, and in the southwestern Zhousong area contains the aeolian sand dune, interdune, aeolian sand sheet, intermittent river, fan delta, and salt lake. The sand dune micro-lithofacies has a scattered distribution. The distribution of the fan delta is tectonically controlled by reverse stair-step faulting and shows a bead-like pattern along the trend of the main fault.

The sand bodies of the sand dune, underwater distributary channel, and estuary bar are thick and have good porosity (average>20%) and permeability. These micro-lithofacies are favorable for oil and gas exploration. The sediments in the interdune are fine grained and the porosity and permeability are both low, on the basis of which the reservoir potential is judged to be low.

ACKNOWLEDGEMENTS

We thank Mr. Zhang Jianliang, Mr. Zhong Siying, Mr. Yan Xuefan, Mr. Wu Yanmei, Mr. Luo Hongfei, Mr. Kong Hong, and others working in the Geology Research Institute, Jiangsu oilfield, for providing data, guidance, and assistance. We are also grateful to Duan Hongliang (Geology Research Institute, Jiangsu oilfield) and Chen Xin (Research Institute, Bureau of Geophysical Prospecting Incorporation, China National Petroleum Corporation) for their guidance.

REFERENCES

[1] Ronglin Ch., "The Plain Research of the Aeolian Sandstones of Chishan group the Cretaceous Period, South Jiangsu Province," *Petroleum Geology & Experiment*, **1982**, *4*, 220–224.

[2] Wenzhe Y., Zhizheng Y., and Yuehua J., "Sedimentary Environment of the Late Cretaceous Chishan Formation in the Jiangsu Province," *Acta Sedimentologica Sinica*, **1997**, *15*, 1–7.

[3] Qinghua Ch., Fei P., and Dongfang Q., "Depositional and Reservoir Haracteristics of the Cretaceous Chishan Formation in Subei Basin and their Significance," *Marine Geology & Quaternary Geology*, **2008**, *28*, 95–100.

[4] Sheng D., Chengyan L., Baojn M., Jian L., et al., "Sedimentary and Diagenetic Characteristics of Aeolian Sandstones of Chenbao Oilfield in Northern Jiangsu and its Petroleum Searching Significance," *Journal of China University of Petroleum*, **2010**, *34*, 8–18.

[5] Wenzhe Y. and Baoliang D., "Study on Cretaceous Stratigraphic Sequence of Continental Basin in Jiangsu," *Olcanology & Mineral Resources*, **1999**, *20*, 287–344.

[6] "Bureau of Geology and Mineral Resources of Jiangsu Province," *Regional geology of Jiangsu Province and Shanghai* Beijing: Geological Publishing House, **1984**, 292–301.

[7] "Bureau of Geology and Mineral Resources of Jiangsu Province," *Geology of Nanking Hills* Nanjing: Jiangsu Science and Technology Publishing House, **1989**, 163–164.

[8] Mader D., "Depositional Mechanisms and Facies Models of Intertonguing Aeolian Environment and Fluvial Milieu in the Middle Buntsandstein in Aspects of Fluvial Sedimentation in the Lower Triassic Buntsandstein of Europe," Lecture Notes in Earth Sciences, Verlag Berlin Heidelberg (Springer), **1985**, 127–164.

[9] Yuwen L., Leyao Ch., and Xinsheng J., "Cretaceous–Eogene Desert Facies in Southern Sichuan and Northern Guizhou and Its Significance," *Sedimentary Facies and Palaeogeography*, **1996**, *16*, 42–51.

[10] Yuan X., Jian W., Xinsheng J., Minghui L., et al., "Sedimentary Characteristics of the Cretaceous Desert Facies in Ordos Basin and their Hydrogeological Significance," *Acta Sedimentologica Sinica*, **2005**, *23*, 73–83.

[11] Rodriguez-Lopez J. P., Melendez N., De Boer P. L., and Soria A. R., "Aeolian Sand Sea Development along the Mid-cretaceous Western Tethyan Margin (Spain): Erg. Sedimentology and Palaeoclimate Implications," *Sedimentology*, **2008**, *55*, 1253–1292.

[12] Hao W., Xiaozhuang C., Guoqing X., Yujie Z., et al., "Earkly Cretaceous Desert and the Reconstruction of Palaeowind Belts in Gaolan-Baiyin-Jingyuan Area, Gansu Province," *Geological Bulletin of China*, **2011**, *30*, 1077–1084.

[13] Li'an L. and Zaixing J., "Depositional Features of the Desert in the Paleogene Sichuan Basin and the Significance for the Reconstruction of Palaeowind Direction," *Geological Science and Technology Information*, **2011**, *30*, 63–68.

[14] Jing Y., Chengling Z., and Shanwen Z., "Genetic Model of the Deep Water Salt Lake of the Paleogene Sha4 Member in Dongying Sag," *Acta Sedimentologica Sinica*, **2000**, *18*, 114–118.

[15] Jing Y., "Study of Tempsetite Sedimentary of Paleogene in Huimin Depression," *China University of petroleum*, **2006**, *30*, 1–6.

[16] Jianbo G., Jianhua Z., Xiaolin X., Zhifeng W., et al., "The Fan Delta in the Altun Slope Bounded by the Altun Rift System in the Qaidam Basin," *Acta Sedimentologica Sinica*, **2004**, *22*, 206–215.

[17] Liqiong Ch., Hao L., Qidong L., and Ming Z., "Control of the Wubao Faulted Structure over the Migration of Petroleum in the Chenbao Oilfield of the Gaoyou Depression, Jiangsu," *Acta Geoscientica Sinica*, **2009**, *30*, 404–412.

[18] Reading H. G., "Sedimentary and Facies," Beijing, Science Press, **1985**, 105–126.

[19] Jiang Xinsheng., "Cretaceous Desert in China and Its Paleoclimate", Chengdu: Chengdu University of Technology Press, **2005**, 5–8.

16

Modeling of Asphaltene Deposition in Pipelines

Behruz Mirzayi[1]*, Seyed Ali Mousavi-Dehghani[2], and Mohammad Behruz -Chakan[1]

[1] Chemical Engineering Department, University of Mohaghegh Ardabili, Ardabil, Iran
[2] Research Institute of Petroleum Industry, Tehran, Iran

ABSTRACT

This paper is concerned with asphaltene deposition in fluid flowing through pipelines. Brownian diffusion and drag, gravitational, thermophoresis, buoyancy, and shear removal are considered as possible mechanisms in the asphaltene deposition process. The thermo-physical properties of the fluid were obtained from Iranian oil fields. A model was used in the pipeline deposition modeling to predict the asphaltene deposition rates under flow conditions. The effects of particle size, temperature gradient, and fluid velocity were studied on asphaltene deposition rate. The results showed that, among the above-mentioned mechanisms, the gravitational and thermophoresis forces played a significant role in the formation of the deposit under the flow conditions. To verify the model, some predictions were compared with the available aerosol deposition data in the literature.

Keywords: Asphaltene deposition, Drag, Thermophoresis, Buoyancy, Shear Removal

INTRODUCTION

The deposition of asphaltene on a pipe surface in oil flow is commonly encountered in petroleum industrial processes. Asphaltene is a polydisperse mixture of the heaviest and most polarizable fraction of the oil. Due to the tendency to precipitate and deposit, asphaltene causes many problem in oil production, transfer, and storage. Several researchers have investigated the asphaltene deposition problems under dynamic conditions, i.e. during the oil flow. Mansoori has studied the factors affecting the heavy organic deposition such as asphaltene in the well-head production of petroleum [1]. He proposed a model which considered four different effects (mechanisms) on asphaltene depositions including polydispersivity, steric colloidal, aggregation, and electrokinetic effects. Only one of these mechanisms, namely electrokinetic effect, is of great importance during oil transportation. The other factors are considerable under the static (stationary) conditions of petroleum fluid. The electrokinetic effect rises due to streaming potential generated by the flow of charged solid particles such as asphaltene. Mansoori showed that when the oil entered pipelines, positively charged asphalttene particles began to become attached to the pipe wall [1]. As the negatively charged oil flows toward downstream, the charge difference between wall and bulk fluid grows and leads to creating a potential difference. This potential difference causes the asphaltene particles to flow backwards and hence the probability of

***Corresponding author**

Behruz Mirzayi

Email: mirzayib@uma.ac.ir

their deposition increases. To demonstrate the existence of electrokinetic effect, Mansoori used the static phase diagram of some oils and did not apply any data for the dynamic deposition of asphaltene [1].

Ramirez-Jaramillo et al. have investigated the asphaltene deposition in production pipelines via a multiphase multicomponent hydrodynamic model [2]. They assumed the radial diffusion of asphaltene particles with two competing mechanisms: molecular diffusion and shear removal. To validate their model, Ramirez-Jaramillo et al. used available data of pressure-temperature-depth production profiles related to asphaltene deposits in two problematic wells from the southwest producing area of Mexico [2]. They used these data to predict the thickness of deposited asphaltene layer as a function of the axial coordinate for different oil flow rates. The results indicated that the wells presented the formation of asphaltene deposit located at various depths.

Escobedo and Mansoori studied the mechanism of the migration of suspended heavy organic particles towards the walls in oil-producing wells and pipelines [3]. They proposed an analytical model for calculating particle deposition coefficient under the conditions of petroleum fluids flowing in oil wells. They used eddy diffusivity, Brownian diffusivity, as well as inertial effects to develop their model. This model was able to predict the rate of particle deposition during various turbulent flow regimes. They compared the predictions of the proposed model with the available experimental data. Since experimental data for particle deposition from petroleum fluid flows were scarce, they used the deposition data from aerosols (deposition rates for aluminum and iron particles in air) to verify their model. The modeling results were in very good agreement with the experimental data. It was also shown that deposition rates decreased with increasing particle size. Eskin et al. analyzed the deposition mechanism of asphaltene in pipelines based on the theoretical and experimental approaches [4]. Their model consisted of three mechanisms for modeling particle transport to the wall, namely Brownian motion, turbulent diffusion, and turbophoresis. They also used the special Couette device to simulate asphaltene deposition in pipeline and tubing wells. In their experimental setup, Eskin et al. measured the mass of asphaltene deposits at the different running times [4]. These results were used to calculate the model parameters, such as critical particle size, particle-particle collision efficiency and particle-wall collision efficiency. This model was then used to determine the distribution of the thickness of the asphaltene deposit layer along vertical production tubing. The results showed that only particles which were smaller than a critical size were able to deposit.

In the previous proposed models, turbulent flow parameters were considered, but the most important mechanisms which had a crucial role in the asphaltene deposition of a laminar flow were not addressed. Also, in those models, the asphaltene deposition rate was not considered. In this paper, a theoretical model is developed to predict the asphaltene deposition velocity during fluid flow in pipelines. The proposed model takes into account the effects of main mechanisms such as Brownian diffusion and drag, gravitational, thermophoresis, buoyancy, and shear removal which influence the asphaltene deposition phenomena.

THEORY

Asphaltene Deposition Model

Many factors affect the asphaltene deposition phenomena in the petroleum fluids flow. Generally, in every flow regime (laminar or turbulent), it is always possible to find a narrow layer (the viscous or laminar sublayer) of fluid near the wall. Within this layer, the dominating transport mechanism is molecular diffusion. In the center, on the other hand, the transport mechanism depends upon flow regime and the distribution form of momentum, heat, and mass species. In the case of a laminar flow regime, the aforesaid distributions are obtained using governing equations of motion, heat, and mass in fluid flow; however, in a turbulent flow, a complex set of equations must be solved for predicting momentum, heat, and mass species profiles [5].

Herein, it is assumed that the turbulence effects on fluid flow are negligible. Under these conditions, the asphaltene deposition occurs under the influence of some significant mechanisms such as Brownian (molecular) diffusion, drag, gravitational, thermophoresis, buoyancy, and shear forces deposition. The role of different mechanisms affecting the asphaltene deposition phenomena depends on many parameters such as temperature, fluid velocity, asphaltene particles size, etc. These mechanisms and their formulations are described below.

Brownian Diffusion

Brownian motion is the apparently random motion of small particles when suspended in a fluid. The molecules strike the suspended particles and exert a small force on them as a consequence. If the particles are small enough, the force on the particles as a result of these collisions is sufficient to produce observable motion. Because of the random motion of the particles, they diffuse from the regions of high concentration to the regions of low concentration. Here, this diffusion is called Brownian diffusion. In the crude oil-asphaltene system, the collisions between the asphaltene particles and oil molecules lead to the random Brownian motion of suspended asphaltene particles and because of this random motion, the asphaltene particles diffuse in the fluid. The asphaltene mass flux due to Brownian diffusion is calculated by Fick's law:

$$J_B = -D_B \frac{\partial C}{\partial r} \tag{1}$$

where, J_B is the Brownian diffusive asphaltene flux in the radial direction, $\partial C/\partial r$ is the radial gradient in asphaltene concentration and D_B is the asphaltene Brownian diffusivity coefficient. The Brownian diffusivity can be calculated by the Stokes-Einstein relation [6]:

$$D_B = \frac{k_b T}{3\pi\mu d_p} \tag{2}$$

where, k_b is the Boltzmann constant (1.38×10^{-23} J/K); T stands for the absolute temperature; μ and d_p represent the absolute viscosity and the asphaltene particle mean diameter respectively. According to the Pilat and Prem's work, the Brownian force can be expressed as the following equation [6]:

$$F_{Br} = \frac{k_b T\, u_p}{D_B C_c} \tag{3}$$

where, u_p is the particle settling velocity and C_c stands for the Cunningham correction factor, which can be estimated by the expression:

$$C_c = 1 + K_n\left[1.257 + 0.4\exp\left(-\frac{1.1}{K_n}\right)\right] \tag{4}$$

where, $K_n = d_p/\lambda$ is the Knudsen number and λ represents the mean free path of molecule. According to the kinetic theory, λ is expressed as [7]:

$$\lambda = \frac{k_b T}{\sqrt{2}\pi d_p^2 P} \tag{5}$$

where, P is pressure.

Gravity Force

The higher mass density of the asphaltene particles with respect to oil (about 1.2 gr/cc) causes the settlement and deposition of the asphaltene particles in a gravity field. The gravitational force acting on asphaltene particles may be given as:

$$F_g = \rho_p g V_p \tag{6}$$

where, V_p and ρ_p are the particle volume and density respectively and g is the acceleration of gravity (9.8 m.s^{-2}).

Shear Induced Lift Force

Due to shear effects, the dispersed asphaltene particles in the oil phase experience a lift force which is applied perpendicular to the direction of flow [8]. The magnitude of this shear-induced lift force on asphaltene particles can be calculated using Saffman equation [9, 10]:

$$F_l = \frac{1.62\mu d_p^2 (du/dr)}{\sqrt{\nu|du/dr|}}\left(u - u_{px}\right) \tag{7}$$

where, u is the fluid velocity; du/dr is the fluid velocity gradient (shear strain) and u_{px} represents the asphaltene particle velocity in the axial or x-direction; d_p, μ, and ν are particle diameter, fluid absolute, and kinematics viscosities respecttively.

Thermophoresis

Thermophoresis, thermodiffusion, or Soret effect (or Ludwig-Soret effect) is a phenomenon observed when a mixture of two or more types of motile particles (particles able to move) is subjected to the force of a temperature gradient and the different types of particles respond to it differently [8]. In the presence of temperature gradient in asphaltene-oil fluid flow in pipelines, thermophoresis can affect the motion of dispersed asphaltene particles. This force is the reason for asphaltene movement from hot toward cool regions. This force can be evaluated by applying the following equation [11]:

$$F_{th} = -\frac{3\pi\mu^2 d_p H}{\rho_a T_p}\frac{dT}{dr} \tag{8}$$

where, $\frac{dT}{dr}$ is the radial temperature gradient; T_p and H are the absolute temperature of particles and the thermophoretic force coefficient, which can be calculated by Equation 9:

$$H = \left(\frac{2.34}{1+3.42\,K_n}\right)\left(\frac{k_f/k_p + 2.18\,K_n}{1+2k_f/k_p + 4.36\,K_n}\right) \tag{9}$$

where, K_f and k_p are the thermal conductivities of the fluid and the asphaltene particle respectively.

Drag Force

Drag refers to forces that oppose the relative motion of an object through a fluid (a liquid or gas). This force acts in a direction opposite to the oncoming flow velocity. In the general case, the drag force exerted on a particle moving along with a fluid is calculated by Equation 12 [12]:

$$F_D = \frac{\pi d_p^2 \rho_f \left(u_r - u_p\right)^2 C_D}{8C_c} \tag{10}$$

where, u_r is the r-component of fluid velocity and C_{Di} stands for the drag coefficient.

Buoyancy Force

According to Archimedes law, when an object is immersed in a fluid, it displaces the same volume of the fluid as the immersed volume of the object and is consequently buoyed up by a force equal to the weight of the displaced fluid. The buoyancy force is obtained by the following equation:

$$F_{buoy} = \rho_f g V_{ip} \tag{11}$$

where, g the gravity acceleration and V_{iP} is the immersed volume of particle.

MODEL EXECUTION

The applicability of the proposed model is examined in a pipe with inner radius of r_i and length of L. The hydrocarbon mixture flowing in the pipe contains asphaltene particles with a concentration of C_0 at the beginning of the pipe (L=0) and C at the arbitrary pipe length of x. The initial temperature, pressure, and velocity of hydrocarbon mixture are T_i, P_i, and u_i respectively. Due to various deposition mechanisms, asphaltene can be deposited on a pipe surface and the asphaltene concentration of hydrocarbon mixture is reduced at the pipe exit. Figure 1 shows deposition asphaltene on the inner surface of a pipe.

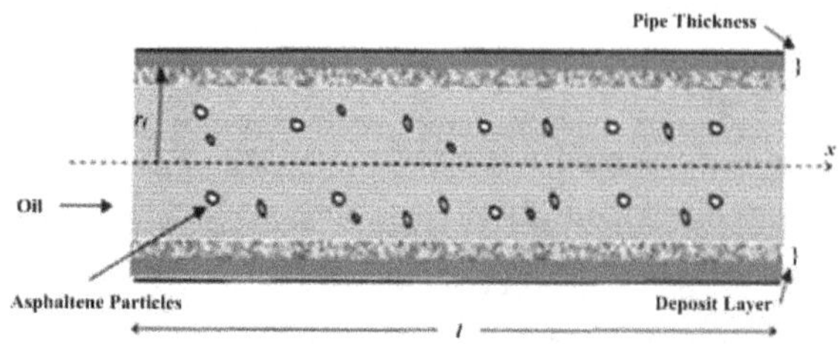

Figure 1: Schematic of asphaltene deposition on a pipeline surface

The original fluid contains some dissolved asphaltene and perhaps some dispersed colloidal particles of the asphaltene. Changes in temperature, pressure, and compositions are the main causes of the asphaltene precipitation. At first, the precipitated particles are dispersed in fluid. Thereafter, other factors such as particle tendency to adhesion, the size of particles, the conduct characteristics, and the fluid flow behavior can control the asphaltene deposition process. The settling (deposition) velocity of asphaltene is obtained via balancing between different forces acting on the flowing asphaltene particles in the pipe. Brownian diffusion is significant for very small particles (d_p<0.1 μm), but has a negligible effect on the large asphaltene particles with the size considered herein (d_p=0.5-100 μm). Hence, herein, the effects of other dominating forces, namely gravity settling, shear induced lift force, thermophoresis force, drag, and buoyancy force are considered.

The motion of asphaltene particles from bulk phase toward pipe surface can be intensified or weaken by different effective forces. For example, forces such as buoyancy, drag, and lift are opposed to asphaltene deposition on surface and hence decelerate the effect of the others. On the other hand, a force such as gravity is in accordance with deposition direction and thus accelerates the deposition process. Some forces such as thermophoresis force have a double effect depending on flowing fluid temperature whether it is higher or less than wall temperature. For a hot wall, heat is transferred toward fluid and as a result decelerates the asphaltene deposition; however, in the case of a cold wall, thermophoresis force and deposition velocity has the same direction and therefore the asphaltene deposition is intensified.

Force Balance on the Whole Deposit

Among the mechanisms previously discussed, gravity deposition and thermophoresis force help the deposit grow while shear removal, buoyancy, and drag forces do the opposite. Therefore, a total force balance on the whole deposit can be written as follows:

$$F_g \pm F_{th} - F_{buoy} - F_l - F_D = m_p \frac{du_p}{dt} \qquad (12)$$

By replacing F_D from Equation 10 into the above equation with an assumption that u_f=0, the following expression is derived:

$$F_g \pm F_{th} - F_{buoy} - F_l - \frac{\pi d_p^2 \rho_f C_D}{8 C_c} u_p^2 = m_p \frac{du_p}{dt} \qquad (13)$$

u_p is the deposition velocity of asphaltene particles, which can be obtained by solving the above equation. The maximum attainable velocity of asphaltene particles is called the terminal velocity and, under these conditions, the particles reach a constant deposition velocity. The terminal velocity of the asphaltene deposition can be evaluated from Equation 14: $\left(\frac{du_p}{dt} = 0\right)$:

$$u_p = \left(\frac{8C_c \left(F_g \pm F_{th} - F_{buoy} - F_l \right)}{\pi d_p^2 \rho_f C_D} \right)^{0.5} \qquad (14)$$

In this equation, u_p is the deposition (settling) velocity of asphaltene particles, which depends on many different variables. To estimate the unknown parameters, we assume that the oil is flowing inside the pipe with a diameter of 50 cm and a length of 100 m. The fluid velocity as well as its temperature profile is supposed parabolic in the radial direction. In addition, it is assumed that the temperature changes linearly during the axial direction.

Table 1 shows the concerned parameters of the fluid and asphaltene used in this work [14].

Table 1: Characteristics of the Fluid under Study

Parameters	Oil	Asphaltene
$\rho\ (kg.m^{-3})$	860	1200
$Cp\ (J.kg^{-1}.K^{-1})$	213	920
$K\ (W.m^{-1}.K^{-1})$	0.120	0.75
$\mu\ (Pa.s)$	0.075	-
$\alpha = \dfrac{K}{\rho\ Cp}\ (m^2.s^{-1})$	6.55×10^{-4}	6.79×10^{-7}

RESULTS AND DISCUSSIONS

Asphaltene Particle Size Effect

As it can be expected, asphaltene particle size plays a main role in the deposition velocity. To establish a fact of this claim, the asphaltene particle size is assumed to be in the range of 0.5 to 100 micron [15-17]. Equation 14 quantifies the effect of asphaltene particle size on its deposition velocity. The results are depicted in Figure 2.

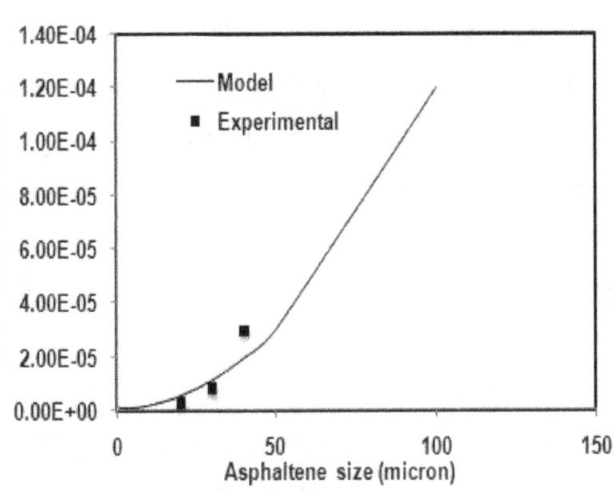

Figure 2: Effect of asphaltene particle size on deposition velocity

As shown in the figure below, increasing the asphaltene particle diameter causes a rise in deposition velocity. Since experimental data for asphaltene deposition from petroleum fluid flow were scare, the verification of the model was accomplished using some available aerosols deposition data [3]. As it is shown, the model predictions are in good agreement with deposition data.

It is also shown that among the mentioned forces, the gravitational force is more affected by particle size variation. This is due to the cubic relation of this force to the particle diameter. The concerned results are shown in Figure 3.

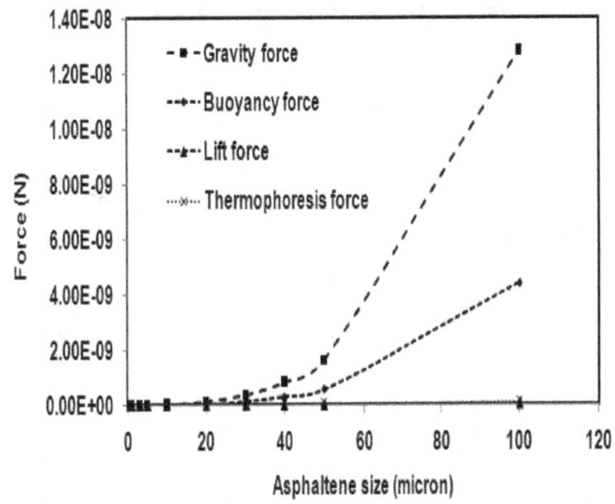

Figure 3: Effect of asphaltene size on the magnitude of different forces

Temperature gradient between fluid and pipe wall causes heat transfer phenomenon. This phenomenon can influence the thermophoresis force which is one of the forces for asphaltene deposition. Depending on temperature gradient direction, deposition velocity may be increased or decreased. While the wall temperature is higher than the fluid bulk temperature, heat flows from the wall toward the fluid bulk. This leads to a decrease in asphaltene deposition rate; this is due to thermophoresis force which acts in the direction of temperature gradient.

On the other hand, in the case of a cold wall the temperature gradient, thermophoresis force, and deposition velocity has the same direction and therefore the asphaltene deposition is intensified. Figure 4 shows the effect of temperature gradient on asphaltene deposition velocity. As depicted in this figure, the deposition velocity of asphaltene particles rises by increasing temperature gradient.

Fluid Velocity Effect

Fluid velocity is one of the main parameters affecting the asphaltene deposition phenomena. Changes in velocity influence the temperature gradient which is the major cause of thermophoresis force. Furthermore, the velocity variations have crucial effects on the lift force which can delay the asphaltene deposition.

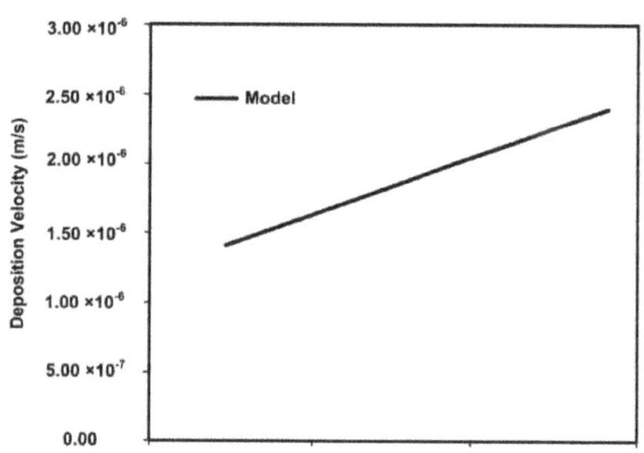

Figure 4: The effect of temperature gradient on deposition velocity

Figure 5 shows the effect of fluid velocity on both thermophoresis and lift force. The results of this work show that the fluid velocity affects the thermophoresis force more effectively than the lift force. Therefore, according to Equation 14, the deposition velocity rises as the fluid velocity is increased. The results obtained from velocity effect investigations are shown in Figure 6. The results obtained from our model are compared with some experimental aerosols deposition data [3]. According to Figure 6 there is a good agreement between the calculated and limited experimental data.

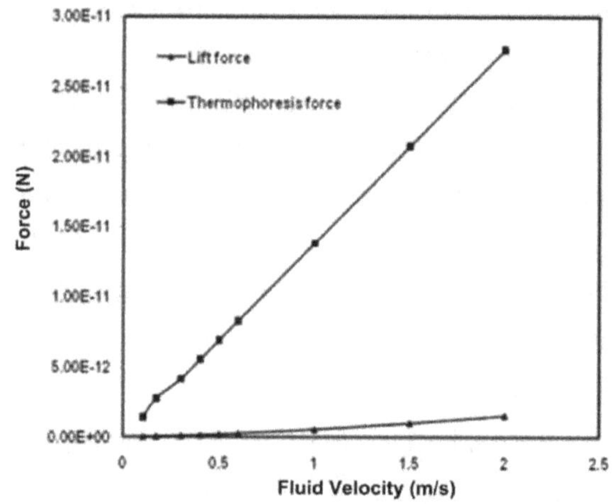

Figure 5: The effect of fluid velocity on forces

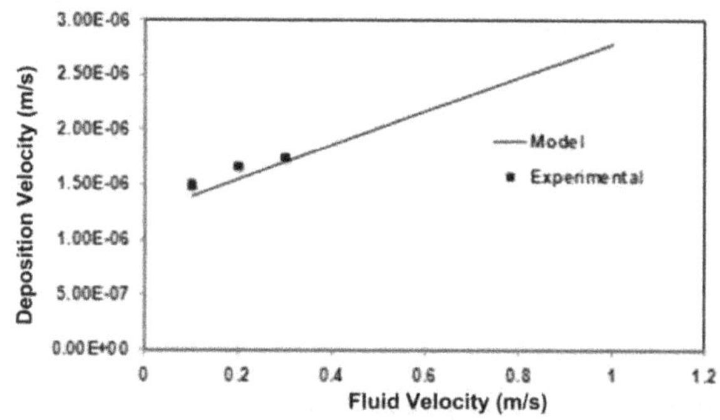

Figure 6: Effect of fluid velocity on deposition velocity

Asphaltene Concentration Changes

The deposition of asphaltene particles on pipe surface causes a significant variation in its concentration in the fluid bulk. The amount of deposited asphaltene can be estimated using a material balance around the inlet and outlet of pipe. For this reason, it is necessary to calculate the amount of asphaltene concentration at the pipe exit. The relation between particle concentration and its deposition velocity in a pipe flow was suggested elsewhere [13]. This equation is applied for the proposed system:

$$C = C_0 \exp\left(-\frac{u_p}{u}\frac{x}{R}\right) \tag{15}$$

where, u_p is the particle deposition velocity; u is the flow velocity; C and C_0 are the particle concentration in the x-direction and the particle concentration at the inlet of the pipe respectively; R stands for the pipe radius. We assume that the hydrocarbon mixture contains an initial concentration of asphaltene particles equal to 0.05 *gr/cc*. Setting *x=L* under different conditions, the asphaltene concentration at the end of the pipe can be calculated using Equation 14. Since the asphaltene deposition velocity is affected by many different parameters such as particle size, fluid velocity, and temperature gradient, the asphaltene concentration is also changed by those parameters. The effects of some mentioned parameters on asphaltene concentration are investigated and the obtained results are depicted in Figures 7, 8 and 9.

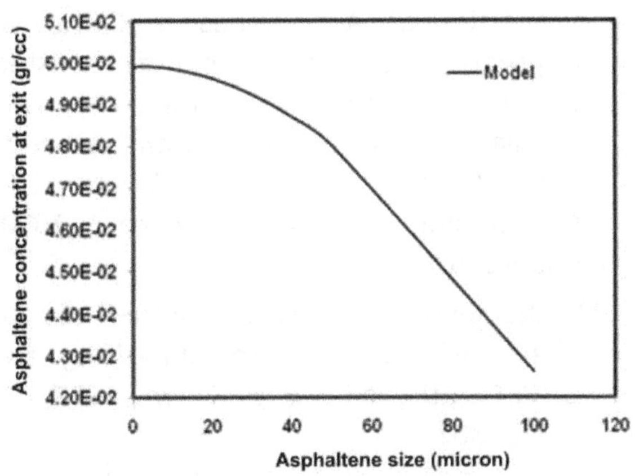

Figure 7: Effect of asphaltene particles size on the exit asphaltene concentration

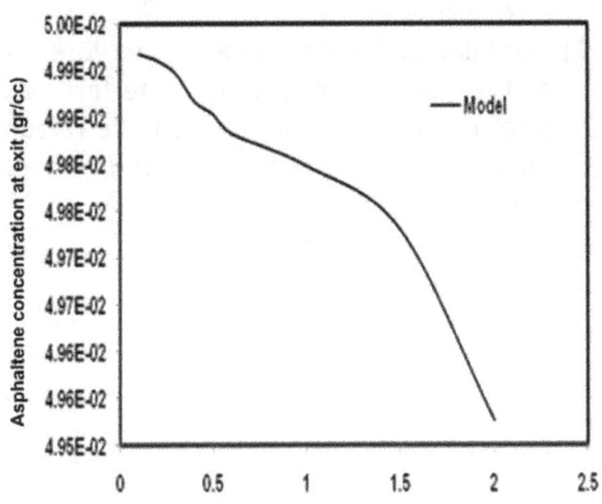

Figure 8: Effect of fluid velocity on asphaltene exit concentration

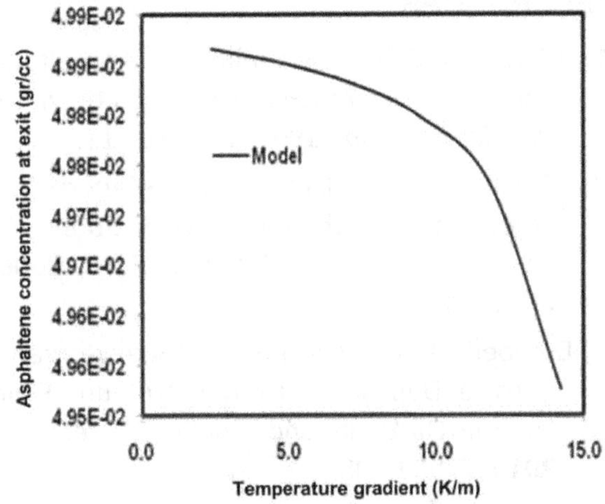

Figure 9: Effect of temperature gradient on asphaltene exit concentration

CONCLUSIONS

A new dynamic model is introduced for asphaltene deposition. The basis of this model is the contribution of different forces which act on the dispersed asphaltene particles in the petroleum fluid flow. The results show that the gravitational force is more affected by the asphaltene particle size in comparison with the other contributed forces; consequently, the deposition velocity rises by increasing asphaltene particles size. Also, the thermophoresis force, which is resulted from the temperature gradient between fluid bulk and the pipe wall, is exerted on the asphaltene particles. The results of this work show that the deposition velocity increases by a rise in the temperature difference between the fluid bulk and the pipe wall; the major reason is the existence of thermophoresis force, which is intensified by the temperature gradient increment. In addition, the results show that both the thermophoresis and lift forces are affected by the fluid velocity. Furthermore, the results show that the deposition velocity increases with a rise in fluid velocity. On the basis of the limited experimental data on aerosol deposition, the proposed model is able to predict the asphaltene deposition velocity in petroleum fluids flowing in pipes.

REFERENCES

[1] Mansoori G. A., "Modeling of Asphaltene and other Heavy Organics Depositions," *J. Pet. Sci. and Eng.*, **1997**, *17*, 101-111.

[2] Ramirez J. E., Lira G. C., Manero O., "Modeling Asphaltene Deposition in Production Pipelines," *Energ. Fuel.*, **2006**, *20*, 1184-1196.

[3] Escobedo J., Mansoori G. A., "Heavy-organic Particle Deposition from Petroleum Fluid Flow in Oil Wells and Pipelines," *Pet. Sci.*, **2010**, *7*, 502-508.

[4] Eskin D., Ratulowski J., Akbarzadeh K., Pan S., "Modeling Asphaltene Deposition in Turbulent Pipeline Flows," *Can. J. Chem. Eng.*, **2011**, *89*, 421-441.

[5] Bird R. B., Stewart W. C., Lightfoot E. N., *Transport Phenomena*, 2nd Ed., New York, John Wiley & Sons Inc., **2002**.

[6] Pilat M. J., Prem A., "Calculated Particle Collection Efficiencies of Single Droplets Including Inertial Impaction, Brownian Diffusiophoresis, and Thermophoresis, Atmos Environ," *Atmos. Environ.*, **1976**, *10*, 13-19.

[7] Chapman S., Cowling T. G., *The Mathematical Theory of Non-uniform Gases*, New York, Cambridge University Press, **1990**.

[8] Sippola M. R., Nazaroff W. W., "Particle Deposition from Turbulent Flow: Review of Published Research and its Applicability to Ventilation Ducts in Commercial Buildings, Berkeley," Berkeley, CA, **2002**, 94720-1710.

[9] Saffman P. G., "The Lift on a Small Sphere in a Slow Shear Flow," *J. Fluid Mech.*, **1965**, *22*, 385-400.

[10] Saffman P. G., "Corrigendum to the Lift on a Small Sphere in a Slow Shear Flow," *J. Fluid Mech.*, **1968**, *31*, 624-624.

[11] Talbot L., Cheng R. K., Schefer R. W., Willis D. R. "Thermophoresis of Particles in a Heated Boundary Layer," *J. Fluid Mech.*, **1980**, *101*, 737-758.

[12] French A. P., *Newtonian Mechanics (The M. I. T. Introductory Physics Series)*, New York, W. W. Norton & Company Inc, **1970**.

[13] Mednikov E. P., *Turbulent Transfer and Sedimentation of Aerosols*, Moscow: Nauka, **1981**.

[14] Diallo M. S., Cagin T., Faulon J. L., and Goddard W. A., "Thermodynamic Properties of Asphaltenes: a Predictive Approach Based on Computer Assisted Structure Elucidation and Atomistic Simulations," in *Asphaltenes and Asphalts* Yen, T. F., Chilingarian G. V. Ed., Netherlands, Elsevier Science, **2000**, Chapter **5**.

[15] Witherspoon P. A., "A Study of the Colloidal Characteristics of Petroleum Using the Ultra-Centrifuge," *J. Phys. Chem.*, **1957**, *61*, 1296-1302.

[16] Kevin A. F., William Y. S., Anil K. M., "Measurement of Asphaltene Particle Size Distributions in Crude Oils Diluted with n-heptane," *Ind. Eng. Chem. Res.*, **1993**, *32*, 955-959.

[17] González G., "Surface Charge and Potential at the Asphaltenes-solution Interface, 8th International Conference on Petroleum Phase Behavior and Fouling," *PAU-France*, France, June **2007**.

A New Screening Evaluation Method for Carbon Dioxide Miscible Flooding Candidate Reservoirs

Li Yanjie*[1], Zhang Yanyu[2], Wang Shichao[3], Chen Huijuan[2], Shi Xiaoying[4], and Bai Xuzheng[5]

[1] Development Research Department, China National Offshore Oil Corporation (CNOOC), Beijing, China
[2] School of Petroleum Engineering, China University of Petroleum (Eastern China), Qingdao, China
[3] Research Institute of Western South China Sea Petroleum Administrative Bureau, CNOOC Zhanjiang Branch, Zhanjiang, China
[4] Exploration and Development Research Institute of Huabei Oilfield Company, Renqiu, China
[5] Gudong Oil Production Plant of Shengli Oil Field Co., Dongying, China

ABSTRACT

Prior to the implementation of CO_2 injection EOR projects, the screening evaluation of candidate reservoirs will promote the economic benefits of CO_2 injection. Currently, a uniform screening method for CO_2 miscible flooding does not exist. Based on more than 112 successfully implemented CO_2 miscible flooding reservoirs, which was referred in 2010 Worldwide EOR Survey, and CO_2 miscible flooding mechanisms, this paper picks out 12 reservoir and fluid parameters affecting CO_2 miscible flooding results as comprehensive evaluation parameters for screening candidate reservoirs. According to investigations on a large number of domestic and international CO_2 miscible flooding projects, the quantitative methods are determined by theoretical analyses, field experience, and probability statistics. By means of calculating the combinational weights by improved analytical hierarchy process (AHP) and entropy method and combining the advantages of technique for order preference by similarity to ideal solution (TOPSIS) with gray relational analysis to construct a new similarity nearness degree, the weighted GC-TOPSIS model is established for screening candidate reservoirs. This screening method was employed for the assessment of five classical candidate reservoirs proposed for CO_2 miscible flooding. The results show that this new method can correctly evaluate and compare the potential of CO_2 miscible flooding.

Keywords: CO_2 Miscible Flooding, Screening Index, Evaluation Criteria, Ideal Solution, Grey Correlation

INTRODUCTION

CO_2 miscible displacement, as a key EOR method, is becoming increasingly important to improve oil recovery efficiency these days. Miscible displacement can recover trapped oil; the main mechanism is efficiency due to the low interfacial tension (IFT) between oil and the injected CO_2. IFT tends towards zero when miscibility is reached, which means that oil recovery can reach the maximum in the swept area. Additional mechanisms such as the assistance of pressure, viscosity reduction, oil

***Corresponding author**
Li Yanjie
Email: liyj65@cnooc.com.cn

swelling, and improved reservoir contact further contribute to a higher oil recovery. Based on a recently updated report from Department of Energy, within Advanced Resource International's Large Oilfield's Database, 1,719 out of 2,790 reservoirs are favorably for CO_2-EOR. These 1,719 reservoirs contain 305 billion barrels of OOIP (R.C. Ferguson et al., 2010). Therefore, CO_2 miscible flooding to enhance oil recovery has become a major potential in petroleum industry.

According to 2010 Worldwide EOR Survey, gas injection is currently the most common EOR process apart from thermal recovery (Figure 1), and CO_2 miscible displacement is also the most widely applied gas injection EOR method. However, with the gradual improvement to exploitation and development technologies, reservoirs prepared for CO_2 miscible displacement need to be carefully studied in order to optimize economic benefits. Based on the statistics of successful CO_2 miscible displacement examples from 2010 Worldwide EOR Survey, theoretical analysis, probability statistics, screening index systems, and evaluation criteria of key parameters, including reservoir characteristics, rock properties, and crude oil properties, were obtained.

The application of CO_2 injection into oil reservoirs started in the 1970's. From then on, some experts and scholars put forward different screening criteria and evaluation methods for CO_2 injection. In order to single out CO_2-EOR candidates (Geffen, 1973; Lewin et al., 1976; NPC, 1976; McRee, 1977; Lyoho, 1978; OTA, 1978; Carcoana, 1982; Taber & Martin, 1983; Mark A Klins, 1989; O. Rivas et al., 1994; Daniel et al., 1996; Taber et al., 1997; Brent, 1998; J. Shaw, 2002; Ahmad et al., 2010), a variety of parameters have been put forward. These parameters include reservoir characteristics (depth, temperature, pressure, and thickness of the reservoir), rock properties (lithology, permeability, porosity), and crude oil properties (API gravity, oil viscosity, oil saturation). However, most of these criteria just give roughly the appropriate range, while unable to determine the applicability of CO_2 miscible flooding for a specific oil field if one or several parameters fail to fall within the right range. Since Louisiana State University, commissioned by the U.S. DOE, investigated CO_2 flooding comprehensive evaluation method for light oil reservoirs after water flooding in 1992, some experts [18-23] have established comprehensive evaluation methods based on fuzzy analytic hierarchy processes. However, due to the use of objective weighting method to determine the index weight, which apparently has some shortcomings, these methods are relatively restricted and limited. Therefore, twelve screening indexes affecting CO_2 injection miscible flooding results were selected in this paper, and their quantitative methods are determined by theoretical analysis, field experience, and probability statistics. Based on improved AHP and entropy method to determine the comprehensive weight of screening indexes, the weighted GC-TOPSIS model was established to screen candidate reservoirs while the accuracy and stability of this method were also evaluated.

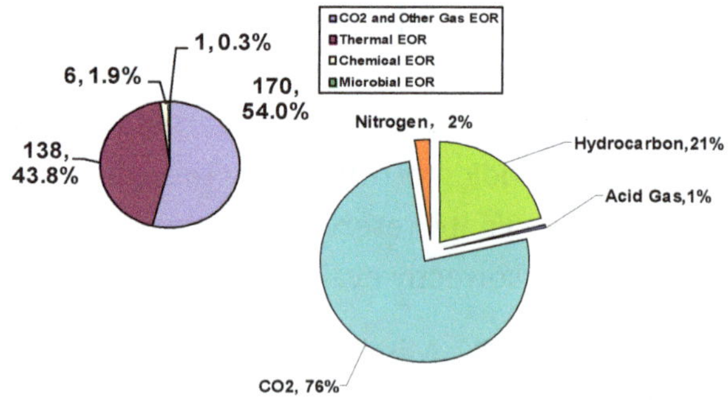

Figure 1: EOR methods distribution based on 2010 Worldwide EOR Survey

The Screening Indexes and Evaluation Criteria

Due to different reservoir conditions, CO_2 injection effect and corresponding flooding mechanism are different. Systematically analyzing the various factors that affect CO_2 injection and selecting the appropriate screening indexes are the basis of the screening model for CO_2 injection miscible flooding.

According to screening indexes mentioned above, and the analysis of factors affecting CO_2 injection, we extract 12 screening indexes, namely reservoir temperature, reservoir depth, reservoir pressure, reservoir thickness, reservoir dip, permeability, porosity, wettability, heterogeneity, API gravity, oil viscosity, and oil saturation. They are divided into three categories, including direct screening indexes, interval screening indexes, and inverse screening indexes.

Direct Screening Indexes

Direct screening indexes refer to that larger evaluation value shows better results. In the identified CO_2 injection miscible flooding screening indexes, the oil wettability, reservoir dip, oil saturation, and oil API gravity are direct screening indexes.

Oil Wettability. Rock wettability directly impacts the displacement efficiency of CO_2 miscible flooding process. In the process of drilling, well completion, workover and production operations, the phenomenon that exotic fluid is retained in porous media will appear. The presence of another immiscible phase in the reservoir or when the saturation of original immiscible phase in porous medium increases will harm the reservoir relative permeability; thus the reservoir permeability and oil and gas relative permeability are significantly reduced, and ultimately affect the reservoir recovery. When the immiscible phase is aqueous phase, this phenomenon is called water lock phenomenon. Theoretical research and core test showed that, in a water-wet medium, the contact area of the oil and CO_2 is reduced due to the presence of water, and water lock phenomenon is more serious, which is not conducive to CO_2 displacement efficiency. If CO_2 is injected into these reservoirs without water injection, this adverse effect can be negligible. But if the reservoirs have been developed by water flooding process, the water lock phenomenon in the water-wet reservoir should be considered. Oil-wet index can be used to reflect rock wettability. Therefore, a higher oil-wet index is better.

Reservoir dip. In slanted reservoirs, gravity could be used as the displacement force in CO_2 miscible flooding. Usually we inject CO_2 into the up dip structure of reservoirs at a low injection rate. The gravity can be used to separate CO_2 and crude oil, in order to suppress the fingering phenomenon and enhance displacement efficiency. The dip angle of candidate reservoirs for gravity flooding is usually at least 15°, and a larger the dip angle is better.

Oil saturation. Bigger oil saturation is definitely more suitable. If oil saturation is very low, it will be difficult to form continuous oil in miscible flooding process, and crude oil will not be easily flooded out. As shown in Figure 2, based on 2010 Worldwide EOR Survey, it is found that initial oil saturation for CO_2 miscible displacement reservoirs ranges between 24% and 89%, or more precisely from 40% to 55%, which accounts for 52% of all the candidate reservoirs.

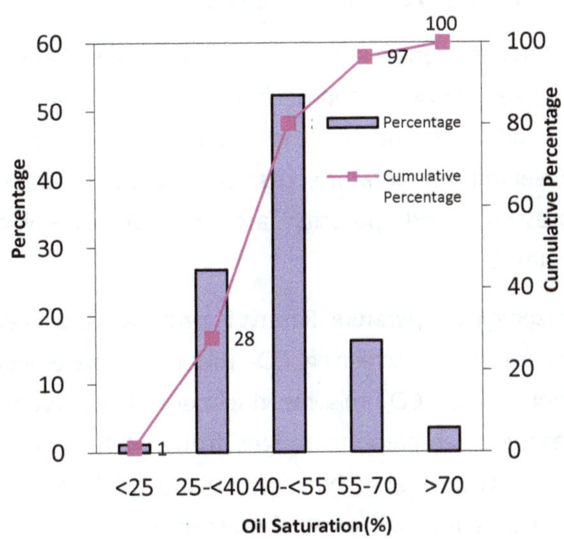

Figure 2: Oil saturation distributions for carbon dioxide miscible displacement reservoirs

Oil API gravity. The lower the API gravity is, the higher the oil viscosity becomes, and CO_2 injection is prone to forming viscous fingering and generating low displacement efficiency. Based on available data, CO_2 miscible displacement reservoirs are characterized by high API gravity, which is usually more than 30°. The proportion of CO_2 miscible flooding reservoirs crude oil API gravity of which is greater than 39° is approximately 37% (Figure 3). Therefore, a smaller API gravity of crude oil is more acceptable.

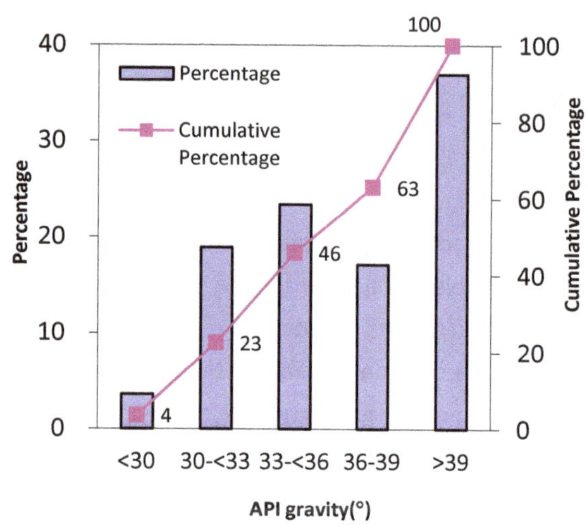

Figure 3: API gravity distributions for carbon dioxide miscible displacement reservoirs

Interval Screening Indexes

Interval screening indexes means that the screening indexes closely approaching a fixed interval (including falling in the interval) are better. Reservoir temperature, reservoir depth, reservoir pressure, and porosity are interval screening indexes.

Reservoir temperature. Reservoir temperature is a key parameter for screening CO_2 miscible displacement process. For CO_2 miscible displacement reservoirs, reservoir temperature ranges from 83 to 257 °F with a typical range between 90 and 130 °F, which accounts for 58% of all the reservoirs (Figure 4). Because distribution parameters are not a continuous function, but discrete screening indexes, the optimal parameters interval for CO_2 miscible flooding can be determined by statistical methods of normal distribution mean. The following equation can be used to calculate the optimal parameters interval:

$$\left[\bar{X} - z_{\alpha/2}\frac{\sigma}{\sqrt{n}}, \bar{X} + z_{\alpha/2}\frac{\sigma}{\sqrt{n}}\right] \tag{1}$$

where, $\bar{X}$ is the mean of the samples; $1-\alpha$ is confidence level, which is calculated at 90% in this paper; $z_{\alpha/2}$ is the quantile; n is the number of the samples and σ stands for standard deviation. The optimal interval of reservoir temperature through calculation using the above formula is [116.35 °F, 130.84 °F].

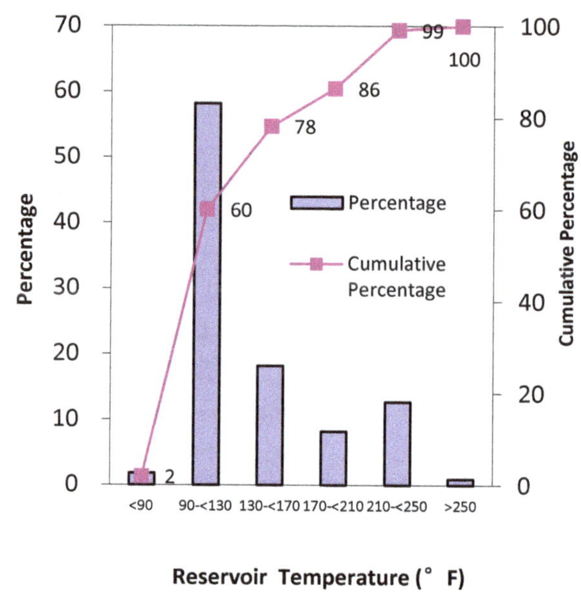

Figure 4: Reservoir temperature distribution for carbon dioxide miscible displacement reservoirs

Reservoir depth. The depth for the CO_2 miscible displacement oil reservoirs in this study ranges from 1150 to 11950 ft below ground level (Figure 5). Approximately 60% of CO_2 miscible displacement reservoirs are at the depth of 4500 to 6500 ft. Very few exceptions of more than 11500 ft are found in Martinville and Soso fields in US, where CO_2 miscible displacement have been successfully applied at depths up to 11550 and 11950 ft respectively. The optimal interval of reservoir depth through calculation using Equation 1 is [5440.06 ft, 6072.08 ft].

Reservoir pressure. The premise of CO_2 miscible flooding is to achieve the miscible phase between

CO_2 and oil. When the reservoir temperature at a certain depth is more than the critical temperature of CO_2, high reservoir pressure is required to ensure CO_2 and crude oil to reach miscibility. When the reservoir pressure becomes greater than the minimum miscibility pressure, the miscible phase can be reached. The reservoir pressure of CO_2 miscible flooding projects is in the range of 1.2 to 22 MPa. Although CO_2 and oil can be easily reached a miscible phase at high reservoir pressure, the high reservoir pressure will increase the level of risk. Therefore, the reservoir pressure is not too high or too low and it is only in a certain range. According to the range of different minimum CO_2 miscible flooding pressures, the optimal interval of reservoir pressure is [16.8 MPa, 22 MPa].

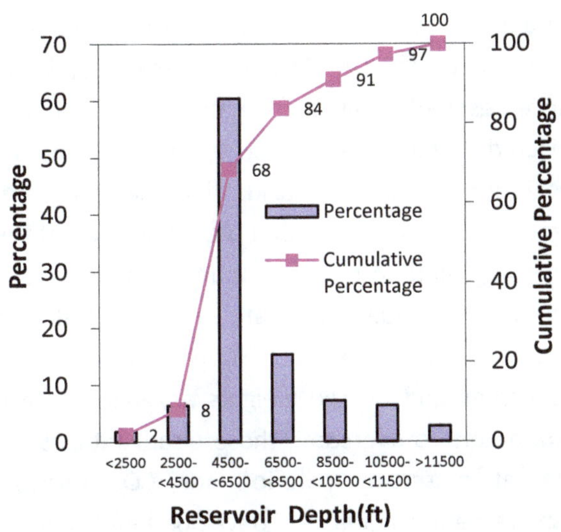

Figure 5: Reservoir depth distribution for carbon dioxide miscible displacement reservoirs

Porosity: CO_2 miscible displacement reservoirs are characterized by low porosity, ranging from 3% to 28% (Figure 6), mainly from 10% to 15%, which accounts for 49% of all the reservoirs. The effective porosity basically obeys normal distribution, and the average effective porosity is 12.33%. The optimal interval of porosity by applying Equation 1 is [11.36%, 13.31%].

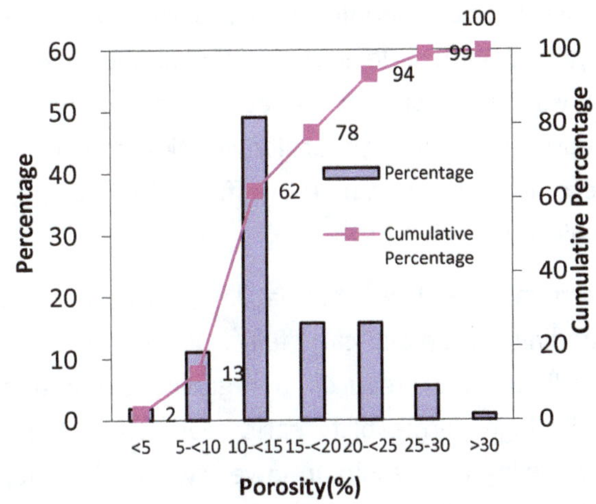

Figure 6: Porosity distribution for carbon dioxide miscible displacement reservoirs

Inverse Screening Indexes

Inverse screening indexes means that the smaller screening indexes are better. Reservoir thickness, heterogeneity, permeability, and oil viscosity are direct screening indexes.

Reservoir thickness. By means of theoretical analysis, it is found that reservoir thickness should not be too large. Because when reservoir thickness becomes larger, the contradiction between layers is more prominent, and thus the overlying effect of the injected gas becomes stronger, which is not beneficial to the expansion of crude oil to reduce viscosity. When reservoir thickness becomes smaller, CO_2 gravity separation becomes weaker; the reservoir is thus less prone to face gas channeling, and CO_2 can spread more evenly to various depths of the reservoir to maximize miscible effect.

Heterogeneity. Vertical heterogeneity has a great effect on CO_2 miscible flooding development, especially for the small plug miscible flooding. Due to the difference in permeability, the volume penetrating into the high permeability layer will be greater than the one into the low-permeability layer. Furthermore, the small plug penetrating into the low-permeability layer becomes diluted due to the transverse and longitudinal dispersion;

hence miscible flooding in low-permeability was reported with little success. The above analysis shows that the smaller the reservoir heterogeneity (represented by permeability variation coefficient) becomes, the better the effects that can be achieved.

Permeability. Low permeability in CO_2 miscible flooding reservoir can fully provide miscible conditions and reduces the gravity separation, while high permeability easily leads to early gas channeling and results in lower sweep efficiency. But for high-dip reservoirs, according to the gravity stable displacement mechanism, the reservoir should have a high vertical permeability. As shown in Figure 7, CO_2 miscible displacement reservoirs are characterized by low permeability, ranging from 1.5 to 3000 mD, mainly less than 10 mD, which accounts for 44% of all the reservoirs. Therefore, the permeability for CO_2 miscible flooding reservoir should be relatively smaller.

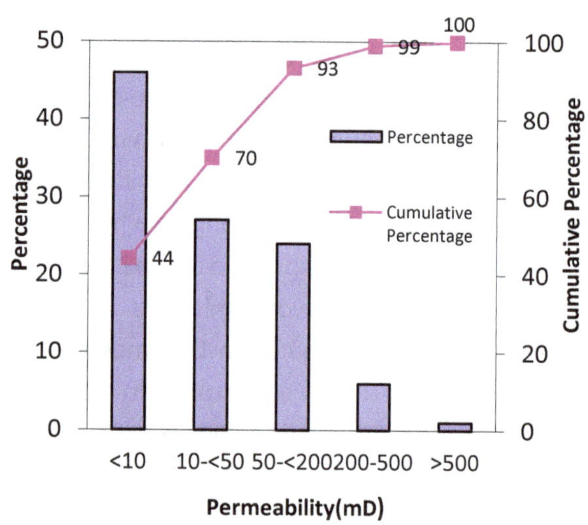

Figure 7: Permeability distribution for carbon dioxide miscible displacement reservoirs

Oil viscosity. The oil for CO_2 miscible displacement reservoirs is characterized by low viscosity. Viscosity is generally less than 4 cp, and typically <1 cp, which accounts for 55% of all the reservoirs (Figure 8). Theoretically, at lower oil viscosity, better effects can be achieved.

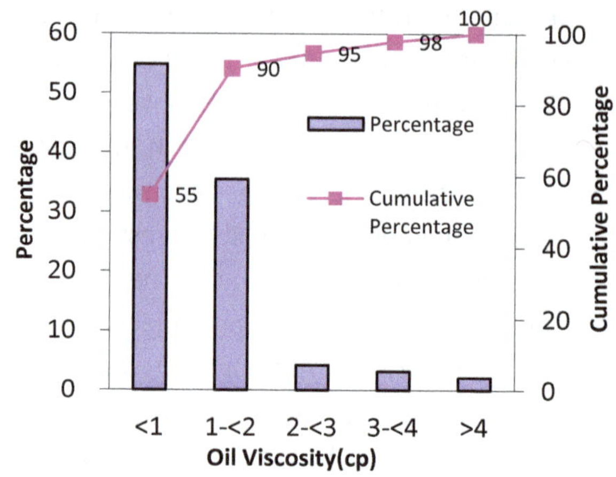

Figure 8: Oil viscosity distribution for carbon dioxide miscible displacement reservoirs

Establishment of the Screening Assessment Model

Conventional screening procedures consist of comparing the properties of reservoirs with certain screening criteria. Those reservoirs matching all the conditions are selected, while those reservoirs failing to match the values are rejected. In studies of selecting the most suitable reservoirs for carbon dioxide flooding in Eastern Venezuela, it was found that the application of the screening method to a large number of reservoirs did not provide adequate results. Different reservoirs were selected depending on the researches and on the weights assigned to each parameter. To overcome those disadvantages, an alternative screen criterion for CO_2 miscible displacement candidates is developed in this paper based on a new GC-TOPSIS model, which combines Euclidean distance and gray relational grade. The combinational weight is calculated by improved AHP and entropy method, giving more accurate results. First, the standardized evaluation matrix should be built. Then, the screening indexes weight is determined. At last, based on the above two steps, the weighted GC-TOPSIS model is established to screen and evaluate reservoirs considering a new "similarity nearness degree" constructed by Euclidean distance and gray relational grade.

Establishing Standardized Evaluation Matrix

The screening method developed in this paper is based on determining each property (i) of the reservoir (j), X_{ij}, and a corresponding normalized parameter r_{ij}. According to three types of screening indexes, r_{ij} is defined by different equations: for "direct indexes" parameters, it is defined by Equation 2; for "inverse indexes" parameters, it is defined by Equation 3; for "interval indexes" parameters, it is defined by Equation 4.

$$r_{ij} = \frac{\max_i x_{ij} - x_{ij}}{\max_i x_{ij} - \min_i x_{ij}} \tag{2}$$

$$r_{ij} = \frac{x_{ij} - \min_i x_{ij}}{\max_i x_{ij} - \min_i x_{ij}} \tag{3}$$

$$r_{ij} = \begin{bmatrix} 1 - \dfrac{q_{1j} - x_{ij}}{\max\left(q_{1j} - \min_i x_{ij}, \max_i x_{ij} - q_{2j}\right)} \\ 1 \\ 1 - \dfrac{x_{ij} - q_{2j}}{\max\left(q_{1j} - \min_i x_{ij}, \max_i x_{ij} - q_{2j}\right)} \end{bmatrix} \tag{4}$$

$$x_{ij} < q_{1j} \qquad x_{ij} \in [q_{1j}, q_{2j}] \qquad x_{ij} > q_{2j}$$

where, x_{ij} is the magnitude of property (i) in the reservoir (j). $\max_i x_{ij}$ is the maximum of property (i) in a fictitious reservoir, while $\min_i x_{ij}$ is the minimum of property (i) in a fictitious reservoir. [q_{1j}, q_{2j}] is the optimal interval for property (i). The variable r_{ij}, as given by Equations 2 to 4, changes between 0 and 1. When r_{ij} is closer to 1, reservoir conditions are better. Through the standardization process of screening indexes, the evaluation matrix of the candidates can be transformed into standardized matrix R. The columns and rows of the matrix, respectively, reflect the suitability magnitude of reservoirs.

$$R = \begin{bmatrix} r_{11} & r_{12} & \cdots & r_{1n} \\ r_{21} & r_{22} & \cdots & r_{2n} \\ \vdots & \vdots & \vdots & \vdots \\ r_{m1} & r_{m2} & \cdots & r_{mn} \end{bmatrix} = (r_{ij})_{m \times n} \tag{5}$$

all evaluation parameters for one reservoir and that of the same parameter for different

Determining Screening Indexes Weight

In the screening index system, the importance of each index to a program is not identical. Furthermore, the weight is used to express relative importance of each index. Some indexes may have considerable effects, while some indexes may have small effects. Thus it is very important and necessary to determine the weight of each index for screening different programs. The scientific and reasonable weights can improve the accuracy and validity of evaluating the results; hence it is one of the most important factors. In the multi-attribute decision problem, if the weights are given, they can directly be used. Otherwise, it is required to determine the weight of each index. Purely subjective weighting methods (Delphi method, the analytical hierarchy process (AHP), efficiency coefficient method, etc.) or objective weighting methods (entropy method, principal component analysis, factor analysis, standard deviation coefficient method, etc.) have certain defects to take into account the relative importance or weight of each reservoir property more reasonably in determining the indexes weight. By comparing the various methods for determining screening indexes weight, the improved analytical hierarchy process and entropy method were used to get screening indexes weight to achieve subjective and objective unity for determining the weights.

The Improved Analytical Hierarchy Process Method

The analytic hierarchy process is a decision making model that aids us in making decisions in our complex world. It is a three-part process which includes identifying and organizing decision objectives, criteria, constraints, and alternatives into a hierarchy; evaluating pair wise comparisons between the relevant elements at each level of the hierarchy; and the synthesis using the solution algorithm of the results of the pair wise comparisons over all the levels. Further, the algorithm result gives the relative importance of the alternative courses of action. The AHP has a special concern with departure from consistency, with the measurement of this departure, and with dependence within and between the groups of the elements of its structure. In comprehensively evaluating programs, the traditional AHP method uses 1-9 scale methods to construct a judgment matrix. When the order of the judgment matrix is large, it is hard to satisfy the consistency requirements of the judgment matrix. It needs to repeatedly adjust the judgment matrix, which is time consuming, and thus the results are not satisfactory. Three-scale AHP method is a simplified algorithm, which is based on the idea of the traditional AHP method. By establishing the judgment matrix, using "optimal transfer matrix" concept, and transforming the matrix, the optimal transforming matrix and consistency matrix are obtained by calculation, which naturally meet the conformance requirements. Therefore, it does not need to be tested for consistency. Comparing with traditional AHP method, this method has better judgment transitivity and a rational scale value, which is conducive for the decision-makers to improve accuracy in pair wise comparison judgment process. The comparison between original AHP consistency check and improved AHP consistency check is shown in Figure 9 and Figure 10.

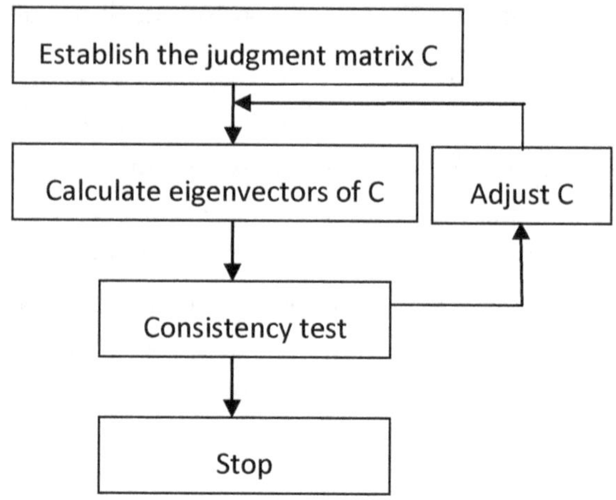

Figure 9: Original AHP consistency check

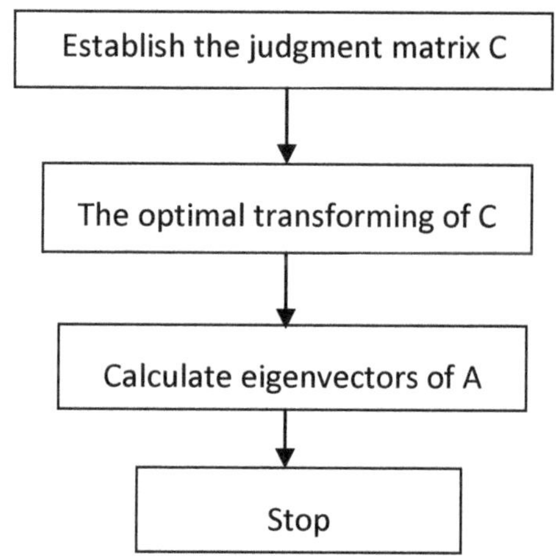

Figure 10: Improved AHP consistency check

The calculation steps of three-scale AHP method are as follows:

(1) Through the pair wise comparisons of screening indexes, the judgment matrix C is established.

$$C = \begin{bmatrix} c_{11} & c_{12} & \cdots & c_{1n} \\ c_{21} & c_{22} & \cdots & c_{2n} \\ \vdots & \vdots & \vdots & \vdots \\ c_{m1} & c_{m2} & \cdots & c_{mn} \end{bmatrix} = (c_{ij})_{m \times n} \tag{6}$$

where, c_{ij} is the ratio of the relative importance of property (i) to property (j), and

$$c_{ij}=\begin{cases}1 & property(i)\ is\ more\ important\ than\ property(j)\\ 0 & property(i)\ is\ as\ important\ as\ property(j)\\ -1 & property(i)\ is\ less\ important\ than\ property(j)\end{cases}$$

$i=1,2,3,\cdots n$

$j=1,2,3,\cdots m$

(2) Calculate the optimal transfer matrix of the judgment matrix C.

$$D=\left(d_{ij}\right)_{nm}=\left(\frac{1}{m}\sum_{k=1}^{m}\left(c_{ik}+c_{kj}\right)\right)_{nm}\qquad \forall i,j \tag{7}$$

(3) Transform the matrix D to the consistency judgment matrix A.

$$A=\left(a_{ij}\right)_{nm}=\left(\exp\left(d_{ij}\right)\right)_{nm}\qquad \forall i,j \tag{8}$$

(4) Calculate the eigenvectors of the matrix A with square root method.

$$\omega=\left(\omega_i\right)_n=\left(\sqrt[n]{\prod_{j=1}^{n}a_{ij}}\right)_n\qquad (i=1,2,\cdots,n) \tag{9}$$

Conduct normalization processing to the eigenvectors.

$$W_1=\left(\overline{\omega_i}\right)_n=\left(\omega_i\Big/\sum_{i=1}^{n}\omega_i\right)_n \tag{10}$$

Then $W_1=\left(\overline{\omega_1},\overline{\omega_2},\cdots,\overline{\omega_n}\right)$ is the weight vector.

The Entropy Method

The entropy method is to determine the index weight based on the judgment matrix constituted by the evaluation indexes, which eliminates anthropogenic interference to the index weight. To some extent, this method avoids the subjective arbitrariness, which makes the evaluation results more in a linear distribution with actual data. The calculation steps of the entropy method are as follows:

(1) Establish standardized evaluation matrix $R=(r_{ij})_{m\times n}$.

(2) Determine the entropy value of property (*i*).

$$H_i=-\frac{1}{\ln n}\sum_{j=1}^{n}f_{ij}\ln f_{ij},\quad i=1,2,\cdots m; j=1,2,\cdots,n \tag{11}$$

$$f_{ij}=\frac{1+r_{ij}}{\sum_{j=1}^{n}\left(1+r_{ij}\right)} \tag{12}$$

(3)Determine difference coefficient of screening indexes.

$$d_i=1-H_i \tag{13}$$

(4) Calculate the entropy weight of screening indexes.

$$\omega_i'=\frac{d_i}{\sum_{i=1}^{m}d_i} \tag{14}$$

where, $0\le\omega_i'\le1$, and $\sum_{i=1}^{m}\omega_i'=1$.

Then $W_2=\left(\omega_1',\omega_2',...,\omega_n'\right)$ is the entropy weight vector.

Comprehensive Weight Calculation

A large number of screening indexes are presented in this paper, and in order to avoid the weight of screening indexes without a big difference, the multiplication synthesis method is used to empower comprehensive weight to the screening indexes. First, the weight coefficients determined by the subjective and objective weighting methods are multiplied respectively, and then the normalization processing of the product is made.

$$\omega_i = \frac{\left(\overline{\omega_i}\right)^\alpha \times \left(\omega_i^{'}\right)^\beta}{\sum_{j=1}^{m}\left(\overline{\omega_i}\right)^\alpha \times \left(\omega_i^{'}\right)^\beta} \tag{15}$$

where, $\overline{\omega_i}$ and $\omega_i^{'}$ are the weight of property (*i*) determined by improved analytical hierarchy process and entropy method respectively. ω_i is the comprehensive weight of property (*i*). α and β are the relative importance of the subjective and objective weight respectively; $0 \le \alpha, \beta \le 1, \alpha + \beta = 1$.

Then, $W = (\omega_1, \omega_2, ..., \omega_n)$ is the comprehensive weight vector.

Establishing the Weighted GC-TOPSIS Mathematical Model

TOPSIS (technique for order preference by similarity to ideal solution) method and the improved gray correlation analysis method both can be applied to the screening evaluation of CO_2 miscible flooding candidate reservoirs. Euclidean distances in TOPSIS method can better reflect the curve position of program data, but there are some defects in shape similarity or reflecting the trend changes between program data curves; however, gray correlation analysis method can precisely reflect the similarity of curve geometry and the trend change between the program data curve. Therefore, they can be organically combined to build "similar approach degree" indexes to reflect the curve position of program data, as well as the trend changes between program data curve and geometry similarity. On this basis, the weighted GC-TOPSIS model was built to improve the accuracy and reliability of the screening evaluation methods.

The calculation steps are as follows:

(1) Establish the evaluation matrix X of the screening indexes to CO_2 miscible flooding candidate reservoirs. Build the corresponding standardized evaluation matrix $R=(r_{ij})_{m\times n}$ by Equations 2-4.

(2) Determine the screening indexes weight. In this paper, we use the above comprehensive weight method to determine the weight of the screening indexes, $W = (\omega_1, \omega_2, ..., \omega_n)$.

(3) Calculate the grey correlation of property(*i*) of the reservoir(*j*) with the positive ideal program and negative ideal program.

First, calculate the weighted standardization matrix U

$$U = \omega^T \times R = \left(\omega_i r_{ij}\right)_{m\times n} = \begin{bmatrix} u_{11} & u_{12} & \cdots & u_{1n} \\ u_{21} & u_{22} & \cdots & u_{2n} \\ \vdots & \vdots & \vdots & \vdots \\ u_{m1} & u_{m2} & \cdots & u_{mn} \end{bmatrix} \tag{16}$$

Then, determine the positive ideal program and negative ideal program U_0^-, which is composed of the maximum value of the screening indexes as the best program and the minimum value of the screening indexes as the worst program respectively.

$$U_0^+ = \left\{\max_{1\le i\le m} r_{ij} \middle| 1 \le j \le n\right\} = \left(u_0^+(1), u_0^+(2), ..., u_0^+(m)\right) \tag{17-a}$$

$$U_0^- = \left\{\min_{1\le i\le m} r_{ij} \middle| 1 \le j \le n\right\} = \left(u_0^-(1), u_0^-(2), ..., u_0^-(m)\right) \tag{17-b}$$

Calculate the grey correlation of R_j^+ and R_j^- property (*i*) of the reservoir (*j*) with the positive ideal program U_0^+ and negative ideal program U_0^- by the following equations.

Calculate the grey correlation coefficient of property (*i*) between the reservoir (*j*) and the positive ideal program U_0^+.

$$r_{ij}^+ = \frac{m + \zeta M}{\Delta_i(k) + \zeta M}, \quad \zeta \in (0,1) \tag{18-a}$$

where, $\Delta_i(k) = \left|u_0^+(k) - u_{ij}\right|$, $1 \le j \le n$, $m = \min_i \min_k \Delta_i(k)$, $M = \max_i \max_k \Delta_i(k)$. ζ is a distinguishing coefficient and $\zeta = 0.5$ in this paper. Due to $0 \le m \le \Delta_i(k)$, then $\frac{1}{3} \le r_{ij}^- \le 1$

Calculate the grey correlation coefficient matrix between the reservoir (*j*) and the positive ideal program U_0^+.

$$r_{ij}^- = \frac{m+\zeta M}{\Delta_i(k)+\zeta M}, \quad \zeta \in (0,1) \tag{18-b}$$

where, $\Delta_i(k) = |u_0^-(k) - u_{ij}|$, $1 \le j \le n$, $m = \min_i \min_k \Delta_i(k)$, $M = \max_i \max_k \Delta_i(k)$. ζ is a distinguishing coefficient and $\zeta = 0.5$ in this paper.

Calculate the grey correlation coefficient matrix between the reservoir (*j*) and the negative ideal program U_0^-.

$$R^+ = (r_{ij}^+)_{m \times n} = \begin{bmatrix} r_{11}^+ & r_{12}^+ & \cdots & r_{1n}^+ \\ r_{21}^+ & r_{22}^+ & \cdots & r_{2n}^+ \\ \vdots & \vdots & \vdots & \vdots \\ r_{m1}^+ & r_{m2}^+ & \cdots & r_{mn}^+ \end{bmatrix} \tag{19-a}$$

Then calculate the grey correlation between the reservoir (*j*) and the positive ideal program U_0^+.

$$R^- = (r_{ij}^-)_{m \times n} = \begin{bmatrix} r_{11}^- & r_{12}^- & \cdots & r_{1n}^- \\ r_{21}^- & r_{22}^- & \cdots & r_{2n}^- \\ \vdots & \vdots & \vdots & \vdots \\ r_{m1}^- & r_{m2}^- & \cdots & r_{mn}^- \end{bmatrix} \tag{19-b}$$

Then, calculate the grey correlation between the reservoir (*j*) and the negative ideal program U_0^-.

$$R_j^+ = \frac{1}{m}\sum_{i=1}^{m} r_{ij}^+, \quad (j=1,2,\ldots,n) \tag{20-a}$$

Similarly, calculate the grey correlation coefficient of property (*i*) between the reservoir (*j*) and the negative ideal program U_0^-.

$$R_j^- = \frac{1}{m}\sum_{i=1}^{m} r_{ij}^-, \quad (j=1,2,\ldots,n) \tag{20-b}$$

(4) Calculate the Euclidean distance D_j^+ and D_j^- of the reservoir (*j*) with the positive ideal program U_0^+ and negative ideal program U_0^-.

$$D_j^+ = \sqrt{\sum_{i=1}^{m}(V_j^+ - u_{ij})^2} \quad j=1,2,\ldots,n \tag{21-a}$$

$$D_j^- = \sqrt{\sum_{i=1}^{m}(V_j^- - u_{ij})^2} \quad j=1,2,\ldots,n \tag{21-b}$$

where, V^+ is the positive ideal solution, and $V^+ = \left\{\max_{1 \le i \le n}(u_{ij}) \mid j=1,2,\ldots,m\right\} = (v_1^+, v_2^+, \ldots, v_m^+)$,

While V^- is the negative ideal solution, and $V^- = \left\{\max_{1 \le i \le n}(u_{ij}) \mid j=1,2,\ldots,m\right\} = (v_1^-, v_2^-, \ldots, v_m^-)$

(5) Make indexes (R_j^+, R_j^-, D_j^+, and D_j^-) being dimensionless by the following equation.

$$M_{new} = \frac{M_j}{\max_{1 \le i \le n}(M_j)} \tag{22}$$

where, M_j is R_j^+, R_j^-, D_j^+, and D_j^- respectively.

(6) Combine R_j^+, R_j^-, D_j^+, and D_j^-

According to the definition, greater values of R_j^+ and D_j^- make the program closer to the negative ideal program, while greater values of R_j^- and D_j^+ make the program closer to the positive ideal solution program. Thus, the combined formula can be determined as:

$$S_j^+ = \alpha_1 D_j^- + \alpha_2 R_j^+, \quad (j=1,2,\ldots,n) \tag{23}$$

$$S_j^- = \alpha_1 D_j^+ + \alpha_2 R_j^-, \quad (j = 1,2,...,n) \tag{24}$$

where α_1 and α_2 reflects the preference of the decision-makers on the position and shape, and $\alpha_1 + \alpha_2 = 1$. The decision-makers can decide these two values according to their preference. In this paper, $\alpha_1 = \alpha_2 = 0.5$. S_j^+ reflects the closeness of the candidate program to the positive ideal program; the greater the S_j^+ value is, the closer the program to the positive ideal program becomes; on the other hand, S_j^- reflects the closeness of the candidate program to the negative ideal program; greater S_j^- values make the program closer to the negative ideal program.

(7) Calculate the "similarity nearness degree" C_j of the candidate program decided by the S_j^+ and S_j^-.

$$C_j = \frac{S_j^+}{S_j^- + S_j^+}, \quad (j = 1,2,...,n) \tag{25}$$

where, C_j is based on the Euclidean distance and the grey correlation, reflecting the position relation and the similar differences of the data curves between the candidate program with the positive/negative program, and the physical meaning is more explicit .

(8) Rank the candidate programs.

Rank the alternatives in accordance with the magnitude of the "similarity nearness degree" C_j, the greater the "similarity nearness degree" is, the more superior the program is; on the contrary, the smaller the "similarity nearness degree" is, the more inferior the program becomes.

Model Validation

On the basis of GC-TOPSIS mathematical model, an example is carried out to verify whether this model is accurate in screening candidate reservoirs for CO_2 miscible displacement. Basic reservoir characteristics and fluid properties for five cases (California CO_2 injection programs) have been put forward in (Table 1). The simplified calculation steps are shown in appendix. Based on the magnitude of the similarity nearness degree, the sequence of the five reservoirs from superior to inferior is: 3 (Gatchell)>1 (Emery)>5 (Potroro)>2 (Stevens)>4 (Westerm21-1), which is entirely consistent with the ultimate recovery. The contrast between the reservoir evaluation results and total ultimate recovery is shown in Table 2. Hence the example analysis verifies that this model is accurate in screening candidate reservoirs for CO_2 miscible displacement.

Table 1: Parameter values of candidate reservoirs

Parameters	1	2	3	4	5
Permeability [mD]	83	200	421	108	75
Reservoir thickness [ft]	302.8	109.9	385.8	309.1	191.9
Oil viscosity [mPa's]	1.05	0.37	0.65	0.45	0.72
Oil API gravity [°]	33.0	37.0	31.9	34.0	35.0
Heterogeneity [fraction]	0.52	0.63	0.46	0.64	0.6
Oil saturation [fraction]	63	62	72.3	53	70
Wettability [fraction]	0.58	0.43	0.78	0.38	0.52
Reservoir dip [°]	32	13	35	7	9
Porosity [fraction]	26.5	22	15	19.5	18
Reservoir temperature [°C]	185.0	209.8	212.0	234.9	169.9
Reservoir pressure [MPa]	17.7	24.2	23.7	27	21.1
Reservoir depth [ft]	5299	8199	7500	8799	7001

Table 2: The comparison of reservoir evaluation results and total ultimate recovery

Reservoirs	1	2	3	4	5
Reservoirs evaluation results [fraction]	0.541	0.482	0.574	0.388	0.519
Total ultimate recovery [fraction]	0.447	0.331	0.585	0.213	0.367

During the process of calculating the comprehensive weights and similarity nearness degree, we assume weighting coefficients α and β and similarity nearness degree coefficients α_1 and α_2 for the given values. Usually, considering the different preferences of the decision-makers on the position and shape similarity between the subjective/objective weights and program data curves, the changes of α and β and α_1 and α_2 may have an impact on the final results of the evaluation. Therefore, we make a stability test on the values of α_1 and α, which varies within the range of 0 to 1, and the evaluation results are shown in Figures 11 and 12.

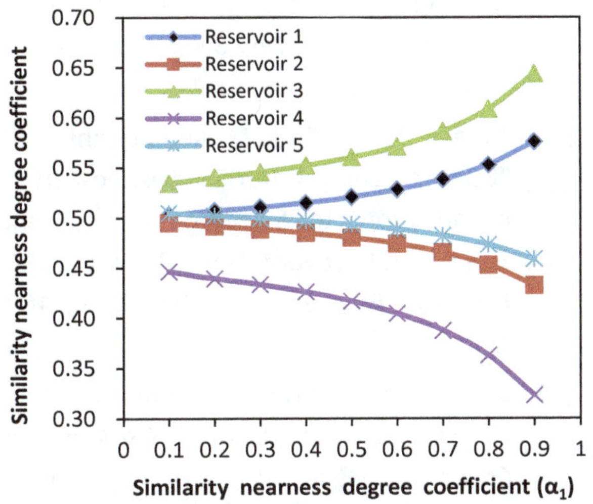

Figure 11: α=0.5 Pictorial diagram of comprehensive evaluation

The results demonstrate that the change of comprehensive weights α and α_1 has a little effect on the relative similarity nearness degree of each reservoir; thus the impact on the final results of the evaluation can be ignored. In addition, the effect of similarity nearness degree because of the changes of weighting coefficients α and β are less obvious than the changes of similarity nearness degree coefficients α_1 and α_2. Through these analyses, it is found that this model is also stable in screening candidate reservoirs for CO_2 miscible displacement to enhanced oil recovery.

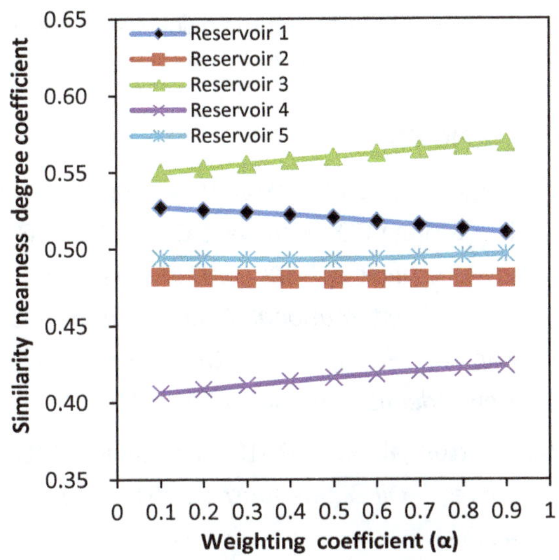

Figure 12: α_1=0.5 Pictorial diagram of comprehensive evaluation

CONCLUSIONS

According to the theoretical analysis, field experience, and probability statistics, 12 parameters affecting CO_2 injection miscible flooding are analyzed and the correct screening indexes system is established. In addition, by respectively applying the improved AHP and entropy method to determine the subjective weight and objective weight, then the comprehensive weight is determined by the multiplication synthesis method. So the comprehensive weight can better reflect expertise, and the inherent law between data to avoid simple use of objective weighting method for determining the index weights. This model is validated in the last part, the accuracy and stability of the screening evaluation model is proved. Therefore, the model can be applied to CO_2 miscible flooding screening studies of other reservoirs.

ACKNOWLEDGEMENTS

This research was supported by the China National Key Development Planning Project for Basic Research "973" Project (Grant No.2006CB705804): Greenhouse Gas EOR Resource Utilization and Underground Sequestration and the Fundamental Research Funds for the Central Universities (Grant No. 11CX06022A). The authors would like to thank the assistance of the petroleum engineering institute staff.

REFERENCES

[1] Ferguson R. C., Kuuskraa V. A., and Remson D., "Storing CO_2 with Next Generation CO_2-EOR Technology," SPE 139717 presented at the *SPE International Conference on CO_2 Capture, Storage, and Utilization held in New Orleans*, Louisiana, USA, **2010**.

[2] Koottungal L., "2010 worldwide EOR survey," *Oil & Gas Journal*, **2010**, *108*(14), 41-53.

[3] Geffen T. M., "Improved Oil Recovery Could Ease Energy Shortage," World Oil, **1977**, *177*(5), 84-88.

[4] Brashear J. P. and Kuuskraa V. A., "The Potential and Economics of Enhanced Oil Recovery," Report US FEA Contract No. CO-0350222-000, Washington DC, **1976**.

[5] NPC (National Petroleum Council)., "Enhanced Oil Recovery-An Analysis of the Potential for Enhanced Oil Recovery from Known Fields in the United States," Washington D.C., **1976**.

[6] McRee B. C., "CO_2: How it Works, Where it Works," *Petroleum Engineering*, **1977**, 52-63.

[7] Lyoho A. W., "Selecting Enhanced Recovery Processes," *World Oil*, **1978**, *187*(6), 61-64.

[8] OTA (Office of Technology Assessment), "Enhanced Oil Recovery Potential in the United States," Congress of the United States, *US Government Printing Office*, Washington D. C., **1978**.

[9] Carcoana A., "Enhanced Oil Recovery in Romania," Proceedings Third Joint SPE/DOE Symposium on Enhanced Oil Recovery, *Society of Petroleum Engineers of AIME*, Dallas, TX, **1982**, 367-379.

[10] Taber J. J. and Martin F. D., "Technical Screening Guides for the Enhanced Recovery of Oil," SPE 12069 presented at the *Society of Petroleum Engineers Annual Technical Conference and Exhibition*, San Francisco, CA, **1983**.

[11] Klins M. A., "CO_2 Flooding Mechanism and Engineering Design [M]," *Beijing: Petroleum Industry Press*, **1989**, 216-234.

[12] Rivas O., "Ranking Reservoirs for CO_2 Flooding Processes," *SPE 23641*, **1992**, 95-103.

[13] Diaz D., Bassiouni Z., and Kimbrell W., "Screening Criteria for Application of Carbon Dioxide Miscible Displacement in Waterflooded Reservoirs Containing Light Oil," SPE/DOE 35431 presented at the *SPE Improved Oil Recovery Symposium* held in Tulsa, OK, USA, **1996**, 21-24.

[14] Taber J. J., Martin F. D., and Seright R. S., "EOR Screening Criteria Revisited-Part 1: Introduction, to Screening Criteria and Enhanced Recovery Field Projects," *SPE Reservoir Engineering*, **1997**, *12*(3), 189-198.

[15] Brent Thomas., "Proposed Screening Criteria for Gas Injection Evaluation," *The Journal of Canadian Petroleum Technology*, **1998**, *37*(11), 14-20.

[16] Aladasani J. A. and Bai B., "Recent Developments and Updated Screening Criteria of Enhanced Oil Recovery Techniques," SPE 130726 presented at the *CPS/SPE International Oil & Gas Conference and Exhibition in China* held in Beijing, China, **2010**.

[17] Qin L. C., Hailong L., and Yin X. Q., "Application of CO_2 Miscible Flooding on Gao 89-1 Low Permeability Reservoir," SPE 144918 presented at the *SPE Asia Pacific Oil & Gas Conference and Exhibition* held in Jakarta, Indonesia, **2011**.

[18] Shuqiu Z., "Application of Fuzzy Optimization

Theory to Optimizing Development Program of Oil Gad Field," *Natural Gas Industry*, **1993**, *13*(2), 45-48.

[19] Yu X., Liangtian S., and Lei S., "A New Integrative Evaluation Way for Candidate of Carbon Dioxide Miscible Flooding Reservoirs Based on Fuzzy Analytical Hierarchy Process," *Acta. Petrolei. Sinica.*, **2002**, *23*(6), 60-62.

[20] Yu X., Yubin X., and Tiangang C., "A Screening Candidate Reservoir for Gas Injection Way Based on Characteristic Parameter's Comprehensive Weight Optimization Method," *Journal of Southwest Petroleum Institute*, **2004**, *26*(2), 22-25.

[21] Yunchuan Z., Yu X., and Tianjiang H., "Fuzzy Optimization Method for Screening Candidate of Gas Injection Reservoirs," *Journal of Southwest Petroleum Institute*, **2005**, *27*(1), 44-47.

[22] Shunpeng Z., Qiuwen Y., and Jie C., "Fuzzy Hierarchy Analysis-based Selection of Oil Reservoirs for Gas Storage and Gas Injection," *Henan Petroleum*, **2005**, *19*(4), 40-46.

[23] Lingtai F., Yueming C., and Jun Y., "Application of Fuzzy Analytical Hierarchy Process on Variable Weight Basis in Profile Control Well Selection in Low Permeability Reservoir," *Special Oil and Gas Reservoirs*, **2007**, *14*(3), 81-84.

[24] Wang G. C., "Microscopic Investigation of CO_2 Flooding Process," SPE/DOE 9788, AGU, **1982**.

[25] Fred I. and Stalkup J. R., "Miscible displacement [M]," *Beijing: Petroleum Industry Press*, **1989**, 40-52, 355-401.

[26] Chengzhi Y., Qingshan Y., and Pingping S., "EOR by Miscible Flooding [M]," *Beijing: Petroleum Industry Press*, **1993**, 51-78.

[27] Bacci G., Korre A., and Durucan S., "An Experimental and Numerical Investigation into the Impact of Dissolution/ Precipitation Mechanisms on CO_2 Injectivity in the Wellbore and far Field Regions," *International Journal of Greenhouse Gas Control*, **2011**, *5*, 579-588.

[28] Jahangiri H. R. and Zhang D., "Ensemble Based Co-optimization of Carbon Dioxide Sequestration and Enhanced Oil Recovery," *International Journal of Greenhouse Gas Control*, **2012**, *8*, 22-33.

[29] Cao M. and Gu Y., "Physicochemical Characterization of Produced Oils and Gases in Immiscible and Miscible CO_2 Flooding Processes," *Energy & Fuels*, **2013**, *27*, 440-453.

[30] Valcarcel M. and Rios A., "The Hierarchy and Relationships of Analytical Properties," *Analytical Chemistry*, **1993**, *18*, 781-786.

[31] Junwen B. and Kebin S., "Improved Hierarchy Analysis Process Used in Optimal Selection of Seepage Control Project of Dam Foundation," *Journal of Water Resources and Architectural Engineering*, **2004**, *2*(4), 11-13, 22.

[32] Yu Y. S., Li Y., Lu H. F., Yan L. W. et al. "Performance Improvement for Chemical Absorption of CO_2 by Global Field Synergy Optimization," International Journal of Greenhouse Gas Control, 2011, *5*, 649-658.

[33] Wanhua Q., "Management Decision and Application of Entropy," *Beijing: China Machine Press*, **2001**, 140-158, 193-196.

[34] Fuyuan Z. and Qinghua C., "The Comprehensive Integration Method of Subjective and Objective Granting Weight and Its Application," *Journal of the Academy of Equipment Command & Technology*, **2003**, *14*(6), 106-110.

[35] Chen S. J. and Hwang C. L., "Fuzzy Multiple Attribute Decision Making:

Methods and Applications," New York, Springer, **1992**.

[36] Pop H. and Sarbu C., "A New Fuzzy Regression Algorithm," *Anal. Chem.*, **1996**, *68*, 771-778.

[37] Zhou Q., Chan C. W., and Tontiwachwuthikul P., "Application of Neuro-fuzzy Modeling Technique for Operational Problem Solving in a CO_2 Capture Process System," *International Journal of Greenhouse Gas Control*, **2013**, *15*, 32-41.

[38] Julong Z., "Grey Theoretical Basis," *Wuhan: Huazhong University of Science Technology Press*, **2002**, 35-76.

[39] Xuanmin Z. and Kang X., "An Application of Grey Correlation Analysis to the Study about the Influencing Factors of Oil-gas Production Cost," *Journal of Xi'an Shiyou University (Social Science Edition)*, **2011**, *20*(6), 5-9.

[40] Yanming P., Hongyan G., and Zhicheng Y., "Examples of the Development of Foreign Oilfield Gas Injection," *Beijing: Petroleum Industry Press*, **2001**, 244-253.

Appendix

(1) Build the standardized evaluation matrix $R=(r_{ij})_{m\times n}$ using Equations 1-3.

$$R=\begin{pmatrix} 0.976879 & 0.638728 & 0.000000 & 0.904624 & 1.000000 \\ 0.666667 & 0.055556 & 1.000000 & 0.000000 & 0.222222 \\ 0.300832 & 1.000000 & 0.000000 & 0.278240 & 0.702735 \\ 0.000000 & 1.000000 & 0.588235 & 0.882353 & 0.485294 \\ 0.221374 & 1.000000 & 0.000000 & 0.419847 & 0.618321 \\ 0.000000 & 0.341219 & 0.872005 & 0.530786 & 0.644525 \\ 0.479322 & 0.240526 & 0.219761 & 0.000000 & 0.624676 \\ 1.000000 & 0.560000 & 0.660000 & 0.000000 & 1.000000 \\ 0.948113 & 0.220156 & 0.476403 & 0.000000 & 0.659264 \\ 0.500000 & 0.125000 & 1.000000 & 0.000000 & 0.350000 \\ 0.892857 & 0.214286 & 1.000000 & 0.000000 & 0.071429 \\ 0.518135 & 0.466321 & 1.000000 & 0.000000 & 0.880829 \end{pmatrix} \begin{array}{|l} \text{Permeability} \\ \text{Heterogeneity} \\ \text{Reservoir thickness} \\ \text{Oil viscosity} \\ \text{Oil API gravity} \\ \text{Porosity} \\ \text{Reservoir temperature} \\ \text{Reservoir pressure} \\ \text{Reservoir depth} \\ \text{Wettability} \\ \text{Reservoir dip} \\ \text{Oil saturation} \end{array}$$

(2) Determine the screening indexes weight.

The judgment matrix C

$$C=\begin{bmatrix} 0 & -1 & 1 & -1 & 1 & 1 & -1 & -1 & -1 & -1 & -1 & -1 \\ 1 & 0 & 1 & 1 & 1 & 1 & 1 & 1 & 1 & 1 & -1 & 1 \\ -1 & -1 & 0 & -1 & -1 & 1 & -1 & -1 & -1 & -1 & -1 & -1 \\ 1 & -1 & 1 & 0 & 1 & 1 & 1 & 1 & 1 & 1 & -1 & 1 \\ -1 & -1 & 1 & -1 & 0 & 1 & -1 & -1 & -1 & -1 & -1 & -1 \\ -1 & -1 & -1 & -1 & -1 & 0 & -1 & -1 & -1 & -1 & -1 & -1 \\ 1 & -1 & 1 & -1 & 1 & 1 & 0 & -1 & -1 & -1 & -1 & -1 \\ 1 & -1 & 1 & -1 & 1 & 1 & 1 & 0 & 1 & -1 & -1 & -1 \\ 1 & -1 & 1 & -1 & 1 & 1 & 1 & -1 & 0 & -1 & -1 & -1 \\ 1 & -1 & 1 & -1 & 1 & 1 & 1 & 1 & 1 & 0 & -1 & 1 \\ 1 & 1 & 1 & 1 & 1 & 1 & 1 & 1 & 1 & 1 & 0 & 1 \\ 1 & -1 & 1 & -1 & 1 & 1 & 1 & 1 & 1 & -1 & -1 & 0 \end{bmatrix}$$

The AHP weight vector

$$\omega_1=(0.046801,0.15029,0.033534,0.127218,0.039616,0.028386,\\ 0.055289,0.077161,0.065316,0.107688,0.177547,0.091155)$$

The entropy weight vector

$$\omega_2=(0.080687,0.111962,0.087050,0.077050,0.082948,0.061909,\\ 0.041477,0.080864,0.077895,0.090564,0.128755,0.078837)$$

The comprehensive weight vector

$$\begin{cases} \text{Permeability} & 0.062938 \\ \text{Heterogeneity} & 0.132856 \\ \text{Reservoir thickness} & 0.055336 \\ \text{Oil viscosity} & 0.101401 \\ \text{Oil API gravity} & 0.058711 \\ \text{Porosity} & 0.042935 \\ \text{Reservoir temperature} & 0.049046 \\ \text{Reservoir pressure} & 0.080902 \\ \text{Reservoir depth} & 0.073054 \\ \text{Wettability} & 0.101145 \\ \text{Reservoir dip} & 0.154853 \\ \text{Oil saturation} & 0.086823 \end{cases}$$

(3) Calculate the grey correlation of property (*i*) of the reservoir (*j*) with the positive ideal program and negative ideal program.

The weighted standardization matrix *U*

$$U = R \times W = \begin{pmatrix} 0.061483 & 0.040200 & 0.000000 & 0.056935 & 0.062938 \\ 0.088571 & 0.007381 & 0.132856 & 0.000000 & 0.029524 \\ 0.016647 & 0.055336 & 0.000000 & 0.015397 & 0.038887 \\ 0.000000 & 0.101401 & 0.059648 & 0.089471 & 0.049209 \\ 0.012997 & 0.058711 & 0.000000 & 0.024650 & 0.036302 \\ 0.000000 & 0.014650 & 0.037439 & 0.022789 & 0.027673 \\ 0.023509 & 0.011797 & 0.010778 & 0.000000 & 0.030638 \\ 0.080902 & 0.045305 & 0.053395 & 0.000000 & 0.080902 \\ 0.069264 & 0.016083 & 0.034803 & 0.000000 & 0.048162 \\ 0.050572 & 0.012643 & 0.101145 & 0.000000 & 0.035401 \\ 0.138262 & 0.033183 & 0.154853 & 0.000000 & 0.011061 \\ 0.044986 & 0.040488 & 0.086823 & 0.000000 & 0.076477 \end{pmatrix}$$

The positive ideal program U_0^+ and negative ideal program U_0^- are given by:

$$U_0^+ = (0.062938\ \ 0.132856\ \ 0.055336\ \ 0.101401\ \ 0.058711\ \ 0.037439\ \ 0.030638\ \ 0.080902\ \ 0.069264\ \ 0.101145\ \ 0.154853\ \ 0.086823)$$

$$U_0^- = (0\ 0\ 0\ 0\ 0\ 0\ 0\ 0\ 0\ 0\ 0\ 0)$$

The grey correlation of R_j^+ and R_j^- property (*i*) of the reservoir (*j*) with U_0^+ and U_0^-

$$R^+ = (0.751135, 0.707539, 0.798245, 0.610789, 0.756061)$$

$$R^- = (0.673621, 0.711659, 0.667081, 0.867118, 0.654708)$$

(4) Calculate the Euclidean distance D_j^+ and D_j^- of the reservoir (*j*) with U_0^+ and U_0^-:

$$D^+ = (0.218320, 0.155006, 0.261795, 0.112298, 0.166966)$$

$$D^- = (0.147878, 0.214508, 0.120632, 0.273548, 0.199551)$$

(5) Make indexes (R_j^+, R_j^-, D_j^+, and D_j^-) dimensionless.

$$R^+ = (0.866243, 0.815966, 0.920573, 0.704390, 0.871925)$$

$$R^- = (0.776850, 0.820718, 0.769309, 1.000000, 0.755039)$$

$$D^+ = (0.170539, 0.247381, 0.139119, 0.315468, 0.230131)$$

$$D^- = (0.251777, 0.178760, 0.301914, 0.129507, 0.192553)$$

(6) Combine R_j^+, R_j^-, D_j^+, and D_j^- ($\alpha_1 = \alpha_2 = 0.5$).

$$S^+ = (0.559010, 0.497363, 0.611243, 0.416949, 0.532239)$$

$$S^- = (0.473695, 0.534050, 0.454214, 0.657734, 0.492585)$$

(7) Calculate the "similarity nearness degree" C_j of the candidate program.

$$C = (0.541307, 0.482216, 0.573691, 0.387974, 0.519347)$$

18

Application of Particle Size Distribution Engineering and Nano-technology to Cement Recipes for some Highly Deviated Wellbores in Iran

Yaser Pourmazaheri* and Hamid Soltanian

Division of Petroleum Engineering, Faculty of Upstream Petroleum Industry, Research Institute of Petroleum Industry (RIPI), Tehran, Iran

ABSTRACT

The design of cement slurry for horizontal wellbores is challenging as it must present very high quality cement slurry properties e.g. zero free water, less than 50 cc API fluid loss value, appropriate rheology and yield point for effective displacement, adequate compressive strength, which is critical for lightweight slurries, and finally stable cement column behind casing. The main objective of these requirements is to provide unique cement sheath in the upper side of annulus as well as the lower side and to guarantee complete zonal isolation.

These requisite conditions could be rarely met especially when lightweight cement is needed. Four cement recipes were made and named as RIPI-1 through RIPI-4 representing standard slurry with a density of 118 pounds per cubic feet (pcf), traditional lightweight slurry with a density of 90 pcf, lightweight slurry with engineered PSD and nanoparticles in a liquid form, and finally another type of PSD and nanopowder respectively. These slurries were tested at the temperature and pressure of Iranian Balal, Soroush, and Siri oilfields and finally were compared with those slurries previously used in the oilfields. According to the results, reductions in yield point up to 84% and in fluid loss value up to 44% were recorded by the recipes. Moreover, an increase in compressive strength up to 82% was measured and the maximum strength of 2500 psi was obtained for slurry RIPI-4. All the slurries designed showed adequate stability. However, significant improvements were achieved by slurries RIPI-3 and RIPI-4 with engineered PSD and nanoparticles.

Keywords: Horizontal Cementing, Nanosilica, Particle Size Distribution, Compressive Strength Development

INTRODUCTION

The history of horizontal wells to increase production from reserves backs to 1920's and during several decades it has been under development. The potential applications of horizontal drilling are numerous. They can be related to the targeted location, reservoir characteristics, the nature and properties of formation fluids, or even to an overall field development plan. Nowadays, there is no doubt in their commercial viability. The productivity of a horizontal well can be up to six times more than what a typical vertical well may produce. This significant difference makes it worthy to study horizontal wellbores cementing.

At present, most horizontal wellbores are designed to be completed without cementing. The horizontal section is often lined with a slotted liner, pre-perforated liner or, in some cases, wire wrapped sand control liners. In such conditions, the formation lithology must show sufficient integrity to prevent collapse or sloughing, particularly when the producing formation is approaching depletion. In some cases, these wells are left open hole without any aforementioned completion methods. The previous highly deviated intermediate casing must have a good cement job. However, there are some circumstances that horizontal wells and production intervals must be cased off and cemented, and some forms of isolation must be initiated. Some of these situations are 1) multi-interval stimulation treatments, 2) where water cones or gas coning problems are likely, and 3) when water or gas breakthrough in producing formation may occur [1].

The design of cement slurry for deviated or horizontal sections is challenging as it must present very high quality cement slurry tested and designed according to tough test conditions and standards. For example, cement slurry to be used in horizontal well must have zero free water in all inclinations since little amounts of free water can cause severe lack of zonal isolation especially in upper sections of annulus (Figure 1). In addition, cement shrinkage, gas or water intake from formation to cement sheath, and high fluid loss from cement slurry to formation may lead to improper zonal isolation. These tough requirements could be rarely met especially when lightweight cement is needed. In this study, all the relevant data of some offshore oil fields of Iran such as Balal, Soroush, and Siri have been reviewed and the properties of cement slurry used in the wells have been collected. Then, the cement slurry, which can meet the requirements of horizontal cementing according to standards, has been designed and finally a comparison is made between the traditional cement system previously used in the fields by service companies and the one designed by RIPI.

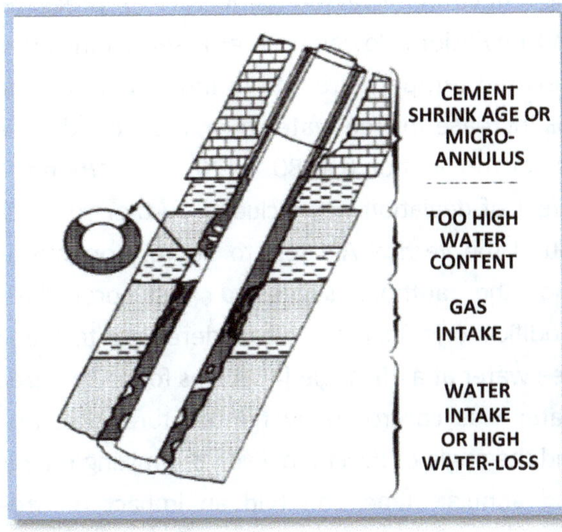

Figure 1: Challenges of cementing in a highly deviated wellbore

Design Challenges of Cement

All cement slurry properties must be kept within the acceptable ranges defined by the API standard for horizontal cementing. For example, the amount of fluid loss for normal cementing conditions (e.g. vertical wells or low inclination) should be less than 100 milliliters per 30 minute of testing, which is considered standard fluid loss but it should be half for horizontal cementing purposes. The pattern of thickening time for slurry should be in the right angle mode to decrease the possibility of excessive gel development and gas migration phenomena. In addition, the compressive strength development pattern in the slurry should be in such a way that the optimum *wait on cement* (WOC) with a reasonable final strength is available. However, the most important design criterion in the lab is the slurry stability. Stability can be divided into two categories: free water content and settling tendency of solid contents (density segregation) in slurry. It was not until 1978's [2, 3] that the industry began to explore the possible relationship between flow-after cementing and changes in hydrostatic pressure due to the accumulation of free water. Most of the research up to this period of time was conducted without considering the effect of increased temperature, pressure, and deviation. However, later research included elevated temperature and found as much as an 8% increase in free water over tests conducted as per API Section 6 at 80 °F (26.7 °C). When the effect of deviation was included, further research found the vertical API test to be very optimistic, and the authors suggested that procedural modifications should be considered to test and free water at a 45° angle [4]. It was found that free water was controlled by temperature, pressure, and the angle of deviation. Even the mixing energy and annular dimension had an impact on free water content. Higher shear stress applied by mixing procedure can decrease the free water. This free water content is directly changing with the angle of deviation, but at deviations higher than 45 degrees, no substantial increase was observed. Pressure shows higher direct effect on solid settling rather than on free water [2,4]. For testing the density segregation, the procedure stated by API Spec 10 A can be used [5]. In this method, the cement slurry is cured under down hole conditions for 24 hours and then the set cement is cut to 3 or 4 similar sections and the density of each section is measured via Archimedes law as follows:

$$\rho_{section}=\frac{m_a}{m_w} \tag{1}$$

where, m_a is the weight of set cement section in the air and m_w stands for the weight of set cement section when hanging in water.

Then, the relative density of each section is compared to the density of slurry before curing and a stability profile is established. The ideal profile is a flat one, which shows similar density between the top and bottom of the set cement column. If lower sections show higher density, then, there is density segregation in the cement column. In lightweight and heavy weight cements, there is a tendency for solid segregation; in fact, in lightweight cement the lightweight additives tend to float on top, while in heavy weight cements, weighting agents tend to settle.

For this study, the cementing program of Iranian hydrocarbon fields such as Balal, Soroush, and Siri was studied carefully and tested in lab again. Based on these data, it was clarified that two cement systems were needed if one would design cementing program for horizontal or highly deviated wellbores in these oil and gas fields. In addition to standard cements with a density of 118-120 pounds per cubic feet (1.89-1.92 S.G), a lightweight cement system with a density of 85-95 pcf (1.36-1.52 S.G) was also desired.

Generally, conventional lightweight cements present high free water and fluid loss. They suffer from low compressive strength and delayed strength development. Definitely, these cement systems are not suitable candidates for cementing horizontal sections as they are very likely prone to formation fluid migration and lack of zonal isolation. Therefore, the main challenge at this stage is to formulate a lightweight cement system, which meets all sever standard requirements for cementing horizontal wellbore intervals.

Lightweight Cement Systems

Three methods, which can be used to prepare lightweight cement systems, are as follows:

1. Using extender additives (additives which adsorb water to reduce the density);

2. Lightweight additives (low density additives to decrease the overall density of the solid content [6]);

3. Nitrogen system (pumping nitrogen to cement and reducing the weight).

The most famous extenders to be used for lightweight systems are bentonite, diatomite, pozzolan, perlite, and microsilica. These additives viscosity the cement slurry and do not pose solid segregation problems, but the cement system does not show adequate compressive strength and free water and fluid loss control properties. In addition, most of these solids represent strength retrogression [6-8].

In RIPI, it took great effort to achieve to a lightweight slurry system, which showed significantly improved test results, using nanotechnology and particle size distribution (PSD) engineering. For PSD purposes, D124 or Litefill system was used. It consists of hollow spherical solid particles filled with nitrogen or CO_2 to be light enough to reduce the density of solid fraction of slurry [9]. The diameters of the spheres range between 60 to 315 microns. These spheres consist of Al_2O_3 (35%) and SiO_2 (65%) and they can be applied to cement powder in dry blend. It does not take additional water, thus the compressive strength is increased and the overall porosity is reduced. With these particles, the slurry weight can vary between 67 to 90 pcf according to the design. Figure 2 shows the SEM image of Litefill additive.

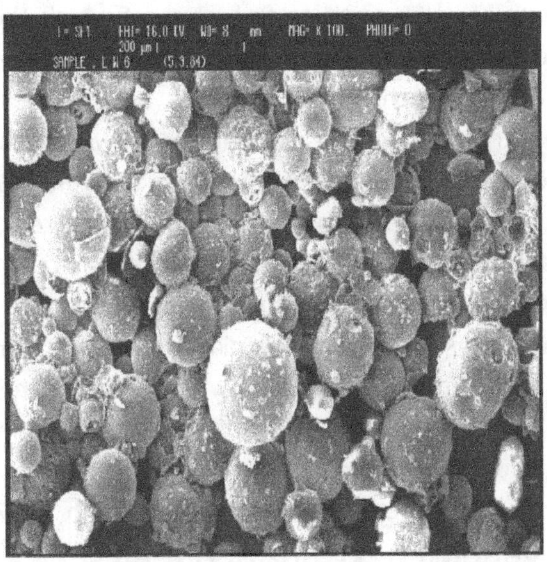

Figure 2: The SEM image of Litefill particles with 100X magnification

Another constituent of the RIPI novel system is a liquid extender which is mainly composed of especially processed amorphous inorganic microsilica with a diameter of 0.15 microns and a specific area of 21 m^2/gr. It reacts with calcium hydroxide released by cement hydration to form calcium silicate hydrate (C-S-H), which results in higher bonding between particles and improved compressive strength. Figure 3 shows the SEM image of this special liquid extender called Microblock™.

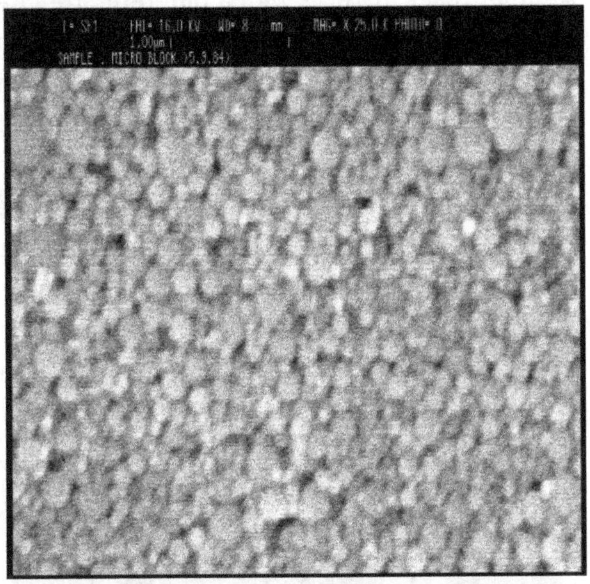

Figure 3: The SEM image of Microblock with 75X magnification

The benefits of using Microblock and the related mechanisms are as follows:

1. Zero free water: as Microblock proposes high surface area, it acts like a strong water-wet material;

2. Reduced fluid loss: since Microblock reduces the overall porosity of the filter cake, the amount of fluid loss is reduced as well;

3. Low friction (viscosity): very fine particles of Microblock reduce the friction between solid particles in such a way a lubricant does and thus the equivalent circulation density is decreased, which is beneficial in long radius horizontal and low-diameter wellbores;

4. High early and final compressive strength: due to high pozzolanic reactivity and silica content and surface area, the compressive strength is

sometimes 30 percent higher than the API recommended compressive strength;

5. Higher stability: due to higher surface tension between particles, Microblock helps maintaining floating forces and prevents solids from settlement or segregation.

Another magic particle with a different size in a nanorange is a kind of hydrophobic silica which is called HSL. This product, which is processed in RIPI, exhibits superior properties in term of compressive strength, free water, fluid loss control ability, and cement stability profile. The mechanism of improvement is similar to what can be seen by Microblock, but in a more effective way. This material is basically a pure non-crystallized silica powder combined with plenty of other constituents with a significant surface area (100 m^2/gr), which is contributing to its remarkable successful applications. Figure 4 shows the SEM image of HSL particles.

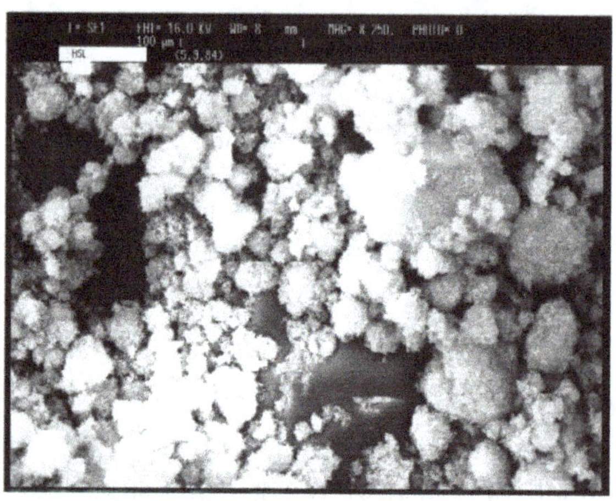

Figure 4: The SEM image of HSL particles with 750X magnification

The Optimum Cement Design

The main idea in cement design by RIPI is based on applying particles with a nonuniform size, which is known as particle size distribution (PSD). Finer lightweight additives fill between cement particles and increase the solid content in the unit volume of cement slurry and reduce the porosity of set cement.

Due to lower water to solid ratio, it results in faster and higher compressive strength progression. The size of HSL particles are around 0.1 micrometer and fill between cement particles with a diameter of 50 microns, Litefill additive with a size of 60 to 315 microns, and Microblock particles with a size of 0.5 microns. With this particle size distribution, the cement designed by RIPI presents lower porosity, permeability, free water, and fluid loss value and higher compressive strength. Figure 5 shows a schematic of particle size distribution technology (PSD) [9].

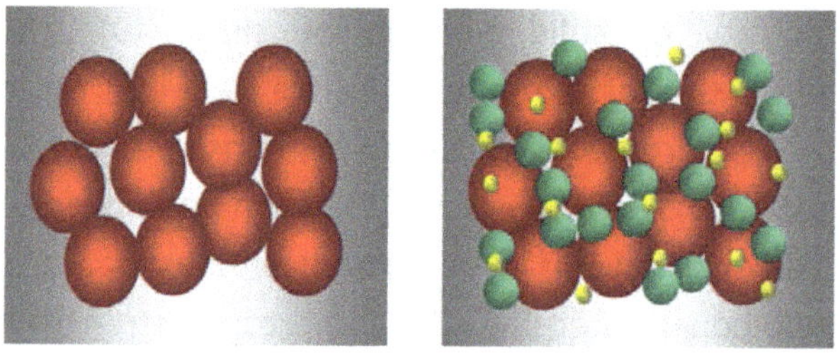

Figure 5: A comparison of void spaces between optimum cement design with a PSD (right) and traditional design (left)

In addition to particle size, another difference between HSL particles and microspheres is their reactivity in cement slurry. Most of lightweight additives such as microspheres are chemically neutral and used just as filler in between other solid particles. Since they increase the volume of water in cement, the final compressive strength is inversely affected. Nanoparticles propose higher surface area and they increase the rate of chemical reactions taking place in cement slurry during hydration. Even at low temperatures, these particles react with the lime content in cement powder to produce a cement phase, and thus they increase the compressive strength [6,7].

EXPERIMENTAL PROCEDURES

Tables 1 and 2 show the slurry composition designed by Schlumberger (Dowell) for Balal and Soroush oil fields respectively. It should be noted that Sor.20-26 refers to well numbers 20 to 26 in Soroush oil field. The first step to design cement slurry is to determine the range of density and applicable pressure and temperature. Based on the data of the aforementioned oil and gas fields, two slurries with a density of 90 and 118 pcf were proposed. The range of bottom hole static temperature (BHST) falls between 70 to 90 °C and the bottom hole circulating temperature (BHCT) is in the range of 50 to 70 °C. In addition to standard cement with a density of 118 pcf (RIPI-1), two types of lightweight cement slurry were designed. One contains D124 as hollow spherical solid particles without any particle size distribution and nanotechnology (RIPI-2), while the other type of cement slurry, based on D124 lightweight additive, uses an engineered PSD and contains nanoparticles (RIPI-3).

Table 1: Slurry composition for Balal oil field wells

Well ID.	CMT G (gr)	Water (cc)	Fluid Loss Controller (gr)	Dispersant (gr)	Retarder (gr)	Antifoam (cc)	Para Gas (cc)
Balal 7I. 7" & 4.5" Liner	1000	417	4	5.5	2.5	1.33	26.6
Balal 8I. 7" Liner	1000	404	3	2	5	1.33	39.95
Balal 9I. 7" Liner	1000	417	4	5.5	3	1.33	26.63
Balal 10I. 7" Liner	1000	404	3	2	4.5	1.33	39.95

Table 2: Slurry composition for Soroush and Siri oil field wells

Well ID.	CMT G (gr)	Water (cc)	Fluid Loss Controller (cc)	Dispersant (cc)	Retarder (cc)	Antifoam (cc)	Anti settling Agent (BWOC)	Para gas (cc)
Sor.20-26	1000	390	-	11.3	4.5	0.8	0.1	-
Sor.27 7" Liner	1000	390	76	11.3	1.5	0.8	0.05	-
Sor. 28 7" Liner	1000	390	4	7.5	7.5	0.8	-	-
Siri DPH-8	1000	460	-	% 0.35 BWOC	% 0.38 BWOC	1.33	-	30

Table 3: Composition and rheological properties of slurry RIPI-1

Slurry Formulation•					
Cement G (Lb)•	FLC (Lb/Sk)•	Dispersant (Lb/Sk)•	Retarder (Lb/Sk)•	Fresh Water (Gal/Sk)•	Slurry Weight (pcf)•
110•	0.44•	0.33•	0.165•	6.07•	118•

Rheological Properties•								
•	•	θ 3•	θ 6•	θ 100•	θ 200•	θ 300•	PV (cp)•	YP (lbf/100ft^2)•
Before Heat•	R.U. Readings•	1•	2•	26•	50•	76•	72•	4•
	R.D. Readings•	2•	3•	30•	54•	76•		
	Average•	**1.5•**	**2.5•**	**28•**	**52•**	**76•**		
After Heat•	R.U. Readings•	1•	3•	40•	70•	98•	85•	13
	R.D. Readings•	4•	5•	43•	74•	98•		
	Average•	**2.5•**	**4•**	**41.5•**	**72•**	**98•**		

Table 4: Test results of slurry RIPI-1

Slurry Properties, W=118 pcf						
Gel 10 sec ($lbf/100ft^2$)	Gel 10 Min ($lbf/100ft^2$)	FL (cc) P=1000 psi T= 158 °F	Thickening Time (min) P=5000 psi• T= 158°F	Compressive Strength UCA (psi) 24-hour P= 3000 psi T=194 °F	Free Water (%)	Permeability (md)
3.5	6.5•	50•	328•	2640•	0•	0.0629

Slurry with a Density of 118 pcf (RIPI-1)

Typical cement slurry with a density of 118 pcf was designed by RIPI. Table 3 shows the composition and rheological properties of the slurry. The lab results of this slurry are presented in Table 4. Figure 6 shows the stability profile of slurry RIPI-1. As can be seen, a relatively flat profile was established.

Lightweight Slurry without a PSD and Nanoparticles (RIPI-2)

The first type of the lightweight slurry was prepared just with D124 spherical particles. Neither particle size distribution nor nanotechnology was used for its design. Tables 5 and 6 show the composition and test results of slurry RIPI-2. The stability profile of this slurry is presented in Figure 7. This profile shows an acceptable density difference between the top and bottom of the cement column.

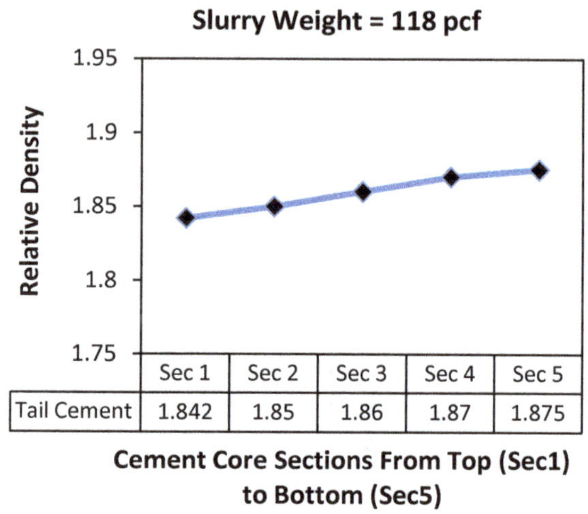

Figure 6: The stability profile of RIPI-1 set cement

Table 5: Composition and rheological properties of slurry RIPI-2

Slurry Formulation•						
Cement G (Lb)•	D124 (Lb/Sk)•	FLC (Lb/Sk)	Dispersant (Lb/Sk)•	Retarder (Lb/Sk)•	Fresh Water (Gal/Sk)•	Slurry Weight (pcf)•
110•	30.58•	0.88•	0.33•	0.08•	8.58•	89•

Rheological Properties•								
•	•	θ 3•	θ 6•	θ 100•	θ 200•	θ 300•	PV (cp)•	YP ($lbf/100ft^2$)•
Before Heat•	R.U. Readings•	4•	8•	67•	128•	183•	177•	6•
	R.D. Readings•	2•	5•	63•	128•	183•		
	Average•	3•	6.5•	65•	128•	183•		
After Heat•	R.U. Readings•	6•	11•	73•	131•	184•	167•	17
	R.D. Readings•	4•	7•	72•	129•	184•		
	Average•	5•	9•	72.5•	130•	184•		

Table 6: Test results of slurry RIPI-2,

Slurry Properties, W=89 pcf						
Gel 10 sec ($lbf/100ft^2$)	Gel 10 Min ($lbf/100ft^2$)	FL (cc) P=1000 psi T= 158 °F	Thickening Time (min) P=5000 psi T= 158°F	Compressive Strength UCA (psi) 24-hour P= 3000 psi T=194 °F•	Free Water (%)	Permeability(md)
5	10•	48•	283•	2500•	0•	0.0537•

Table 7: Composition and rheological properties of slurry RIPI-3

Slurry Formulation, Slurry Weight = 89 pcf•								
Cement G (Lb)•	D124 (Lb/Sk)•	HSL (Gal/Sk)	FLC (Lb/Sk)	Disp. (Lb/Sk)•	Retarder (Lb/Sk)•	M.B Gal/Sk	Gas Block (Gal/Sk)•	Fresh Water (Gal/Sk)•
110•	30.58•	0.8•	0.55•	2.2•	0.55•	1.90•	0.79	6.6•
Rheological Properties•								
•	•	θ 3•	θ 6•	θ 100•	θ 200•	θ 300•	PV (cp)•	YP ($lbf/100ft^2$)•
Before Heat•	R.U. Readings•	4•	7•	49•	88•	135•	130.5•	4.5•
	R.D. Readings•	4•	6•	47•	88•	135•		
	Average•	4•	6.5•	48•	88•	135•		
After Heat•	R.U. Readings•	13•	21•	67•	122•	160•	145•	15•
	R.D. Readings•	3•	6•	60•	115•	160•		
	Average•	8•	13.5•	63.5•	118.5•	160•		

Table 8: Test results of slurry RIPI-3

Slurry Properties, W=89 pcf						
Gel 10 sec ($lbf/100ft^2$)	Gel 10 Min ($lbf/100ft^2$)	FL (cc) P=1000 psi T= 158 °F	Thickening Time (min) P=5000 psi T= 158°F	Compressive Strength UCA (psi) 24-hour P= 3000 psi T=194 °F	Free Water (%)	Permeability (md)
4	15	28•	120•	1970•	0•	0.0211•

Lightweight Slurry Using a PSD and Nanoparticles (RIPI-3)

The other type of the lightweight slurry was prepared by implementing particle size distribution engineering and nanotechnology. The additive called Microblock was used in conjunction with nano-HSL particles. Tables 7 and 8 present the composition and properties of slurry RIPI-3. Figure 8 shows the stability profile of slurry RIPI-3

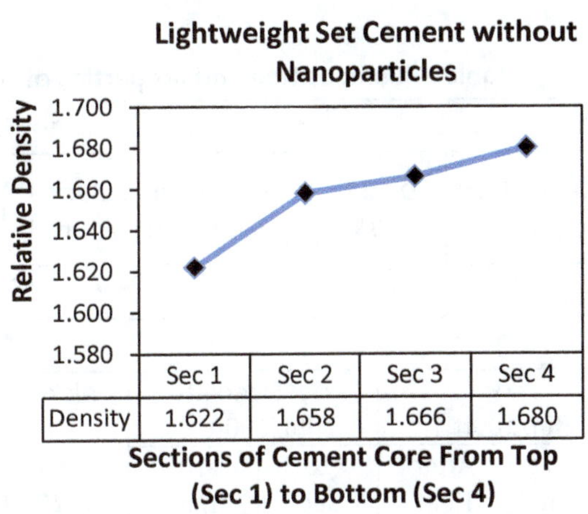

Figure 7: The stability profile of lightweight set cement without nanoparticles

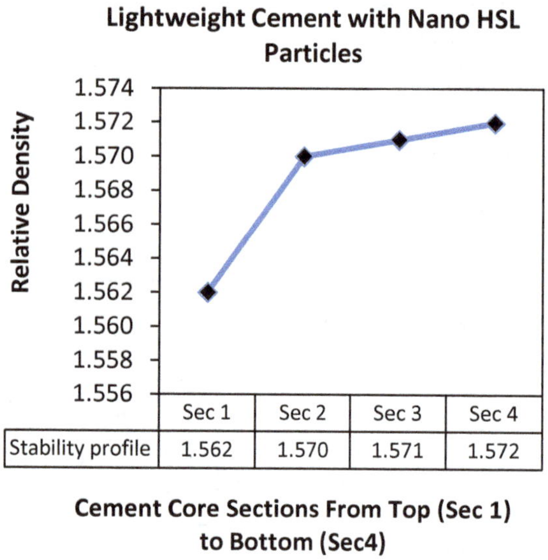

Figure 8: Stability profile of slurry RIPI-3

Lightweight Slurry Using New Type of Nanoparticles and PSD for High Temperature Wells (RIPI-4)

Another adjustment for the lightweight slurry to be used in high-temperature wells was made by RIPI and the slurry was named as RIPI-4. The idea of implementing PSD engineering and nanotechnology was also used. However, the type and quantity of the additives were changed as listed in Table 9. As can be seen, different fluid to solid ratios can be obtained, which results in improved cement properties.

Table 10 presents the summary of test results for each cement slurry compositions.

RESULTS AND DISCUSSION

Based on the results obtained, different cement properties could be evaluated. Thus each category has been discussed separately.

Rheological Properties

The first important slurry property is rheology. The ramp up/down method of reading rheology helps establish a rheology profile, which is informative of slurry tendency to settle or segregation at a glance.

If the ratio $\frac{\text{R.U Reading}}{\text{R.D Reading}}$ is close to 1 for all reading velocities, the slurry does not show the settling tendency and rheological properties are independent of time. If this ratio is higher than 1, there is a settling tendency in the slurry in the test pressure and temperature conditions. Finally, if the ratio is less than 1, gelation behavior has been developed in the cement slurry [5]. For slurry RIPI-1 with a density of 118 pcf, this ratio is around 1 and no settling is observed as well. For slurries RIPI-2 and RIPI-3, this ratio is a little over 1. To make sure that no segregation has happened, we should take a look at the results of free water test and the stability profile of the set cements. According to Figures 7 and 8, the density difference between the top and bottom of the set cement for slurry RIPI-2 is higher (0.06) compared to RIPI-3 (0.01), in which nanoparticles was used.

Table 9: Composition and properties of slurry to be used in high-temperature wells (RIPI-4)

Slurry Formulation								
Cement G• (lb)	D-124 (lb/sk)	HSL (lb/sk)•	Microblock (gal/sk)	Anti Gas Migration (gal/sk)•	Fluid loss controller• (lb/sk)	Retarder (lb/sk)	Dispersant (lb/sk)•	Water• (gal/sk)
110	50	4.5	2.17	0.72	0.48	1.2	2.4	6.5
Slurry Properties								
YP ($lbf/100ft^2$)		Gel Strength ($lbf/100ft^2$)		Thickening Time• (min) T=200°F P=5000psi	Compressive Strength• (psi)• T=230°F, P=3000 psi	Free Water• (%)	Fluid Loss• (cc) T=200°F P=1000psi	Permeability• (md)
Before Heat	After Heat	10 sec	10 min					
14	18	19	35	160	2550•	0	18	0.01

Table 10: The summary of test results of all cement slurry compositions

Well ID.	Casing/Liner	Inclination Angle(°)	Slurry Weight (pcf)	BHT (°C)	API Fluid Loss (cc)	T.T. (min)	YP ($lbf/100ft^2$)	Free Water (%)	Compressive Strength (psi/24 hours)
Balal **7**	7	57	118	75	32	292	40	0	2375
Balal 7	4 1/2	71	118	77	34	259	95	0	-
Balal 8	7	59	118	71	24	265	76	0	500
Balal 9	7	59	118	71	32	260	47	0	50
Balal 10	7	61	118	71	24	260	86	0	0
Soroush 20-27	9 5/8	87	118	82	358	250	10	0	3411
Soroush 28	9 5/8	84	118	40	376	300	6.5	1.6	0
Siri DPH-8	7	88	115	88	87	267	10	0	4600
RIPI-1	-	90	118	90	50	328	4	0	2640
RIPI-2	-	90	90	90	48	283	6	0	2500
RIPI-3	-	90	90	90	28	120	4.5	0	1970
RIPI-4	-	90	90	110	18	160	14	0	2550

However, for both slurries the segregation is negligible. It seems that nanoparticles can fill between larger constituents and prevent them from being packed. In addition, these particles could increase the viscosity of the slurry and provide enough suspending force to prevent solid particles from settling. The other important rheological property is the yield point, which represents the amount of force needed to make the slurry to flow. It is recommended that the slurry for horizontal sections should be designed with a yield point below 20 $lbf/100ft^2$ [8]. Those slurries designed by RIPI and the one used for Siri and Soroush oilfields meet this requirement. However, the slurry used for Balal oilfield presents a very high viscous rheology, which is not suitable for the low side annulus of horizontal sections.

Stability Test

To make judgment about the stability of slurries, one should take into account the free water content, fluid loss value, and the stability profile of them. For RIPI-1 with a density of 118 pcf, fluid loss value is 50cc / 30 min of the test (Table 9), which meets the requirements for horizontal sections. The free water content in 90 degrees inclination is zero as needed. Figure 6 shows a density difference of 2 pcf between the top and bottom, which is acceptable by industry standards.

For RIPI-2 with a density of 90 pcf and without nanoparticles, the fluid loss value and free water content are similar to RIPI-1. According to Figure 7, the density difference between the top and bottom of the set cement is 3.7 pcf, which was predictable based on rheological readings and the ratio explained in section 6-1. Slurry RIPI-3 and RIPI-4 use nanoparticles and PSD engineering. This has led to a lower fluid loss value (28 and 18cc/30min) and zero free water. The stability of this system shows a significant improvement since the density difference between the top and bottom is as low as 0.12 pcf. This situation presents totally stable cement slurry. Although, the temperature of slurry RIPI-4 is higher than RIPI-3, the stability profile of these two slurries were similar to each other. Figure 9 compares the stability profiles of slurries with and without nanoparticles.

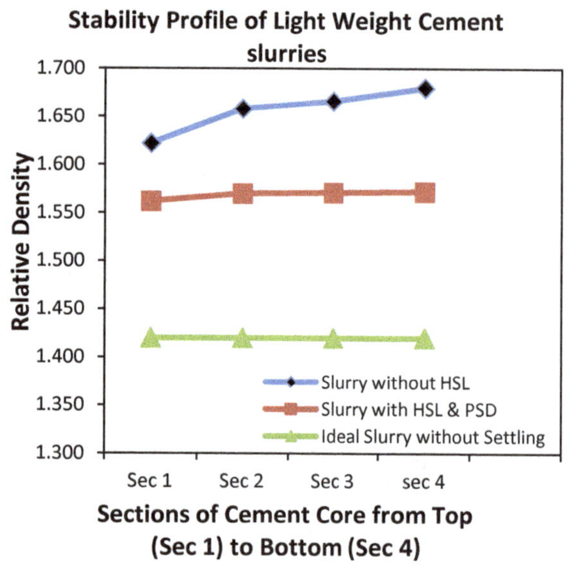

Figure 9: Comparison of stability profiles of slurries with and without nanoparticles

Based on Figure 9, the densities of the tested slurries (red and blue curves) are higher than the original designed slurry (ideal slurry without settling). When the slurry with lightweight additives (hollow spheres) undergoes the curing pressure of 3000 psi, these spheres may collapse to some extend and, as a result, the total volume of slurry is reduced and the density is increased.

The application of nanoparticles and particle size distribution engineering has led to better performance of the cement slurry under down hole conditions. As can be seen in Figure 9, the amount of density change in slurries RIPI-3 and RIPI-4 (red curve) is less compared to slurry RIPI-2 (blue curve), which is just using the lightweight additive. This phenomenon is affected by the presence of nanoparticles and PSD technique, by which finer particles of nanomaterials and Microblock fill the pore spaces and prevent the solid content from collapse and contraction during compressive strength development (see Figure 5 as well). This improved performance of the cement slurry when located behind casing or liner strings will result in better cement/pipe or formation bonding, which is definitely critical to provide zonal isolation.

Compressive Strength Development

Figure 10 compares the compressive strength development of different slurry systems. Those slurries designed during this study have been compared to Balal well No.9 cement slurry.

According to Figure 10, the compressive strength of all the lightweight cement slurries (RIPI-2, 3, and 4) are competing with the standard slurry with a density of 118 pcf (RIPI-1). Since traditional slurries with a density of 90 pcf rarely show significant compressive strength, the improvement to strength is directly related to the application of nano and silicate-based additives (such as Microblock and hollow spheres).

As previously mentioned, the higher compressive strength of slurry RIPI-2 in comparison with RIPI-3 might be related to the performance of nanoadditives and particle size distribution engineering employed to prevent the solid constituents from contraction and avoid the collapse of hollow spheres (Figure 9). The more the compaction of hollow spheres in the lightweight slurry systems is, the higher the density is and the higher the compressive strength, measured by sonic waves, becomes. Both slurries RIPI-2 and RIPI-3 developed a compressive strength after 8 hours, and during the first 24 hours the ultimate strength has been reached. Since slurry RIPI-4 has implemented different PSD and type of nanomaterial, its compressive strength started to rise after 6 hours until the final strength after 24 hours of test run. One of the advantages of these three novel slurries is the rapid compressive strength development, which greatly reduces the wait on cement time (WOC) and decreases the chance of fluid migration through cement sheath. This process happens faster when nanoparticles have been used.

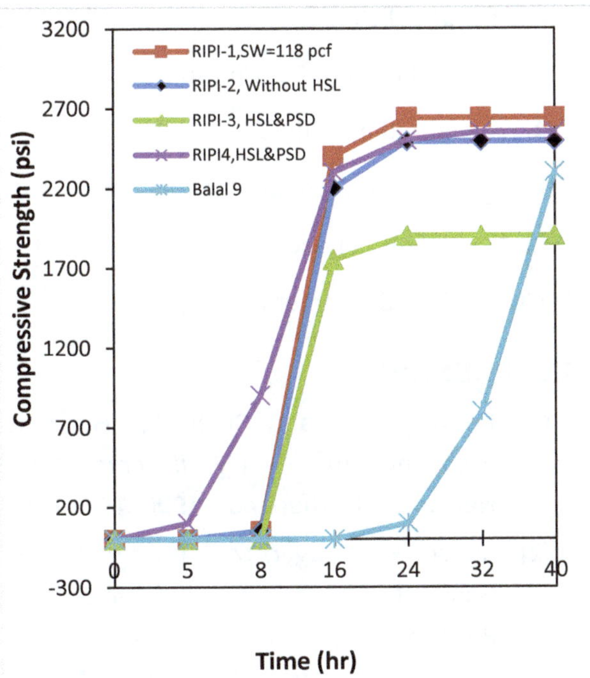

Figure 10: Comparison between compressive strength (UCA) of all the slurry systems

Figures 11, 12, and 13 present respectively comparative information of fluid loss value, total compressive strength development, and the time needed to develop strength from 50 psi to 500 psi. The later one is a representative of WOC and setting off the next phase drilling and can be used as one of the criteria for accessing fluid migration through cement sheath behind casing.

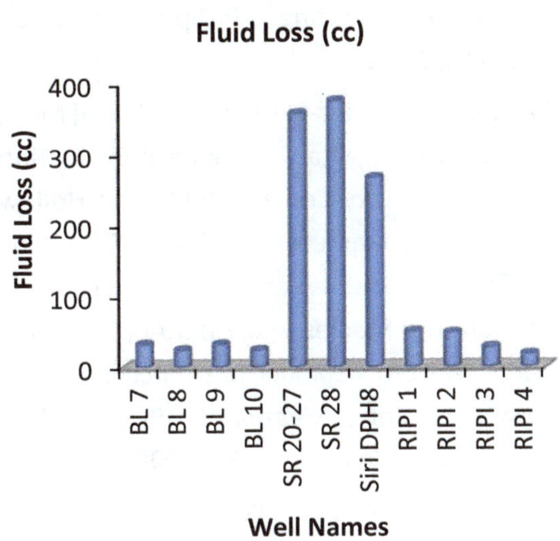

Figure 11: Fluid loss value of all the cement slurry systems

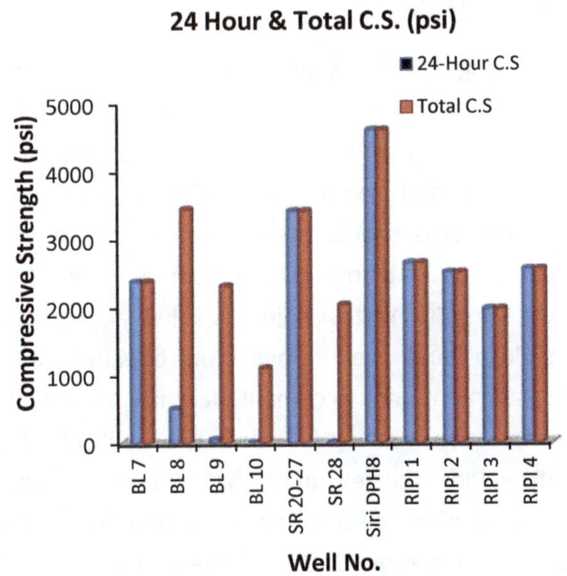

Figure 12: Compressive strength development for all the cement systems

As can be inferred from Figure 13, the best slurry is the one with an earlier start compressive strength (50 psi) and a faster development of compressive strength until the next phase drilling could be commenced (500 psi) as slurries RIPI-3 and RIPI-4 show. Note that by applying nanoparticles and particle size distribution engineering to the lightweight cements, the rapidest compressive strength growth has been obtained.

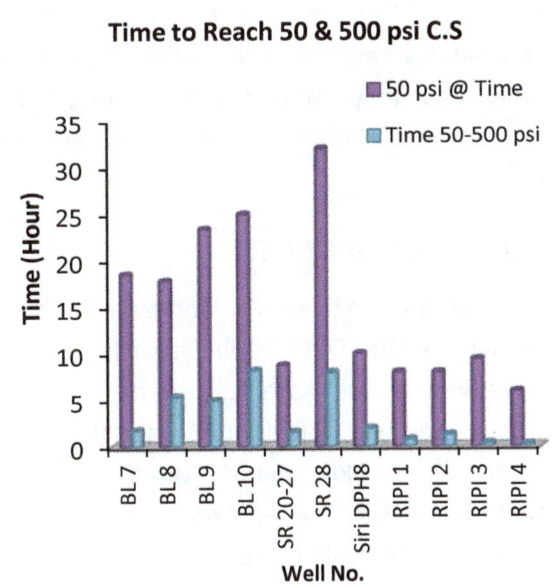

Figure 13: Early strength development and minimum WOC time

CONCLUSIONS

According to the study carried out on some cement slurry recipes, the best slurry for horizontal wells in Iran oilfields has been selected. The proposed recipe for horizontal or highly inclined wells should meet tough standard and operation conditions e.g. zero free water, less than 50cc/30 min test fluid loss value, appropriate rheology and yield point for effective mud displacement prior to cement placement, adequate compressive, strength which is critical for lightweight slurries, and finally stable cement column behind casing. Three lightweight slurries have been designed. One of them was designed as traditional slurry (RIPI-1), for the other one (RIPI-2), nanoparticles in the liquid form and particle size distribution engineering were implemented and finally the slurry with a different PSD (compared to RIPI-2) and nanoparticles in a powder form with different physical and chemical properties was made (RIPI-4). Comparing these slurries with those used in Balal, Soroush, and Siri oilfields, which were basically standard weight slurries, a reduction in yield point up to 84% and a drop in fluid loss value up to 44% were recorded. Besides, an increase in compressive strength up to 82% was measured in some cases. All the slurries designed in RIPI showed adequate stability and they were totally confirmed for application in horizontal wells in Iran oil and gas fields.

ACKNOWLEDGEMENTS

The authors wish to thank Mr. Mortazavi, the head of Drilling and Well Completion Research Group of RIPI as well as Dr. Kazemzadeh, the head of Petroleum Engineering Division. We also thank Mr. Bahmani, Mr. Babaloie, and Mr. Miri for providing oilfield data. Finally, we would like to appreciate the great efforts of Mr. Fathi, Mr. Saiedi, Mr. Tajik, and Mr. Karimi in the cement laboratory of RIPI.

NOMENCLATURES

PSD : Particle Size Distribution

pcf : Pound per Cubic Foot

psi : Pound per Square Inch

Gal/Sk : Gallon per Sack

WOC : Wait On Cement

REFERENCES

[1] Cooper R. E. and Birch G., "Horizontal Well Cementing," in *Well Cementing*, B. Nelson, E., Netherlands, **1990**, 15.1-15.14.

[2] Matson R. P., Rogers M. J., Boncan V. C., and Gandy R. G., "The Effects of Temperature, Pressure, and Angle of Deviation on Free Water and Cement Slurry Stability," *SPE Annual Technical Conference and Exhibition*, SPE 22551-MS, Dallas, Texas, **1991**.

[3] Webster W. W. and Eikerts J. V., "Flow After Cementing-A Field and Laboratory Investigation," *SPE Annual Technical Conference*, SPE 8259-MS, Las Vegas, **1979**.

[4] Wilson M. A. and Sabins F. L., "A Laboratory Investigation of Cementing Horizontal Wells," *SPE Drilling Engineering*, **1988**, *3*(3), 275-280.

[5] Specifications for Cements and Materials for Well Cementing, API Spec. 10 A, 23rd Edition, **2002**.

[6] Al-Yami A. S., Nasr-El-Din H. A., Al-Arfaj M. K., Al-Saleh S. H., et al., "Long-term Evaluation of Low-Density Cement, Based on Hollow Glass Microspheres, Aids in Providing Effective Zonal Isolation in HP/HT Wells: Laboratory Studies and Field Applications," *SPE Western Regional and Pacific Section AAPG Joint Meeting*, SPE 113138, Bakersfield, California, USA, **2008.**

[7] Rae Ph. and Di Lullo G., "Lightweight Cement Formulations for Deepwater Cementing: Fact and Fiction," *SPE Annual*

Technical Conference and Exhibition, SPE 91002, Houston, Texas, **2004**.

[8] Kulakofsky D. and Vargo R., "New Technology for the Delivery of Beaded Lightweight Cements," *SPE Annual Technical Conference and Exhibition*, SPE 94541, Dallas, Texas, **2005**.

[9] Williams H., Khatri D., Voughan M., Landry G., et al., "Particle Size Distribution-Engineered Cementing Approach Reduces Need for Polymeric Extenders in Haynesville Shale Horizontal Reach Wells," *SPE Annual Technical Conference and Exhibition*, USA SPE 147330, Denver, Colorado, **2011**.

19

A Study on the Mechanism of Urea-assisted Steam Flooding in Heavy Oil Reservoirs

Changhua Yang[1], Yiqing Lin[2], Zhang Zhang[3], Ruijian Deng[1], Xinrong Wu[1], Baolun Niu*[1], Zhaomin Li[4], and Shaoran Ren[4]

[1] Zhongyuan Oilfield Branch Company, Sinopec, Puyang, Henan Province, 457001, P.R. of China

[2] School of Foreign Languages, Xi'an Shiyou University, Xi'an, Shaanxi Province, 710065, P. R. of China

[3] Oil and Gas Development Corporation, Zhongyuan Petroleum Engineering Ltd. Sinopec, Puyang, Henan Province, 457001, P. R. of China

[4] School of Petroleum, China University of Petroleum, Qingdao, Shandong Province, 266580, P. R. of China

ABSTRACT

The Biqian-10 block, located in Henan Oilfield of Sinopic, contains many thin and interbedded reservoirs, which have been operated by cyclic steam stimulation for 20 years or more. Therefore, it is a challenge to implement the conventional steam flooding. In order to improve the recovery of steam flooding, urea was used to assist steam flooding. Urea can decompose into CO_2 and NH_3, which are beneficial to enhance oil recovery (EOR). For the sake of exactly quantifying the mechanism of urea assisted steam flooding (UASF), the UASF model was built according to the experimental results. The simulation results show that CO_2 is the key point for EOR, and its pressurization function is more effective than the decrease in oil viscosity by dissolving CO_2. The emulsification of crude oil for NH_3 is weak in interfacial tension tests; thus the effect of emulsification can be ignored in the simulation. The UASF can improve the recovery by about 17.4%, which is 6.8% higher than steam flooding.

Keywords: heavy oil, urea assisted steam flooding, CO_2, mechanism, EOR

INTRODUCTION

Gas-assisted steam flooding has become an effective technology to further EOR after steam flooding [1-4], but it is limited by the gas source. Therefore, gas-assisted steam flooding [5, 6] has not seen a large scale promotion. In order to further enhance the effect of steam flooding, the gas can be generated by chemical agents, which can effectively solve the problem of the gas source and gas generation can also be controlled by a chemical reaction generation rate.

Urea assists steam flooding through the decomposition of urea in the formation of CO_2 and NH_3 to achieve indirect gas-assisted steam flooding. The mechanisms of UASF are the reduction in the viscosity of heavy oil by CO_2 and the emulsification of the heavy oil in theory through NH_3. Meanwhile, the gas can increase the steam spread range, which can further improve heavy oil recovery. In order to describe the mechanism of urea-assisted steam flooding, a real reservoir model from Biqian 10 reservoir in the Henan oilfield of Sinopec was adopted for reservoir simulations using an advanced thermal simulator.

Urea-assisted Steam Flooding Model

In UASF modeling, in addition to conventional heavy oil viscosity-temperature model (Table 1) and relative permeability-temperature model (Figure 1), the main researches were on urea decomposition model, CO_2 dissolved in oil model, and NH_3 emulsifying oil model.

Table 1: Oil viscosity at different temperatures used in the simulator

Temperature (°C)	Oil viscosity (mPa·s) calculated value	experimental value	Error (%)
30	51949	50720	2.4
50	6166	6505	5.2
70	1219	1190	2.4
90	348	350	0.57
130	60	--	--
150	32	--	--
170	19.4	--	--
200	10.7	--	--
250	5.4	--	--
300	3.4	--	--

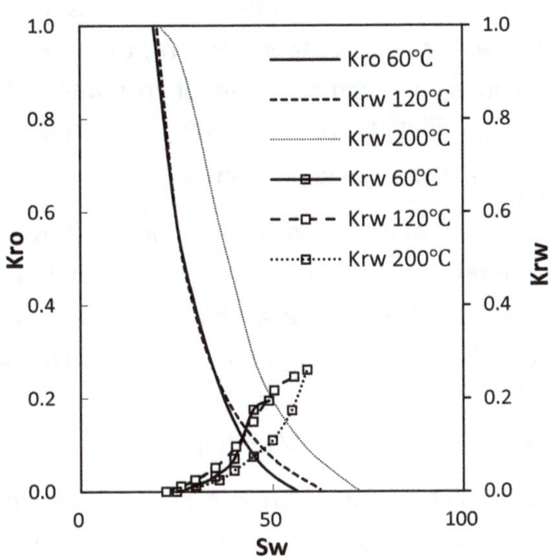

Figure 1: Variations of oil-water relative permeability used in the simulator

Urea Decomposition Model

The decomposition rate of urea is the key factor in controlling the reaction time and the volume of the decomposition gases of CO_2 and NH_3. A high temperature high pressure reactor was employed to simulate the urea decomposition reaction, and the experiment diagram is shown in Figure 2. The experimental conditions and results are shown in Table 2 and Table 3.

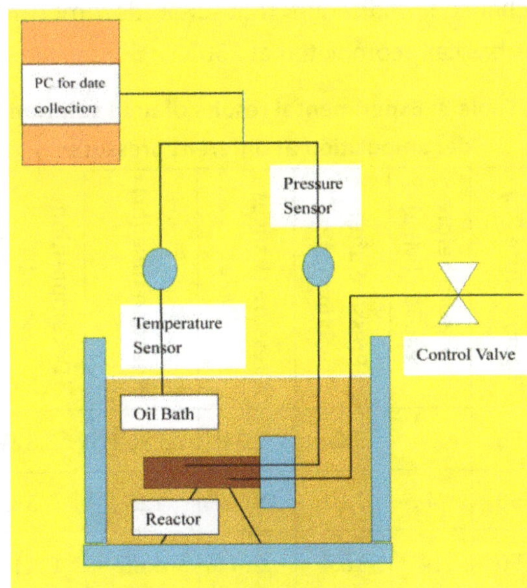

Figure 2: A schematic of urea decomposition reactor

Table 2: Experimental results of urea solution decomposition at different temperatures

Temperature (°C)	Pressure before reaction (MPa)	Pressure after reaction (MPa)	Pressure changes (MPa)	Reaction time(hr)	Decomposition rate ×10^{3} (mol/(l.s))
50	0	0	0	0	0
80	0	0.07	0.07	0.37	0.054
110	0.07	0.3	0.23	0.52	0.116
150	0.3	2.95	2.65	1.87	0.357
180	2.95	3.08	0.13	0.6	0.033

It can be summarized that with increasing temperature, the pressure in the reactor was gradually increased in varying degrees. The reactor pressure was essentially unchanged at low temperatures (50 -80°C), which indicated that no substantial decomposition of urea happened and the reaction rate was low. When the temperature was raised to 110°C, the reactor pressure rose to 0.3 MPa and the reaction rate gradually increased. When the temperature was increased to 150°C, the pressure in the reactor was rapidly increased, and the decomposition rate reached 0.357×10^{-3} mol/(l.s), which was 3 times as much as the one obtained at 110°C. When the temperature reached 180 °C, the pressure remained essentially unchanged, indicating that urea decomposition was basically completed at 150°C.

Table 3: Experimental results of urea solution decomposition at different pressures

Temperature (°C)	Pressure before reaction (MPa)	Pressure after reaction (MPa)	Pressure changes (MPa)	Reaction time (hr)	Decomposition rate $\times10^3$ (mol/(l.s))
150	0	3.04	3.04	0.869	0.829
150	4	7.01	3.01	1.92	0.332
150	14	16.4	2.4	4.86	0.117

For gas generating reactions, the pressure changed with the time to reach equilibrium. At atmospheric pressure, the pressure reached balance after 0.869 hr, and the decomposition rate was 0.829×10^{-3}mol/(l.s). When the pressure rose to 4 MPa (the shallow reservoir pressure), the reaction time was extended to 2 hrs after the pressure reached an equilibrium, and the decomposition rate was reduced to 0.332×10^{-3} mol/(l.s). When the system pressure increased to 14 MPa (the deep reservoir pressure), the average rate of the decomposition was reduced to 0.117×10^{-3} mol/(l.s). Therefore, increasing the reservoir pressure raised the urea decomposition reaction rate, which was not conducive to the rapid decomposition of the urea.

The decomposition of urea belongs to the zero order reaction, because the concentration of urea does not affect the decomposition reaction [8]. Based on the above experimental results, the urea decomposition kinetic parameters used in the thermal simulator were calculated by the Arrhenius equation (Equation 1). The activation energy is 9.83 KJ/mol, and the pre-exponential factor is 1×107 l/(s.kPa).

$$\frac{dp_x}{dt} = k_0 \exp\left(-\frac{E}{RT}\right) \tag{1}$$

where, p_x is a gas partial pressure; *t* is the time and k_0 represents pre-exponential factor; *E* is the activation energy and *R* stands for the molar gas constant equal to 8.314; *T is* the model temperature.

Model of CO_2 Dissolved in Oil

CO_2 is easily dissolved in oil and reduces viscosity. The widely used equation describing CO_2 solubility in heavy oil was proposed by Chung [9] (Equation 2).

$$R_s = \frac{1}{\left[a_1\gamma^{a_2}T^{a_7} + a_3T^{a_4}\exp\left(-a_5P + {a_6}/{P}\right)\right]} \tag{2}$$

where, R_s is the CO_2 solubility in heavy oil and *Y* is the severe of oil; *P* stands for the pressure and *T* represents the temperature; a_1, a_2, a_3, a_4, a_5, a_6, and a_7 are constants equal to 0.4934×10^{-2}, 4.0928, 0.571×10^{-6}, 1.6428, 0.6763×10^{-3}, 781.334, and -0.2499 respectively.

The CO_2 solubility in heavy oil at different temperatures and pressures was calculated by Chung equation. Table 4 shows the calculated data (CD) and the experimental date (ED) used in the thermal simulator. The CO_2 solubility in heavy oil was test by PVT equipment.

The calculated average error in Table 4 is less than 16.67%, which complies with the requirement of engineering computing requirements. Therefore, the CO_2 solubility in the thermal simulator can be calculated by Chung formula.

Table 4: Comparison of the experimental and calculated data on CO_2 solubility in heavy oil at different pressures and temperatures

Parameter	Solubility [ml(gas)/100ml(liquid)]														
Pressure (MPa)	40 °C			60 °C			80 °C			100 °C			120 °C		
	CD	ED	Error (%)	CD	ED	Error (%)	CD	ED	Error (%)	CD	ED	Error (%)	CD	ED	Error (%)
3	27	28	3.7	20	21	5	11	10	9.09	8	7	12.5	6	7	16.67
10	85	87	2.3	65	71	9.23	59	55	6.78	49	52	6.12	38	41	7.89
15	104	110	5.77	85	89	4.71	77	76	1.3	69	72	4.35	62	61	1.61
20	116	119	2.59	97	102	5.15	93	94	1.08	89	91	2.25	81	84	3.7

Model of NH_3 Emulsifying Oil

NH_3 solubility in water is high, which is easy to generate weak alkali ammonium hydroxide (NH_3-H_2O). In theory, ammonium hydroxide reacts with the active ingredient in oil by saponification reactions to form a surfactant. The reaction is given by Equation 3:

$$RCOOH+NH_4OH \longrightarrow RCOONH_4+H_2O \quad (3)$$

The generated surfactant can reduce the oil-water interfacial tension to achieve emulsification, and this mechanism can improve the oil displacement efficiency. The interfacial tension between ammonium hydroxide and oil at different concentrations was tested by tension apparatus of TX-500, but the emulsifying effect was not obvious. The experimental results showed that the interfacial tension was between 50×10^{-3} N/m and 58×10^{-3} N/m. Also, King used capillary force method to calculate the interfacial tension between NH_3 and pure water and showed that the interfacial tension decreased from 72.55×10^{-3} N/m to 22×10^{-3} N/m, i.e. a drop of 69% [10]. Therefore, NH_3 can reduce the interface tension between NH_3 and water, but it is not able to reduce the tension between NH_3 and heavy oil, because ammonium hydroxide is a weak alkaline, and the acids in heavy oil belong to weak acids; thus both react a lesser extent [11]. In conventional alkaline flooding, NaOH is the most important component, while ammonium hydroxide is rarely used for its weak alkaline; NaOH can reduce the interfacial tension to values between 0.1×10^{-3} N/m and 0.01×10^{-3} N/m [12]. Furthermore, the accumulation and diffusion of surface is a dynamic process at the interface. If the concentration of surfactant cannot reach critical micelle concentration (CMC), its effect on reducing interfacial tension is not obvious.

Therefore, the role of NH_3 in decreasing the interfacial tension cannot be considered in the numerical simulation.

Mechanism of Urea-assisted Steam Flooding in Reservoir

Reservoir Geology and Production History

Biqian 10 block is a shallow layer of ultra-heavy oil and the basic reservoir parameters are as follows: reservoir depth is 294.6 m; dip is 12 º; the effective thickness is 9.4 m; the porosity is 34%; the permeability is 2.28 D; the original oil saturation is 75%; the original reservoir pressure is 2.9 MPa; the temperature is 30.3°C; dead oil viscosity under reservoir temperature is 50719.6 mPa.s, and the density is 966.8 kg/m^3.

This block has been operating with the CCS process since 1989. Up to now, the recovery has reached 22%, with an average period of 8.1 cycles per single well, and the reservoir pressure has decreased to 1.2 MPa. Especially, the CCS process has reached the economic limit.

Based on the reservoir parameters, four inverted nine-spot well patterns were established as shown in Figure 3.The well spacing is 70 m×100 m, in which there are four injection wells and 21 production wells. The grid of fine geological model is 45×55×9. Moreover, the results of matching the CCS production history of 20 years are shown in Figure 4.

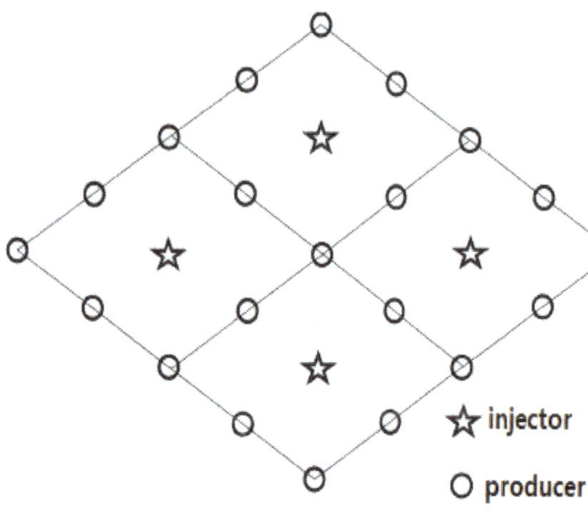

Figure 3: A schematic of 4 inverted nine-spot well patterns

Based on matching the CCS production history of 20 years, steam flooding (SF) and UASF for 30 cycles were compared. The structure of one cycle is shown in Figure 5 and the calculated results are summarized in Table 5. The steam injection pressure was 5.5 MPa and the steam temperature was 200°C. The steam quality was 70%.

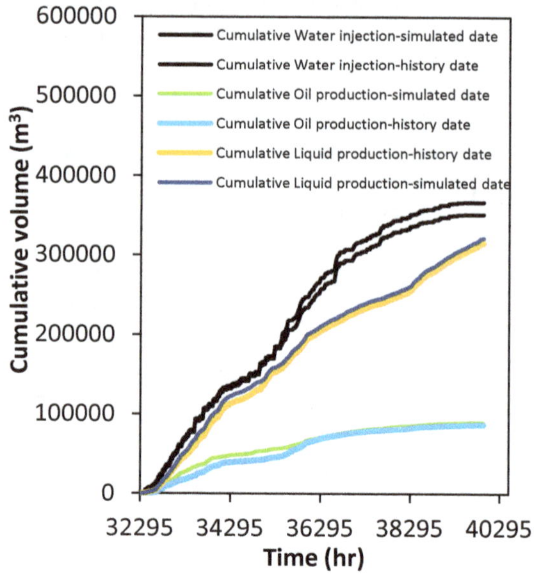

Figure 4: Results of history matching

Table 5: Development results for UASF and SF in 30 cycles

Driving mode	Cumulative cycle	Rate of steam flooding (m^3/d)	Weight of urea in one cycle (t)	The cumulative oil steam ratio (m^3/m^3)	Recovery (%)
Steam flooding	30	80	0	0.14	10.6
Urea-assisted steam flooding	30	80	30	0.24	17.4

Enhancing the Formation Pressure

The decomposition of urea can produce large amount of decomposition gas; reaction of 1 ton urea with 0.3 ton water can generate 373 m^3 CO_2 and 746 m^3 NH_3, which can significantly increase the volume of production. Figure 6 is the average pressure of steam flooding and urea-assisted steam flooding.

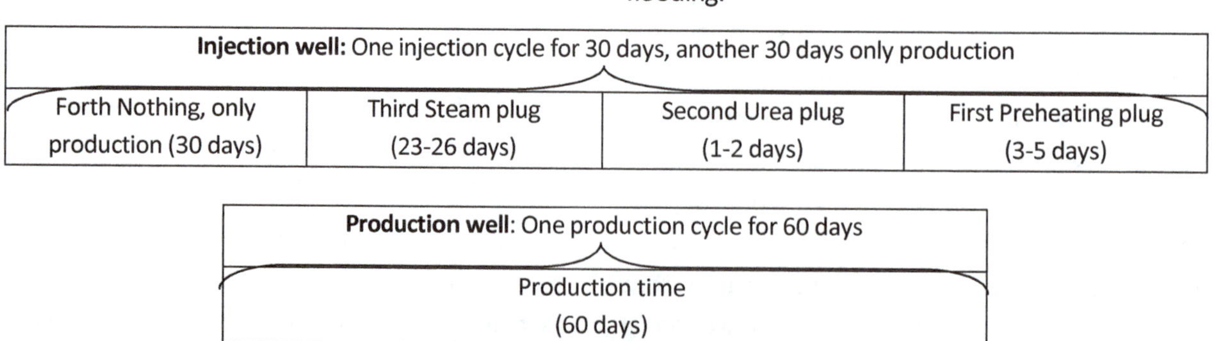

Figure 5: A scheme of single-cycle injection and production for UASF

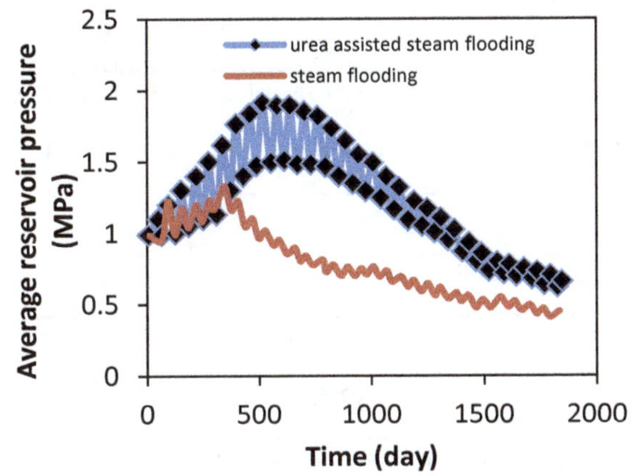

Figure 6: Variations of average pressure of SF and UASF

The reservoir pressure rose to 1.3 MPa by steam flooding, but the pressure declined to 0.5 MPa after steam channeling. UASF boosted the reservoir pressure to 1.9 MPa and the pressure decreased slower than steam flooding after steam channeling. Moreover, the intermittent injection created the jagged form of pressure.

Furthermore, as the reservoir pressure increases, the CO_2 solubility in oil will be enhanced, which can reduce the viscosity of heavy oil. In addition, the increased pressure is beneficial to maintain the steam dryness.

Reducing the Viscosity of Heavy Oil

The viscosity-reducing rate is an important indicator for enhanced heavy oil recovery. The viscosity-reduction rate reached 30% or more in the simulation as tabulated in Table 6.

Expanding Steam Sweep Volume

Figure 7 is the oil saturation field after matching history. Figure 8 displays the oil saturation field for steam flooding and urea-assisted steam flooding after 30 cycles. The red dot positions are the injection well in Figures 7 and 8.

Steam flooding can spread the scope and improve the oil production, but due to the small steam viscosity, high water-oil mobility ratio (which makes it easy to form steam channeling), it results in a lower heat utilization rate.

The gas generated by adding urea to steam flooding can rapidly enhance reservoir pressure and oil production rate.

Table 6: Variation of oil viscosity between injector and producer after UASF

Distance from injector (m)	Mole fraction of CO_2 in heavy oil (%)	Original oil viscosity (mPa·s)	Oil viscosity after UASF (mPa·s)	Viscosity-reducing rate (%)
10	57.57	77.2	11	85.7
30	3.19	741	516	30.4
40	0.1	1129	1114	1.3
50	0	786	786	0

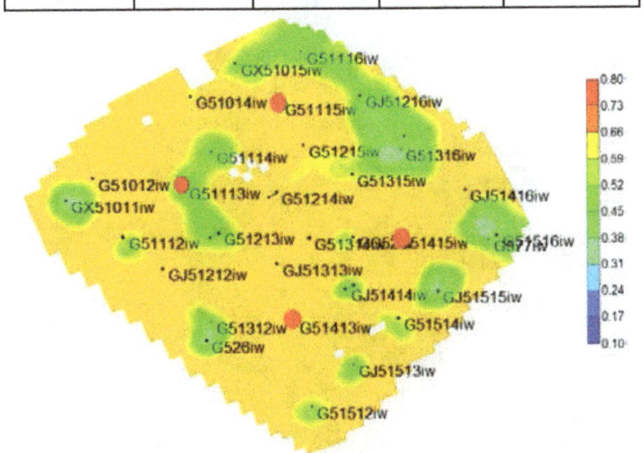

Figure 7: Distribution of oil saturation in the end stage of steam stimulation

The Constitution of EOR by UASF

In order to analysis the contribution of CO_2 and NH_3 to the oil recovery, the gas after the decomposition of urea was selected as the injection fluid; the proportion of CO_2 to NH_3 was 373 m^3 to 746 m^3 for 1 ton of urea. For the sake of the contribution of CO_2 in the viscosity reduction and raised pressure mechanisms, a smaller solubility of N_2 was selected as a counterpoint in the simulation. In order to remove the impact of the intermittent injection on pressure, the mode of single-cycle injection and production was changed as shown in Figure 9. Table 7 and Figure 10 report the results of EOR using different approaches.

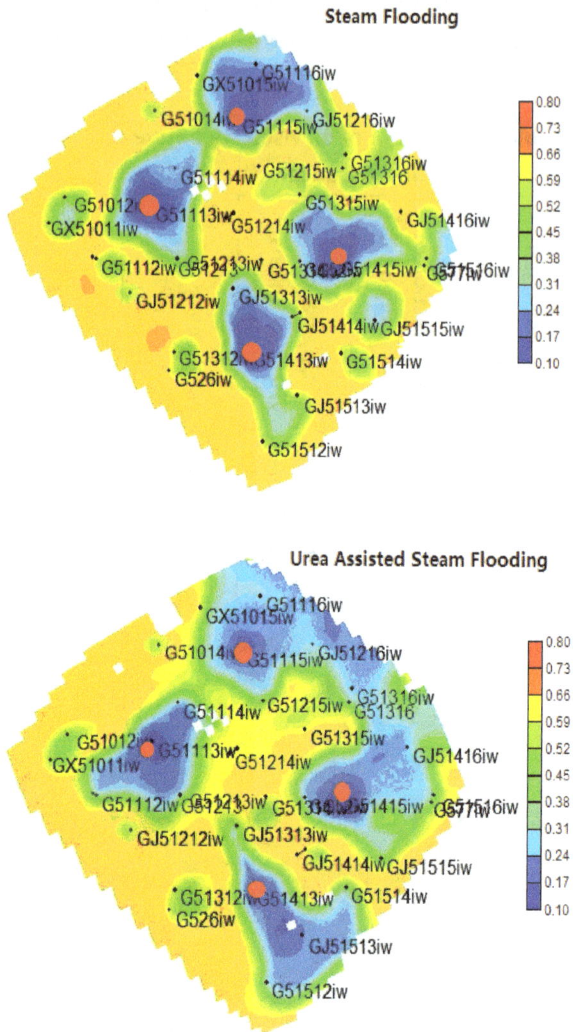

Figure 8: Distribution of oil saturation in 30th cycle after steam flooding (left) and UASF (right)

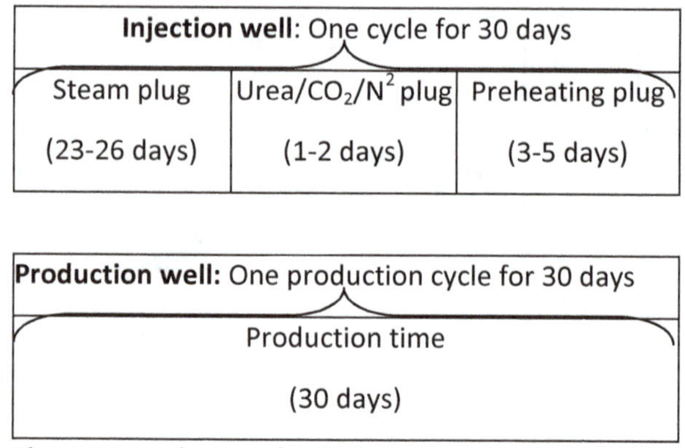

Figure 9: A scheme of a single-cycle injection

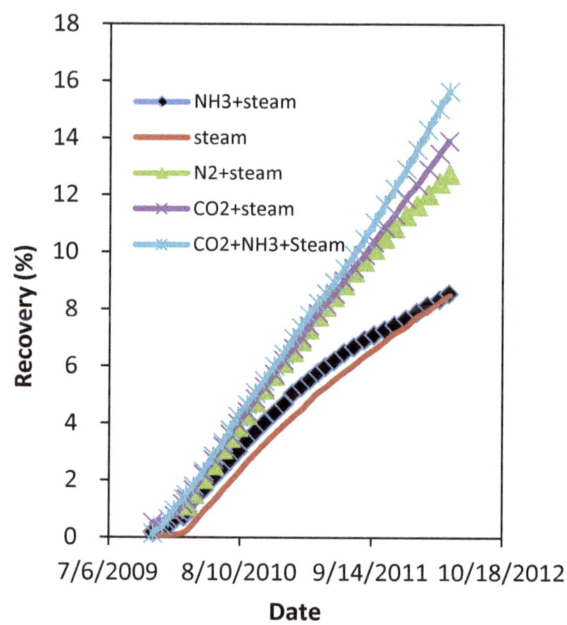

Figure 10: Variations of EOR using different approaches

Table 7: Results of EOR using different approaches

Injection mode	Rate of steam flooding (m^3/d)	Volume of gas in one cycle (m^3/day)		Cumulative volume of steam (m^3/day)	Cumulative volume of gas (m^3/day)		Cumulative volume of oil production (m^3/day)	EOR (%)	Growth based on steam flooding (%)
		CO_2	NH_3		CO_2	NH_3			
Steam flooding	[80	0	0	70879	0	0	6442	8.77	0
N_2 assisted steam flooding	80	11200	0	70879	335995	0	9605	13.07	4.3
CO_2-assisted steam flooding	80	11200	0	70879	335995	0	10568	14.39	5.62
NH_3-assisted steam flooding	80	0	22400	70879		671990	6479	8.88	0.11
CO_2-and NH_3-assisted steam flooding	80	11200	22400	70879	335995	671990	11877	16.17	7.4

Simulation results showed thatNH_3 and CO_2 were beneficial to enhance steam flooding recovery. The final recovery of steam flooding was similar with NH_3-assisted steam flooding, but the production rate of NH_3-assisted steam flooding was faster than steam flooding at early stages, which was due to the pressurization of NH_3. However, the impact of enhanced pressure on recovery was limited for NH_3 owing to its high solubility in water.

The recovery of CO_2-assisted steam flooding reached 14.39%, which was 5.62% higher than steam flooding. Therefore, CO_2 was the key point to enhance heavy oil recovery.

However, the recovery of CO_2-assisted steam flooding was only 1.32% higher than that of N_2-assisted steam flooding, which demonstrated that the dissolution of CO_2 in heavy oil, and thereby viscosity reduction thereof, was not the most important factor in oil recovery. Furthermore, the pressurization mechanism was a major parameter for CO_2 to enhance heavy oil recovery in immiscible flooding.

Meanwhile, the recovery of CO_2- and NH_3-assisted steam flooding reached 16.17%, which was 7.4% higher than that of steam flooding.

CONCLUSIONS

1- Based on the urea decomposition experiments, CO_2 was dissolved in oil and NH_3 emulsified oil; the urea-assisted steam flooding model was built by an advanced thermal simulator. Urea-assisted steam flooding model included urea decomposition model, CO_2 dissolved in oil model, NH_3 emulsifying oil model, and other conventional thermal models such as heavy oil viscosity-temperature model and relative permeability-temperature model. The results of urea-assisted steam flooding model was close to the experiment data.

2- CO_2 was the key factor in EOR and its pressurization function was more effective than the decrease in oil viscosity by the dissolution of CO_2. However, the NH_3-emulsified oil was weak in interfacial tension tests, and thus the effect of emulsification could also be ignored in the simulation.

3- Urea-assisted steam flooding outperforms conventional steam flooding after high stimulation cycles, and this technology has a fine adaptability for heavy oil reservoirs.

REFERENCES

[1] Sedaee Sola B., "Feasibility Experimental Investigation of Steam/Methane Flooding in a Heavy Oil Reservoir," SPE 91968-MS, presented at *the SPE International Petroleum Conference* in Mexico, Puebla Pue., Mexico, **2004.**

[2] Naderi Kh. and Babadagli T., "Experimental Analysis of Heavy Oil Recovery and CO_2 Storage by Alternate Injection of Steam and CO_2 in Deep Naturally Fractured Reservoir," SPE 146738-MS, presented at *the SPE Heavy Oil Conference Canada*, Calgary, Alberta, Canada, **2012**.

[3] Yongrong G., Shangqi L., and Dehuang Sh., "Improving Oil Recovery by Adding N_2 in SAGD Process for Super-heavy Crude Reservoir with Top-Water," SPE 114590-MS, presented at *the SPE Russian Oil and Gas Technical Conference and Exhibition*, Moscow, Russia, **2008**.

[4] Zhong L., Dong Z., and Hou J., "Investigation on Principles of Enhanced Offshore Heavy Oil Recovery by Co-injection of Steam with Flue Gas," SPE 165231-MS, presented at *the 2013 SPE Enhanced Oil Recovery Conference*, Kuala Lumpur, Malaysia, **2013**.

[5] Shen D., Zhang Y., and Zhang X., "Study and Application of Adding Additive to Improve Heavy Oil Recovery," *Petroleum Journal*, *6*, **2004** (in Chinese).

[6] Zhang X., Zhang Y., and Yue Q., "Conformance

Control of CSS and Steam Drive Process with a Carbamide Surfactant," JCPT, presented *at the 7th Canadian International Petroleum Conference* (the 57th Annual Technical Meeting of the Petroleum Society), in Calgary, Alberta, **2006**.

[7] De-huang SH., Jian-jun X., and Xiao-chun W., "Experimental Study and Application of Urea in Steam Flooding of Heavy Oil Reservoir," *Special Oil and Gas Reservoirs*, **2005**, *12*(2), 85-87 (in Chinese).

[8] Hong-kun L., Wei-juan Y., and Jun-hu Z., "Investigation on Thermal Decomposition Characteristics of Urea Solution under High Temperature," *Proceedings of the CSEE*, **2010**, *30*(17), 35-40 (in Chinese).

[9] Chung Frank T. H., Jones Ray A., and Nguyen Hai T., "Measurements and Correlations of the Physical Properties of CO_2/Heavy-crude-oil Mixtures [J]," *SPE 15080-PA.*, **1988**, *3*, 822-828.

[10] King H. H., Hall J. L., and Ware G. C., "A Study of the Density, Surface Tension and Adsorption in the Water-ammonia System at 20 ℃," *Journal of American Chemical Society*, **1930**, *52*(12), 5128-5135.

[11] Tong-yao Z., Wei-yu F., and Guo-zhi N., "Study on Interaction of Heavy Oil Polar Component and Emulsifier in Heavy Crude Oil-in-water Emulsion," *Journal of China University of Petroleum*, **2006**, *30*(5), 101-105 (in Chinese).

[12] Ping J., Ji-jiang G, and Gui-cai Z., "Influence Factor on Oil Recovery Efficiency for Chemical Flooding of Heavy Oil Reservoir," *Journal of China University of Petroleum*, **2011**, *35*(2), 166-169 (in Chinese).

A New Approach towards Precise Planar Feature Characterization Using Image Analysis of FMI Image: Case Study of Gachsaran Oil Field Well No. 245, Southwest of Iran

Mosayeb Shafiezadeh*[1], Mansour Ziaee[2], and Behzad Tokhmchi[3]

[1] Refuling Unit, NIOPDC, Kermanshah Region, Iran
[2] Department of Petroleum Engineering, University of Shahroud, Shahroud, Iran
[3] Department of Petroleum Engineering, University of Shahroud, Shahroud, Iran

ABSTRACT

Formation micro imager (FMI) can directly reflect changes of wall stratums and rock structures. Conventionally, FMI images mainly are analyzed with manual processing, which is extremely inefficient and incurs a heavy workload for experts. Iranian reservoirs are mainly carbonate reservoirs, in which the fractures have an important effect on permeability and petroleum production. In this paper, an automatic planar feature recognition system using image processing was proposed. The dip and azimuth of these features are detected using this algorithm to identify more precise permeability and the career of fluid in reservoirs. The proposed algorithm includes three main steps; first, pixels representing fractures are extracted from projected FMI image into location matrices *x* and *y* and the corresponding value matrix *f*(*x*, *y*). Then, two vectors X and Y as the inputs of CFTOOL of MATLAB are produced by the combination of these three matrices. Finally, the optimum combination of sine function is fitted to the sine shape of pattern to identify the dip and azimuth of the planar feature. The system was tested with real interpretation FMI rock images. In the experiments, the average recognition error of the proposed system is about 0.9% for the azimuth detection and less than 3.5% for the dip detection and the correlations between the actual dip and azimuth with the determined cases are more than 90% and 97% respectively. Moreover, this automatic system can significantly reduce the complexity and difficulty in the planar feature detection analysis task for the oil and gas exploration.

Keywords: FMI, Stratums, Carbonate Reservoirs, Dip, Azimuth

INTRODUCTION

As the oil and gas exploration becomes gradually complicated, the traditional well logging methods have many problems. With limited information, recognizing effective layers and estimating reserves parameters become more and more difficult. In contrast, the formation micro imager (FMI) technology can provide rich information on fractured reservoirs, and most importantly it can be applied to identifying fractured reservoirs qualitatively and can help explain them quantitatively [1,2].

***Corresponding author**

Mosayeb Shafiezadeh

Email: m.shafiezadeh@niopdc.ir

The full-bore formation micro imager provides micro resistivity formation images in water-base mud. Micro resistivity changes are related to lithological and petrophysical variations in the rock, which are conveyed mainly by the high-resolution current component and are interpreted on the image in terms of rock texture, stratigraphic and structural features, and fractures. The physics of the FMI measurement makes it a geological and reservoir characterization tool which produces complete and reliable answer products. These real-time answers are used to characterize the structural dip and faults, identify and evaluate sedimentary features, measure the rock texture, and complement the information obtained from coring programs [3].

To our knowledge, we rarely find automatic structural feature recognition systems with FMI images in the literatures. Consequently, for promoting oil and gas exploration, it is of great significance to develop an efficient and intelligent recognition system using FMI images. For oil and gas reservoirs with FMI images, relative image analysis technologies are mainly for reservoir fracture detection and extraction. Ginkel et al. used the generalized Radon transform to detect fractures in FMI images, which mapped the original 2D image to a 3D orientation space. These methods focus on robust fracture detection and prediction, which are not directly related to rock classification. Furthermore, these fracture analysis techniques with some complicated algorithms are also based on a large number of image samples. Although, Yun et al. proposed an algorithm in 2009, which could be used for limited samples, this algorithm only was used for the classification of the limited type of volcanic rocks (Tuff, Lava, and Breccia).

The interpretation of the FMI log image can be used to estimate planar feature dip and strike. The projection of a planar feature on borehole images will be represented by a sine wave as seen in Figure 3. Because the borehole is very small in relation to stratigraphic and structural features, when they intersect the borehole, they are nearly planar and assumed to be a planar feature; thus a sine function was used herein for fitting the best curve on the fracture pattern.

We analyze rock FMI images with very limited samples in a different way by using image processing. There are several key issues in such recognition systems for FMI images. Firstly, the training samples are limited in most cases due to sensitive company data source. Secondly, FMI images are usually of great noises. In practice, even with manual processing, it is very difficult to analyze and classify rock structures without enough expertise.

The rest of the paper is organized as follows: in section 2, basic information about FMI image is represented. Section 3 simply describes the recognition system and methodology. Conclusions and discussions are presented in section 4. Finally, the experimental results are shown in section 5.

EXPERIMENTAL PROCEDURES

Placement

The Gachsaran oil field is in the southwest of Iran containing several wells (Figure 1). The FMI image used in this paper is related to well 245 of this oil field.

FMI Image

The formation micro imager (FMI) tool (Figure 2) provides a detailed view of stratigraphic, lithologic, and structural information along the wall of the well bore. The FMI tool provides an image with 80% coverage of the borehole wall in an 8 inch diameter borehole.

The spacing of the FMI tool electrodes offers a maximum resolution of 0.2 inches in the azimuthal and vertical directions and a detection limit of 50 microns (about 0.002 inches) in the horizontal direction.

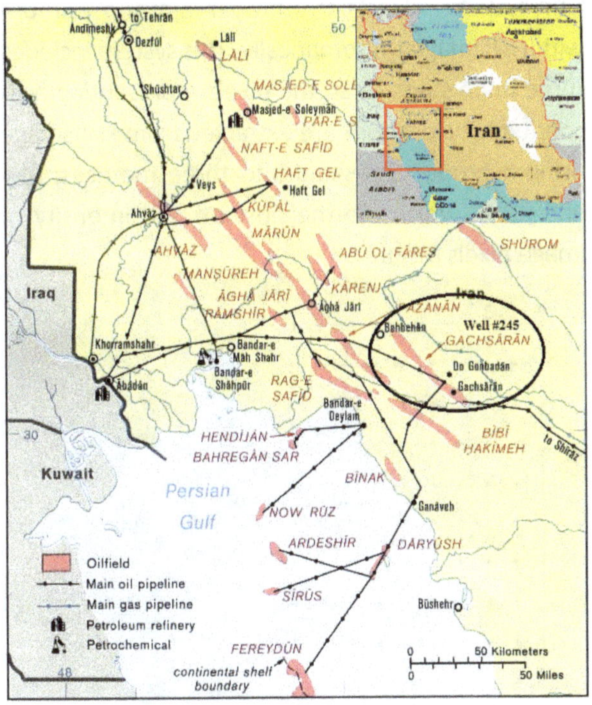

Figure 1: Gachsaran oil field location

The FMI tool uses an alternating current (AC) to produce a current map that minimizes the effects of a formation spontaneous potential (SP) and the direct current (DC) flow between the electrodes on the surface of the borehole (Schlumberger, 2002). The current is emitted through the lower electrodes and received by the upper electrodes after it has passed through the formation. The amount of current received is recorded and digitized [4].

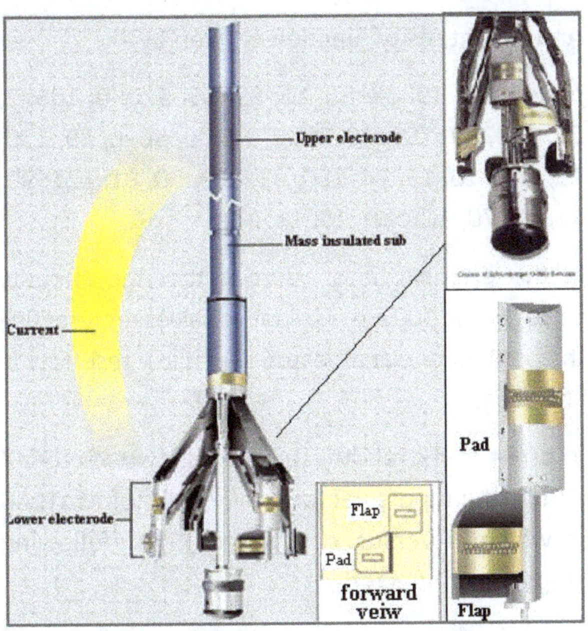

Figure 2: The FMI tool

Current values are recorded at points along the borehole wall. The variations of the measured current flow are portrayed as variations in the borehole color scheme to produce a resistivity image of the borehole (Figure 3). The current map color ranges from white to black with the lighter colors corresponding to relatively low conductivity zones and the darker colors denoting higher conductivity zones. The FMI tool also uses direct current induction to determine formation resistivity.

A typical FMI log header is shown in Figure 3. This header is divided into 7 vertical columns labeled in red. Column 1 provides depth and caliper readings. Column 2 displays a static current map image. The static current map is displayed in grey scale and represents the relative current at any positions in the borehole compared to the absolute highest and lowest current values obtained in the logging session. The static image is used to determine the major lithologic features surrounding the borehole. Column 3 shows the output of multiple induction logs. Generally, a shallow induction, medium induction, and a deep induction log are run with the FMI tool.

Column 4 is a depth and lithology indicator. The lithology indicator is obtained by combining the gamma ray, induction, and photoelectric effect of the rock being tested. This section provides a first approximation of formation lithology. The neutron and density porosity readings are plotted in Column 5. The log display also shows the amount of mudcake present in the borehole in Column 5. Column 6 displays the dynamic current map. Unlike the static current map, the dynamic current map recalculates the discrete current values every inch. This allows small scale features to stand out in the FMI dynamic current map image. Finally, column 7 illustrates the tadpole plots for the planar features interpreted in the FMI log. The circle of the tadpole plot is located at the depth and dip of the planar features. The dip scale is nonlinear and increases from 0 degrees (on left) to 90 degrees (on right).

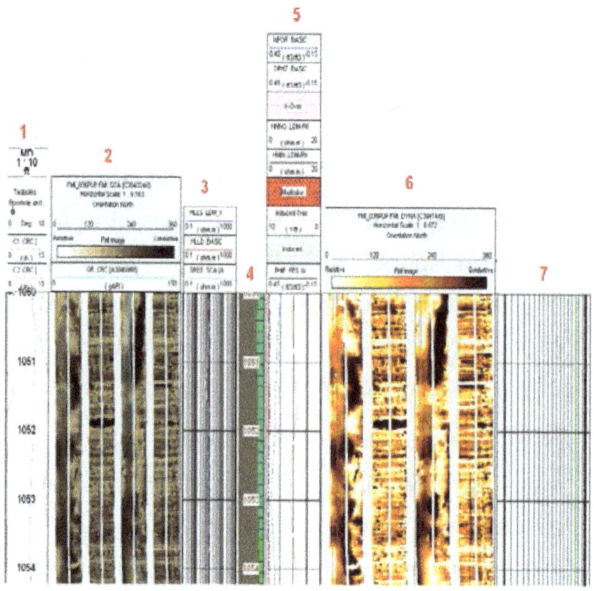

Figure 3: Typical FMI header information and log tracks; this log is from well 67-1-X-10 at Teapot Dome, Wyoming [4].

Methodology

In FMI image, the trough of the sine wave, which represents the planar feature, indicates the azimuthal angle (dip trend angle) of this feature. Figure 4 portrays how the trough value of the sine wave corresponds to the azimuthal angle of the planar feature on the FMI image. The dip of the planar feature can be directly calculated from the amplitude of the sine wave and the diameter of the borehole. However, the difference between the maximum and minimum values of sine wave is used instead of the amplitude.

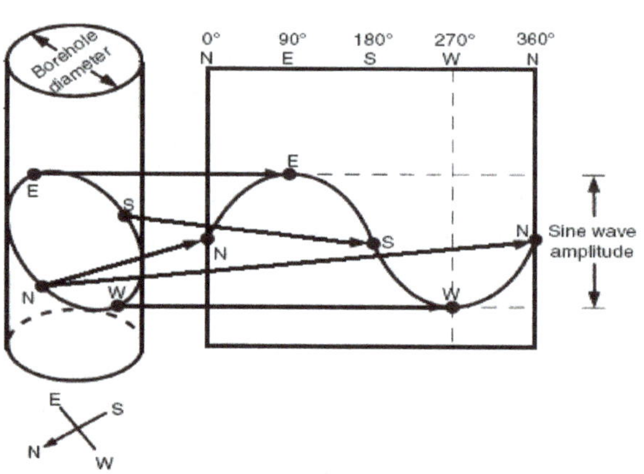

Figure 4: Displaying of directional information on the FMI log [4]

Our system extracts the proper dip and azimuth with FMI images automatically. This system includes three main steps. First, the pixels representing high conductivity locations are extracted from FMI image using a proper threshold into three matrices; *x*, *y* and *f(x,y)*, which are the position and intensity of image pixels (Figure 5).

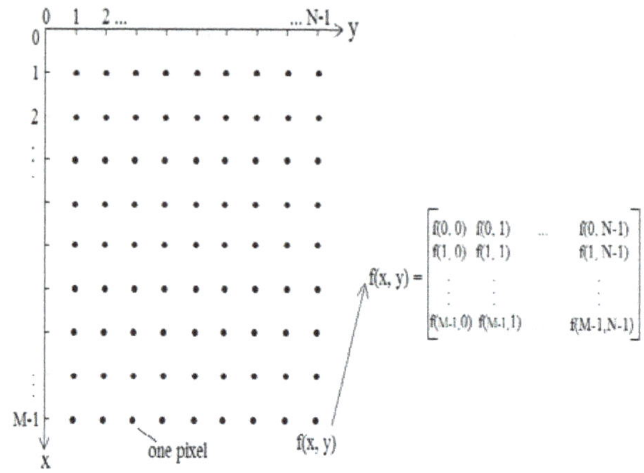

Figure 5: The position and intensity of image pixels

Then, the corresponding matrices are combined and two vectors are made: vectors X and Y. The size of these vectors must be equal.

Suppose that the image is composed by 5×9 pixels; *x* and *y* matrices (vectors) are:

x= [0, 1, 2 ... 4]; size *x* = 5

y= [0, 1, 2 ... 8]; size *y* = 9

And the intensity function is given by:

f(x,y)=[30, 19, 28, 56, 16, 89, 15, 1, 0; 0, 108, 2, 50, 112, 89, 15, 180, 20; 0, 1, 200, 56, 6, 89, 150, 19, 200; 10, 15, 24, 150, 61, 189, 50, 81, 121; 90, 11, 42, 70, 116, 89, 15, 11, 2];

In this example, it is assumed that the intensity values above 100 are expectative values representing the position of planar features with low resistivity in FMI images.

In order to extract these pixels as two vectors X and Y (the inputs vectors of CFTOOL of MATLAB in which Xi versus Yi is plotted), the following steps are conducted:

Use the following threshold:

$f(x, y) => 100$

Since the size of intensity value matrix, $f(x, y)$, is 5×9=45 (size x by size y) and X and Y vectors as the input data of CFTOOL representing pixels with intensity value greater than 100, the size of these vectors is 45 as follows:

Suppose;

m=size(x) = image length,

n=size(y) =image widths,

X= (0: (m×n)-1); for i=1:n

X (1+ (i-1)×m: m+ (i-1) ×m) =X (0: m-1); so the vector X is given by:

X= [0, 1, 2, 3, 4, 0, 1, 2, 3, 4, 0, 1, 2, 3, 4, 0, 1, 2, 3, 4, 0, 1, 2, 3, 4, 0, 1, 2, 3, 4, 0, 1, 2, 3, 4, 0, 1, 2, 3, 4, 0, 1, 2, 3, 4,];

size(X)=45.

Y= (0: (m×n)-1); Y= 0×Y; for i= 0: m, j= 0: n

If f (i, j) =1, Y (m×j+i+1) = j; so the vector Y is given by:

Y= [0, 0, 0, 0, 0, 0, 0, 0, 0, 0, 1, 0, 0, 4, 0, 0, 7, 0, 0, 0, 2, 0, 0, 0, 6, 0, 8, 0, 0, 0, 3, 0, 5, 0, 0, 8, 0, 0, 0, 0, 4, 0, 0, 0, 0];

size(Y)=45.

Figure 6 shows the output of CFTOOL, in which we can see a sine pattern and several noises.

The noises are removed by *imfilter* function or excluding the panel of CFTOOL and several index points, which are in the sine pattern of X-Y plotting, by graphical panel of CFTOOL.

Using the CFTOOL toolbox of MATLAB software, the best curve of sine wave, which is obtained from the linear combination of two sine functions (or three sine functions in some cases), is fitted to the data (Figure 6). Finally, the dip and azimuth of the fracture is determined by analyzing this curve and its property.

Linear combination of two sine functions is obtained by:

$$Y=a_1\sin(b_1X+c_1) + a_2\sin(b_2X+c_2) \quad (1)$$

where, a_1, b_1, c_1, a_2, b_2, and c_2 are six constant parameters. In fitting the best curve, at least six points must be chosen on sine pattern using CFTOOL toolbox.

It should be noted that the analyses panel of CFTOOL can show the points in which the first derivative is zero; hence the maximum and minimum points are extracted. Thus dip and azimuth can be determined (Figure 4).

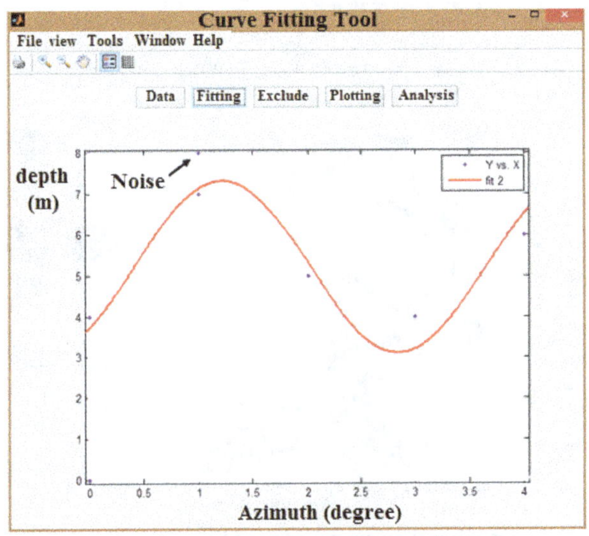

Figure 6: Fitted curve on sine pattern of X-Y

RESULTS AND DISCUSSION

The FMI image used in this paper is related to Gachsaran oil field in the southwest of Iran and the algorithm is tested by six planar features of this image log as shown in Figures 7 and 10. Furthermore, in the following figures, the process can be seen step by step for these tow fractures, including curve fitting on fracture pattern and critical point characterization.

The results of dip and azimuth detection for all the features and the average amount of error are proposed in Tables 1-3.

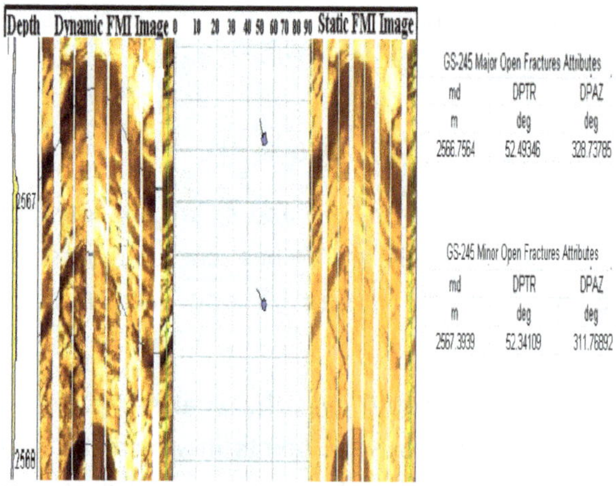

Figure 7: Fractures at depths of 2566 and 2567 m and their characterizations

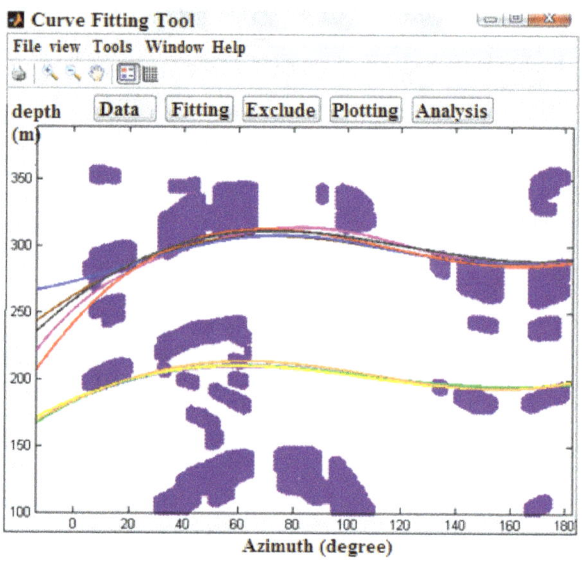

Figure 8: Fitted curve on planar feature pattern

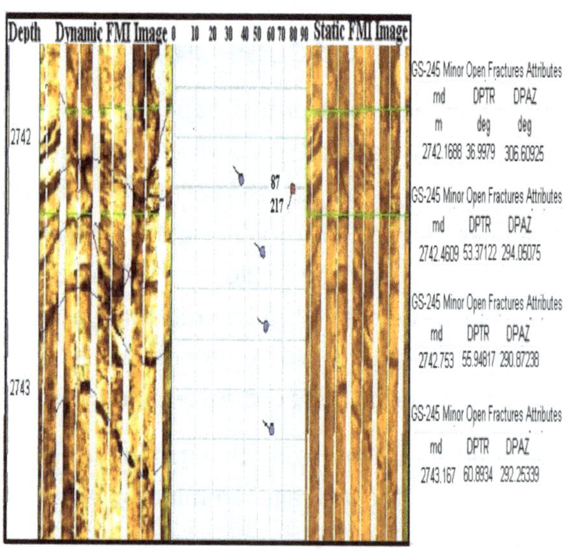

Figure 9: Conductive seam and planar features at depths of 2742 and 2743 m and their characterizations

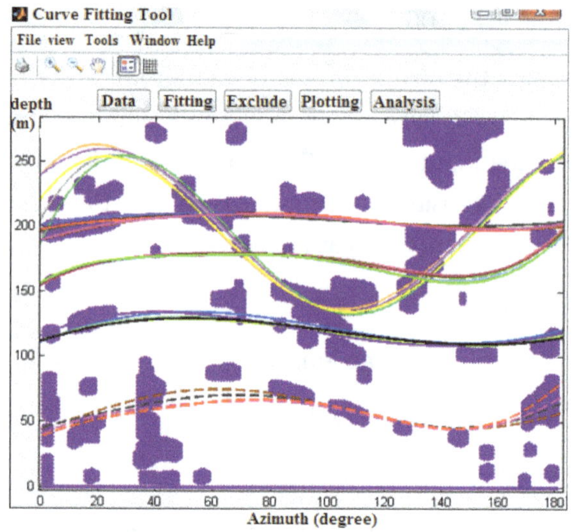

Figure 10: Fitted curve on planar feature pattern

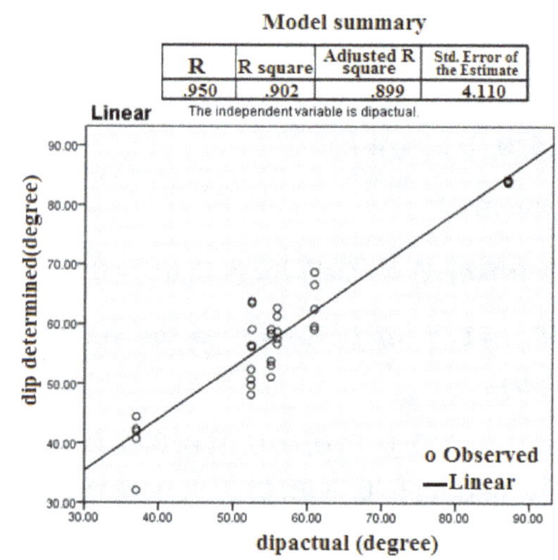

Figure 11: Amount of correlation between the algorithm results and the actual cases for dip detection using SPSS software

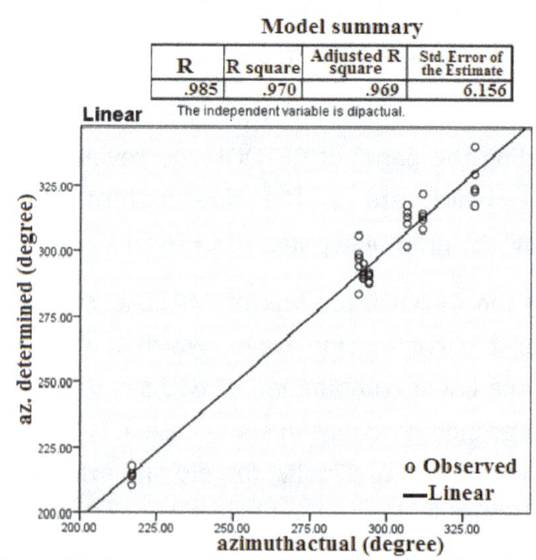

Figure 12: Amount of correlation between the algorithm results and the actual cases for azimuth detection using SPSS software

Table 1: Extracted parameter (x and $f(x)$ of critical points) from curves shown in Figures 7 and 10 for all the features

			Step 1		Step 2		Step 3		Step 4		Step 5	
Planar features	Depth	Parameter	Min.	Max.	Min.	Max.	Min.	Max.	Min.	Max.	Min.	Max.
1	2742.2	*x*	156.6	78.5	158.8	62.8	157.5	76.3	150.9	58.2	155.3	71.4
		f(x)	197.4	209.9	198.2	209.8	198.8	209.8	200.7	208.7	197.9	209.4
2 (Conductive Seam)	2742.3	*x*	107.5	22.7	105.2	19.1	107.1	26.1	106.9	22	108.9	29.5
		f(x)	134.6	253.2	136.8	262.3	134.8	254.3	134.5	258.9	132.8	254.2
3	2742.5	*x*	144	66.3	145.3	78.3	146	70.4	144.2	66.2	145.6	76
		f(x)	162.3	179.3	162.6	178.4	157.1	178.4	158.6	179.4	160.9	178.4
4	2742.7	*x*	148.7	49.8	141.9	41.4	147.2	52.7	149.4	50.1	153	52
		f(x)	108.7	128.8	109.1	133.9	110.9	134.3	109.6	130.6	109.8	129.3
5	2743.2	*x*	144.8	47.9	144.9	78.4	147.8	62.6	145.8	69.3	145.4	75.4
		f(x)	43.3	76.1	45.1	66.9	45.3	74.8	45.7	70.3	44.6	65.9
6	2566.7	*x*	164.7	69.7	158.8	75.3	161.5	71.4	170	73.1	162	82.9
		f(x)	285.3	312.8	287.9	308.1	287.7	307.8	288.9	311.4	287.1	314.2
7	2567.4	*x*	154.2	60.4	156.7	63.3	157.3	62.9	156.1	61.8	161	61.4
		f(x)	193.7	210.2	193.4	213.7	193.8	211.3	193.8	209.7	195.2	210.2

Table 2: General comparison between the determined dip with actual cases and the percentage of error

Planar feature	Depth	Parameter	Step 1	Step 2	Step 3	Step 4	Step 5	Average	Actual	Percentage of error
1	2742.2	Dip	44.4	42.3	40.7	32.1	41.9	40.28	37	3.6
		Az.	313.2	317.6	315	301.8	310.6	311.64	306.6	1.4
(Conductive Seam)	2742.3	Dip	83.8	84.2	83.9	84.2	84	84.1	87	3.3
		Az.	215	210.4	214.2	213.8	217.8	214.3	217	0.8
3	2742.5	Dip	53.1	51.1	59.1	58.4	53.8	55.1	53.4	1.9
		Az.	288	290.6	292	288.4	291.2	290.1	294.1	1.1
4	2742.7	Dip	57.6	62.7	61.3	58.7	56.7	59.4	55.9	3.9
		Az.	297.4	283.8	294.4	298.8	306	296.1	290.8	1.5
5	2743.2	Dip	68.7	59.6	66.6	62.5	59.1	63.3	60.9	2.7
		Az.	289.6	289.8	295.6	291.6	290.8	291.5	292.3	0.2
6	2566.7	Dip	63.8	56.3	56.2	59.1	63.5	59.8	52.5	8.1
		Az.	329.4	317.6	323	340	324	326.8	328.7	0.5
7	2567.4	Dip	50.7	56.4	52.4	49.7	48.1	51.5	52.4	1
		Az.	308.4	313.4	314.6	312.2	322	314.2	311.7	0.7

Table 3: Average percentage of error in dip and azimuth detection

	Parameter	1	2	3	4	5	6	7	Average Error(%)
Error(%)	Dip	3.6	3.3	1.9	3.9	2.7	8.1	1	3.5
	Azimuth	1.4	0.8	1.1	1.5	0.2	0.5	0.7	0.88

CONCLUSIONS

According to the result of the algorithm in dip and azimuth detection, this algorithm can determine these two important parameters of planar features (bedding, fault, or fracture) precisely. Since the average recognition error of our proposed system is about 0.9% for azimuth detection and less than 3.5% for dip detection, the correlations between actual dip and azimuth with the determined cases are more than 90% and 97% respectively. Moreover, this automatic system can significantly reduce the complexity and difficulty in the fracture detection analysis task for the oil and gas exploration. The algorithm can be improved to detect other parameters of fracture such as aperture.

REFERENCES

[1] Gonzalez R. C. and Woods R. E., *Digital Image Processing* (2nd ed.), Prentice-Hall, Inc. Upper Saddle River, New Jersey, **2002**, 69-164.

[2] Gonzalez R. C., Woods R. E., and Eddins S. L., *Digital Image Processing using MATLAB*, Prentice-Hall, Inc. Upper Saddle River, **2004**, 1-104.

[3] Schlumberger, "FMI Borehole Geology, Geomechanics and 3D Reservoir Modeling," Mark of Schlumberger, **2002**.

[4] Schwartz B. C., "Fracture Pattern Characterization of the Tensleep Formation, Teapot Dome, Wyoming," Master Thesis, Department of Geology and Geography Morgantown, West Virginia, **2006**, 115-130.

21

Hydrocracking of Jojoba Oil for Green Fuel Production

Salwa. A. El Khatib[1], Samia Abas Hanafi[1], Mohamed M. H. Arief[2] and Eslam Fawzi Al-Amrousi*[1]

[1] Egyptian Petroleum Research Institute, Nasr City, Cairo, Egypt

[2] Benha University, Faculty Of Science, Department of Chemistry, Behna, Egypt

ABSTRACT

The fast depletion of petroleum crude oil reserves and the serious consideration of environmental issues make the implementation of sustainable energy sources a crucial issue worldwide. Biofuel derived from vegetable oils is receiving a great attention as one of the most suitable and logical alternatives of fossil fuels. Therefore, the production of petroleum-like fractions from nonedible oil extracted from jojoba seeds is the main objective of this study.

Hydrocracking technique has been applied here, using a fixed bed high pressure flow reactor system and using zeolite as a catalyst. A series of experiments were carried out in the temperature ranges from 350-425°C at the pressure of 1.0-5.0 MPa, liquid hour space velocity (LHSV) of 1-5 hr^{-1}, and H_2 to oil feed ratio of 1600 l/l in order to elucidate the influence of operating conditions on the quantity and quality of the biofuel attained.

The major products obtained from hydrocracking of jojoba oil feedstock were organic liquid products contaminated with some water and gases .The yield and conversion were estimated. The cracked products were analyzed and compared with that of the feed via FTIR, GC, ASTM distillation, and some routine testes. The organic liquid layers were fractionated to gasoline, kerosene, and gas oil. The results reveal that the main constituent lies in the gasoline range, which was favored at high temperature and pressure and low LHSV.

Keywords: Biofuel, Hydrocracking, Jojoba Oil, Zeolite, Nonedible Oil

INTRODUCTION

Due to the diminishing petroleum reserve, a worldwide trend is concentrated on searching for alternative and renewable energy sources [1-3]. The biofuels derived from vegetable oils are attracting increased interest in this respect [4,5]. They are nontoxic, biodegradable, and environmentally friendly, since they show great reduction in the emission of harmful matters such as carbon monoxide, sulfur compound, polyaromatics, smoke, etc. [6,7].

Biodiesel production from the transesterification of vegetable oils is currently the primary route for the production of biofuels from plant oils [8]. The transesterification process is the reaction of a lipid with a short chain alcohol, mainly methanol, to form fatty acid methyl ester (biodiesel) and byproducts [9-11]. Although this process has many benefits, the usage of FAME is connected with numerous problems as compared to the use

***Corresponding author**

Eslam Fawzi Al-Amrousi

Email: eslamalamrousi@yahoo.com

of petroleum-based fuel. Some of the disadvantages ensuing employing FAME are deposit formation in the fuel system and combustion chamber, poor storage (low oxidation and heat stability) due to the double bond in the molecules, water adsorption and shortened lifetime of the engine oils, cost, high pour point versus petrodiesel, etc.

It is possible to eliminate or even to minimize the majority of these drawbacks, if vegetable oils are converted to a paraffin-rich mixture of hydrocarbons, which then makes it appropriate to be used in vehicles engines. Some modern technologies such as thermo-catalytic processing of vegetable oils appeared for producing alternative motor fuels, namely gasoline and diesel [12-14]. Three main concepts are described in literature, namely hydroprocessing, thermal, and catalytic cracking [15-17]. These technologies have significant potential as the produced oils have better fuel properties than that obtained via transesterification; in addition, they employ the existing infrastructure of petroleum refineries.

Considerable efforts have been devoted to the promotion of cracking techniques for the production of pet-like fractions from vegetable oils, which is regarded as one of the most reasonable and promising technologies for high yield of liquid products [17-19]. Since the introduction of catalytic cracking, the initial catalysts were clays and amorphous Si-Al and evolved with time to zeolites, particularly the use of rare earth–Y (REY), and finally to the ultra stable–Y(USY, REUSY). Zeolites have shown excellent performance as solid acid cracking catalysts due to their higher selectivity [14,19].

Hydrotreating and hydrocracking of vegetable oils such as soybean sunflower and rapeseed has been reported [20-22]. Considering the problem of biofuels competing with food sources, nonedible plant oil was used as a feedstock to produce green transportation fuel. The aim of this study is to investigate the feasibility of producing high quality petroleum-like fraction by hydrocracking the nonedible jojoba oil and to elucidate the optimum operating conditions for the attainment of the maximum yield of the desired components.

EXPERIMENTAL PROCEDURES

In this study, hydrocracking of jojoba oil purchased from Egyptian Company for Natural Oils has been employed as a feedstock. HYDEX-L, a novel zeolite-based catalyst was developed by the Sud-Chemie Group for dewaxing the lube oil. It selectively hydrocracks waxy molecules to short chain products. HYDEX-L is extruded square in shape with a size of 1.5×2.5 mm, a surface area of 289.75 m^2/gm, a total pour volume of 0.3331 cc/gm, and an average pore radius of 22.99 Å. It was used here to catalyze the hydrocracking reaction. The catalyst was dried at 120 °C for 3 hrs, then calcined at 450 °C for 6 hrs in air, and reduced at 450 °C under 1 MPa hydrogen pressure for 2 hrs.

A series of experiments were conducted to explore the influence of the operating conditions on the quantities and qualities of the obtained products. The reaction has been performed using a continuous high pressure microreactor (cata-test unit) Figure 1.

The apparatus consists mainly of 50 cm length stainless steel reactor, with an internal and external diameter of 19 and 27 mm respectively. It is divided into three zones, each of which has its individual heating element and temperature controller. 30 gm (equivalent to 46 ml) of the catalyst mixed with the same volume of ceramic was charged into the middle zone of the reactor. Hydrogen gas was supplied to the unit from an H_2 cylinder, while the liquid feed was pumped to the top of the reactor by means of a piston pump having an adjusting knob.

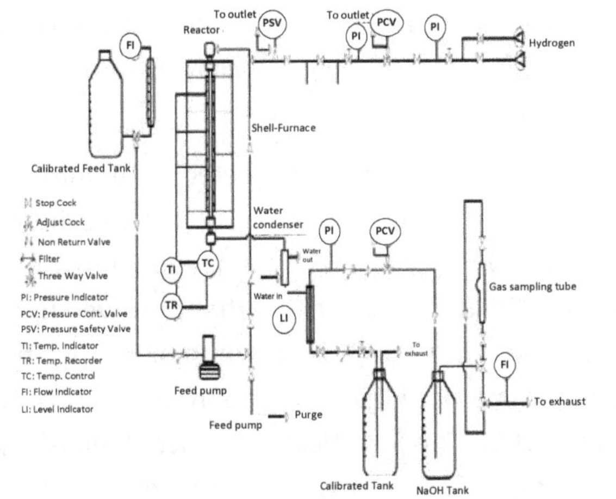

Figure 1: High pressure catalyst activity testing unit

The hydroconversion reaction of jojoba was carried out in various operating conditions. Firstly, the unit was flushed with hydrogen gas before the feedstock oil was fed by the pump; then it was kept under hydrogen gas pressure to check any leakage. Thereafter, the operating conditions in the hydroconversion runs were adjusted as required. Once a steady state had reached, the feed was pumped and the reactor effluent was cooled in a water condenser before separating its liquid and gas phases. The liquid product (organic and aqueous fractions) was collected in a receiver, while the gases were passed through a gas meter to the outlet. The contaminated water involved in the liquid yield was removed after settling in a separating funnel for 24 hrs. Anhydrous sodium sulphate was then added to the organic product to eliminate any traces of water still existed. The effluent gases and liquid yields were estimated.

Weight of gases (gr) = Weight of jojoba oil feed (gr) - Weight of total liquid product (gr).

$$Total\ conversion(wt.\%) = \frac{gases(gr) + water(gr) + biofuel\ yield(gr)}{Feed(gr)} \times 100 \quad (1)$$

The weight of biofuel yield, separated from the organic liquid product, was taken as the weight of fraction distilled below the boiling point of jojoba oil (< 323 °C).

$$Biofuel\ selectivity(\%) = \frac{Biofuel\ yield(gr)}{Converted\ feed} \times 100 \quad (2)$$

The organic fraction was analyzed and evaluated according to the different standard ASTM methods, including density (D-287), kinematic viscosity (D-455), sulfur content (D-4294) using x-ray fluorescence sulfur meter, and distillation (D-86). In addition, the virgin oil and the degraded organic product obtained at 425 °C were analyzed via GC and FTIR techniques. The gas chromatographic apparatus, Agilent 6890, equipped with flame ionization detector (FID) was employed. Separation was performed on HP-1 column with a length of 30 meter and an ID of 0.25 mm. The temperature of injector and detector was set at 300 °C. The FTIR spectroscopy was determined using Nicolet-LOFT-IR, Nicolet IS-10.

RESULTS AND DISCUSSION

The optimization of the operating conditions to assess the highest yield of the desired product will be beneficial. Therefore, a series of experiments were conducted to explore the influence of each operating factor on the product distribution of the hydrocracked jojoba oil and to determine the best conditions for the attainment of the maximum yield of biofuel fraction.

Effect of Liquid Hour Space Velocity (LHSV)

The liquid hourly space velocity (LHSV) is an important operating parameter for regulating catalyst performance. The efficiency of this factor on hydrocracking reaction is one of the main premises of the present investigation. In this respect, hydrocracking of jojoba oil was carried out at three various LHSV's of 1, 2.5, and 5 hr^{-1} in the same experimental conditions, i.e. temperature of 400 °C, pressure of 2.5 MPa, and H_2/oil ratio of 1600 l/l, and using a zeolite catalyst. The quality and yields of the obtained products were evaluated, and the results are summarized in Table 1.

Table 1: Distribution of the hydrocracked products obtained at various LHSV's; temperature: 400 °C, pressure: 2.5 MPa, and H_2/feed: 1600 l/l

Hydrocracked Products	1 hr^{-1}	2.5 hr^{-1}	5 hr^{-1}
Total conversion (wt.%)	99.44	99.17	99
Hydrocracked products (wt.%)			
Gases	18.04	17.84	2.34
Liquid	81.96	82.16	97.66
Water content	6.74	4.49	3.30
Oil content	75.22	77.67	94.36
Density at 20 °C	0.8110	0.8120	0.8215
Kinematic viscosity at 40°C	1.0714	1.1588	2.5250
Sulfur content (ppm)	0.029	0.011	0.016
ASTM distillation, (vol. %)			
IBP (°C)	50	50	50
10	70	70	70
20	80	80	110
30	100	100	140
40	110	120	150
50	120	130	180
60	150	140	273.2
70	150	154.9	294.5
80	237.7	219.7	305.2
90	250	250	305.2
FBP	250	250	305.2
Recovered (vol.%)	**90**	**91**	**97**
Unconverted and residue (vol.%)	**1**	**1**	**1**
Loss (vol.%)	**9**	**8**	**2**
Biofuel yield (wt %)	**74.62**	**76.84**	**93.36**
Biofuel selectivity (%)	**75.04**	**77.48**	**94.3**

The data demonstrate that approximately the whole feedstock was entirely converted to hydrocracked products, at all the LHSV applied. Although the conversion is a key assessment of catalyst activity, higher conversion may not result in a higher yield of the desired components. Beside gases, the effluent liquid products contain organic and aqueous fractions, some residue, and unconverted oil. Among these, the most demandable component is the hydrocracked liquid organic fraction which is mainly composed of alkanes (biofuel).

The depicted results reflected the distinct diminution of the attainable liquid yield and oil content by reducing LHSV from 5 down to 1 hr^{-1}. The overall decreasing trend that observed upon lessening the LHSV is accompanied by the augmentation of gas formation from 2.34 to 18.04 wt.%. This is anticipated since diminishing LHSV implies a longer residence time of the feed in the catalyst section in the reactor, which gives more opportunities for the intensive cracking to proceed, leading to the predomination of gases on the account of the liquid yield.

On the other hand, the concentration of the aqueous layer in the cracked products is also enhanced with minimizing LHSV, since 3.3 wt.% of water was achieved at an LHSV of 5 hr^{-1} and increased to be 6.74 wt.% at an LHSV of 1 hr^{-1}. This may be ascribed to the predomination of hydrodeoxygenation reaction which was favored at longer contact times.

In this study, ASTM distillation was implemented to determine the boiling range distribution of the organic liquid products obtained under the various conditions applied. In addition, the boiling point of crude jojoba oil was determined and found to be, mainly, above 323 °C. Heavy molecules, with boiling points higher than 323 °C, were considered to be the unconverted part of the feed and residue which cannot be utilized as biofuels. It is evident from the accessible data declared in Table 1 that a good yield of petroleum–like fraction, which is the most significant constituent in request, can be obtained at an LHSV of 5 hr^{-1}, at which the highest yield and selectivity of 93.36 wt.% and 94.3%respectively were accomplished. Decreasing the LHSV to 2.5 and 1 hr^{-1} leads to an enhancement of the undesirable components such as water and gasses at the expense of the biofuel attained.

Moreover, the produced biofuels were fractionated and separated into three portions, according to their boiling ranges, i.e. (50-150°C), (150-250°C), and (250-323°C), for gasoline, kerosene, and gas oil respectively. The influence of LHSV on the amount of each of the previously mentioned fractions in the produced biofuel can be deduced from Figure 2. It could be noticed that the substantial conversion to biofuel, frequently composed of gas oil and considerable amount of gasoline distillates, is accomplished at an LHSV of 5 hr^{-1}. Upon reducing the LHSV to 2.5 hr^{-1} or less, the augmentation of gasoline fraction was occurred owing to the degradation of the heavier fractions, which transform to gasoline at such a high residence time. The most suitable LHSV value in getting a good yield of biofuel distillates involving the best concentration of gas oil is 5 hr^{-1}.

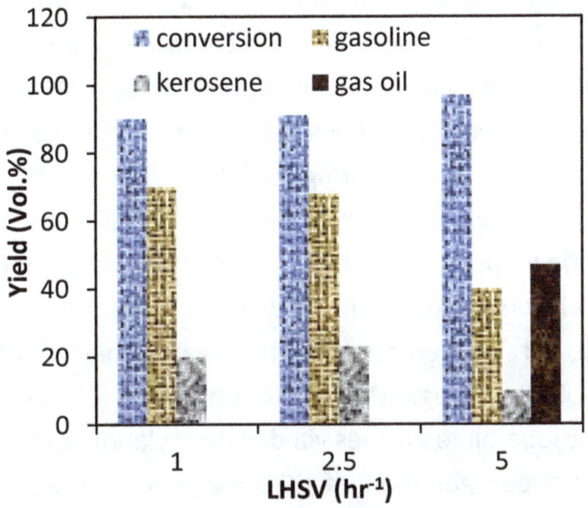

Figure 2: Distribution of the individual fractions of the biofuel obtained as a function of LHSV

Effect of Hydrogen Pressure

The influence of hydrogen pressure, under constant experimental conditions, can be deduced from the results listed in Table 2. These data reveal a significant enhancement of the converted yield from 66.26 to about99 wt.%, accompanied by a pronounced exhibition in the quantities of the produced biofuel fraction as the applied H_2 pressure in the reactor increased beyond 1 MPa. Simultaneously, both density and kinematic viscosity of the obtained oil are also reduced. This situation is expected since hydrocracking reaction can be promoted at sufficiently high pressures, which leads to the predomination of lighter components in the obtained products.

Table 2: Distribution of the hydrocracked products obtained at various H_2 pressures; temperature: 400°C, LHSV: 5 hr^{-1}, and H_2/feed: 1600 l/l

Hydrocracked Products	1.0 MPa	2.5 MPa	5.0 MPa
Total conversion (wt.%)	66.26	99	99.065
Hydrocracked product (wt.%)			
Gases	0.70	2.34	4.64
Liquid	99.30	97.66	95.36
Water content	3.11	3.30	7.45
Oil content	96.19	94.36	87.90
Density at 20 °C (gr/m^3)	0.8295	0.8215	0.8131
Kinematic viscosity at 40 °C	3.9577	2.5250	1.0495
Sulfur content (ppm)	0.012	0.016	0.02
ASTM distillation (vol. %)			
IBP	60	50	40
10	80	70	60
20	120	110	90
30	160	140	110
40	280.3	150	120
50	301.5	180	130
60	301.5	273.2	150
70	-	294.5	170
80	-	305.2	273.2
90	-	305.2	273.2
FBP	301.5	305.2	273.2
Recovered (vol.%)	**65**	**97**	**92**
Unconverted and residue (vol.%)	**34**	**1**	**1**
Loss (vol.%)	**1**	**2**	**7**
Biofuel yield (wt.%)	**62.45**	**93.36**	**86.965**
Biofuel selectivity (%)	**94.25**	**94.3**	**87.79**

At the same time, increases in the amount of H_2O content in the liquid yield are observed too. This may be attributed to the elimination of O_2 from jojoba molecules as water, which is favored at higher H_2 concentration through the hydrodeoxygenation reaction, since decarbonylation and decarboxylation reactions are more favored at lower H_2 pressure according to the following equations [23].

Hydrodeoxygenation reaction:

$$C_nH_{2n+1}COOH + 3H_2 \longrightarrow C_{n+1}H_{2n+4} + 2H_2O$$

Hydrogenation:

$$C_nH_{2n+1}COOH + H_2 \longrightarrow C_nH_{2n+2} + H_2O + CO$$

Decarboxylation:

$$C_nH_{2n+1}COOH \longrightarrow C_nH_{2n+2} + CO_2$$

At an H_2 pressure of 1.0 MPa, only 62.45 wt.% from the virgin oil was transformed to biofuel. Upon increasing the pressure to 2.5 MPa, the recovered petroleum-like fractions increased to be 93.36 wt.%. Beyond which, i.e. at an H_2 pressure of 5.0 MPa, the conversion to biofuel was reduced to 86.97 wt.% due to the losses resulted from the relative abundance of the liberated gases owing to the enhancement of cracking reaction at such a high pressure.

As it is evident from Figure 3, the light fractions are more predominant at an H_2 pressure of 5.0 MPa, indicating that this technology is more suitable for gasoline production at relatively higher pressures rather than gas oil production and vice versa. The best yield of the desired liquid product can be achieved at a hydrogen pressure of 2.5 MPa. On the other hand, the data confirm the unsatisfactory results obtained for hydrocracking and biofuel yield in case of minimizing the reaction pressure to 1 MPa.

Effect of Reaction Temperature

Reaction temperature has been identified as the most crucial parameter for catalyst effectiveness. The influence of this factor on the yield and quality of the hydrocracked products of jojoba oil, using zeolite catalyst, was evaluated at three different reaction temperatures, i.e. 350, 400, and 425 °C. The remaining operating conditions were kept constant (pressure: 2.5 MPa, LHSV: 5 hr^{-1}, and H_2/oil ratio: 1600 l/l). The physicochemical characteristics and distribution of the main products are represented in Table 3.

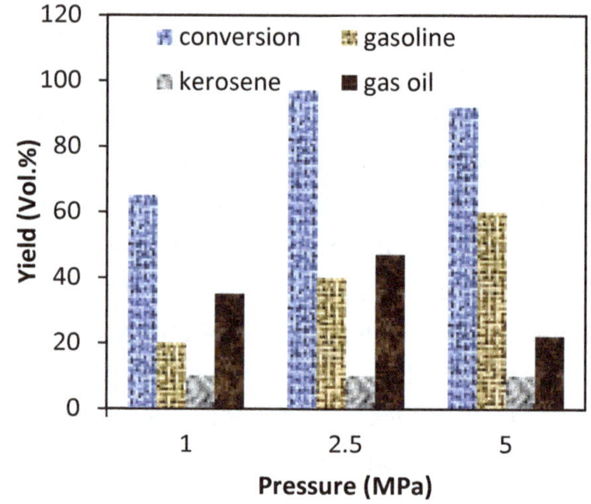

Figure 3: Distribution of the individual fractions of the biofuel obtained as a function of pressure

As can be observed from the depicted data, the amount of the converted crude jojoba is 61.36 wt.%, based on feed at 350°C. Beyond this temperature, i.e. at 400 and 425°C, almost a complete conversion of the feed to hydrocracked constituents was occurred. Nevertheless, at 425°C, the obtained liquid included somewhat lesser content of oil, amounting to 90.6 wt.%, compared to 94.36 wt.% achieved at 400°C. This can be ascribed principally to the contamination of the produced liquid with more water at the expense of the oil concentration. This water has been resulted through the hydroconversion of some of jojoba oil to alkanes via decarbonylation and/or hydrodeoxygenation of COOH group involved in the structure of jojoba molecule. The formation of water is more favored at elevated temperatures owing to the acceleration of the rate of such reactions at high temperatures.

ASTM distillation has been performed to determine the boiling range distribution of the organic hydrocracked product and also to elucidate the amount of the unconverted oil and that transformed to biofuel (Table 3). The fractions of the individual fractions in the produced biofuel at different reaction temperature are represented in Figure 4.

Table 3: Distribution of the hydrocracked products obtained at various temperatures; pressure: 2.5 MPa, LHSV: 5 hr^{-1}, and H_2/ feed: 1600 l/l

Hydrocracked products	350 °C	400 °C	425 °C
Total conversion (wt.%)	61.355	99	99.035
Hydrocracked product (wt.%)			
Gases	2.05	2.34	2.77
Liquid	97.95	97.66	97.23
Water content	2.07	3.30	6.63
Oil content	95.88	94.36	90.60
Density at 20 °C	0.8372	0.8215	0.8125
Kinematic viscosity at 40 °C	4.8540	2.5250	1.0496
Sulfur content (ppm)	0.045	0.016	0.016
ASTM distillation (vol. %)			
IBP(°C)	60	50	40
10	70	70	70
20	110	110	90
30	150	140	120
40	273.2	150	130
50	283.9	180	140
60	305.2	273.2	150
70	-	294.5	150
80	-	305.2	150
90	-	305.2	250
FBP	305.2	305.2	269.7
Recovered (vol.%)	60	97	93
Unconverted and residue (vol.%)	39	1	1
Loss (vol.%)	1	2	6
Biofuel yield (wt.%)	57.245	93.36	89.635
Biofuel selectivity (%)	93.30	94.3	90.508

It is obvious from the figure that the produced petroleum-like fraction composed of 50 vol.% gasoline, 16.7 vol.% kerosene, and 33.3 vol.% gas oil can be obtained at 350 °C, but the yield is unsatisfactory. By increasing the reaction temperature to 400 °C, the major fraction was distilled in the gas oil range and comprised 48.46 vol.% of the produced biofuel, whereas 41.23 vol.% was gasoline and the remainder (10.31 vol.%) represented the kerosene fraction. Upon raising the reaction temperature up to 425°C, relatively lighter products were attained involving a high percentage of gasoline (86.02 vol.%) associated with some kerosene comprising 12.9 vol.% and an insignificant amount of the heavier gas oil fraction (not exceeding than 1.07 vol.%). This can definitely be attributed to the promotion of the intensive hydrocracking reaction at such a high temperature.

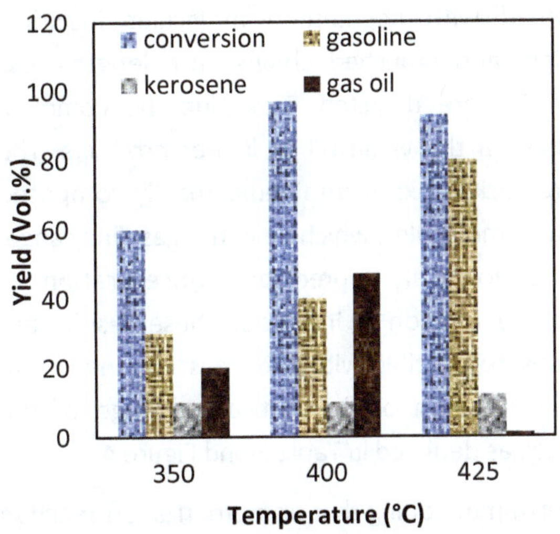

Figure 4: Distribution of the individual fractions of the biofuel obtained as a function of temperature

Therefore, the reaction temperature of 400°C possesses the highest selectivity to petroleum-like fraction formation. Moreover, it could be regarded as the most suitable temperature for gas oil production mixed with a considerable amount of gasoline. On the other hand, the best yield of light fraction, particularly gasoline, can be realized at a reaction temperature of 425°C.

Gas Chromatography

In order to investigate the quantity and quality of the hydrocracked products, gas chromatographic analysis was carried out for both the segmented as well as the virgin jojoba oil for comparison.

It is apparent from the GC spectrum of the crude oil that it is principally composed of heavy, straight chain constituents including more than 16 carbon atoms per molecule, particularly in the range of C_{40}-C_{44} as identified according to the retention time of the recorded peaks declared in Figure 5a.

The chromatogram of the products obtained after catalytic hydrocracking process, performed at 425 °C, 2.5 MPa hydrogen pressure, and an LHSV of 5 hr^{-1}, is represented in Figure 5b. It is evident from this chart that the peaks corresponding to the high boiling jojoba feed is completely vanished. Instead, lighter components enriched with fragmented molecules including straight and branched chains with lengths less than C_{16} are detected, indicating the complete splitting of the virgin oil to lighter products. The hydrocracked bio-oil attained is mainly composed of short molecules, which lie in the gasoline range. Furthermore, an appreciable concentration of kerosene fraction is involved. These results are directly compatible with the ones obtained from the distribution of the distillation range of the biofuel, as depicted in Table 3 and Figure 4.

It worth mentioning that an isomerization reaction has also occurred as indicated from the appearance of numerous peaks in the chromatogram (Figure 4b), corresponding to the isomerized molecules. The substantial presence of branched chain components in the product is definitely leading to a significant improvement in the physicochemical properties of the acquired biofuel such as octane number, viscosity, pour point, etc. The promotion of both hydrocracking and isomerization reactions can be undoubtedly imputed to the intrinsic acidity of the used zeolite catalyst.

FTIR Spectra

The FTIR spectra of jojoba oil and the organic liquid product obtained at 425 °C, 2.5 MPa H_2 pressure, a LHSV of 5 hr^{-1}, and an H_2:oil ratio of 1600 l/l are shown in Figure 5a and b respectively.

The spectrum of the virgin jojoba oil appears to be prevalently aliphatic hydrocarbons as evidenced by the appearance of the intense C-H stretching bands of alkane in the 2857-3000 cm^{-1} region and aliphatic C-H bending of methyl and methylene groups at 1359 and 1459 cm^{-1} respectively. The observation of an absorption peak at 727 cm^{-1} suggested the out of plane bending of an alkene. Since jojoba oil is mainly monoalkyl ester, the intense C=O stretching band of the ester appears at 1740 cm^{-1} and those for the stretching vibration of C-O are observed at 1174 and 1255 cm^{-1}. The FTIR spectrum showed also two weak absorption peaks at 2676 and 3410 cm^{-1} for the OH of carboxylic acid and the H- bonded of alcohol respectively (Figure 6a).

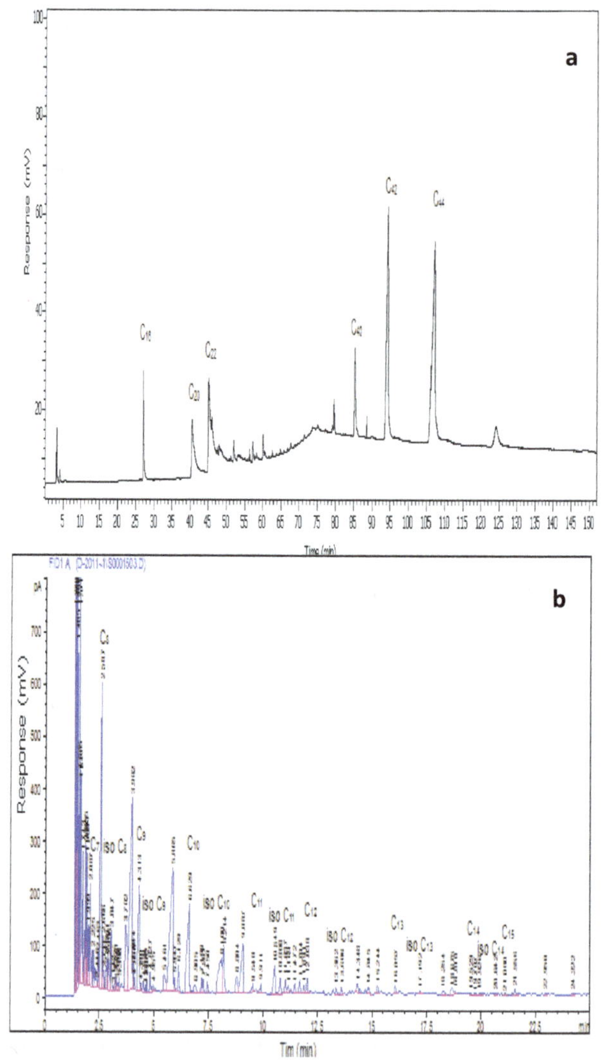

Figure 5: Gas chromatograph for (a) virgin jojoba oil and (b) cracked jojoba oil at 425 °C

The IR analysis, therefore, characterizes the raw jojoba oil as being composed of aliphatic ester, most of which are saturated and associated with some alcohol and carboxylic functional groups.

Figure 6b reveals the IR spectrum of the organic liquid product obtained from hydrocracking of jojoba oil using a zeolite catalyst.

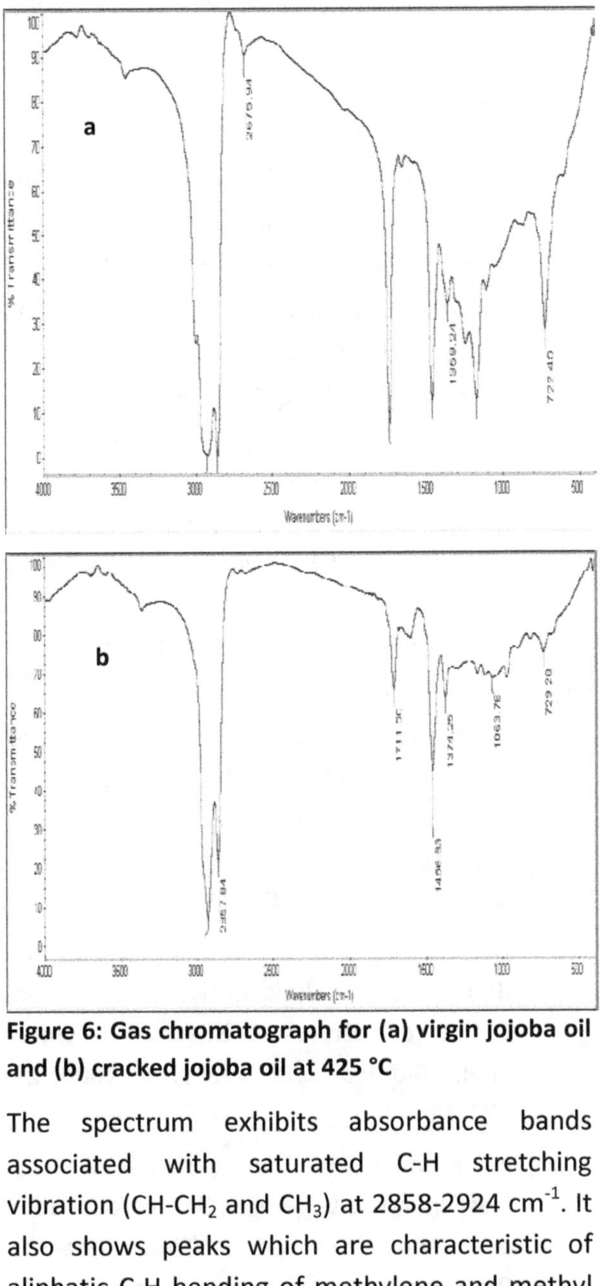

Figure 6: Gas chromatograph for (a) virgin jojoba oil and (b) cracked jojoba oil at 425 °C

The spectrum exhibits absorbance bands associated with saturated C-H stretching vibration (CH-CH_2 and CH_3) at 2858-2924 cm^{-1}. It also shows peaks which are characteristic of aliphatic C-H bending of methylene and methyl groups at 1467 and 1374 cm^{-1}. Therefore, it bores close resemblance to that of the crude jojoba oil; however, there are some main differences between them. It is obvious that the intensity of the stretching bands of C=O at 1711.5 and C-O at 1174 cm^{-1} are significantly reduced in the cracked spectrum compared to that of jojoba oil with the lack of the absorption bands of OH group at 2676 cm^{-1}, which can be ascribed to the occurrence of the deoxygenation reaction. In addition, the pronounced diminution of the intensity of the alkene band (C=C out of plane bending vibration at 729 cm^{-1}) confirms the procession of the saturation of the double bond involved in the structure of jojoba molecule.

CONCLUSIONS

It is concluded from this study that:

1. Hydrocracking process is a promising technique for producing biofuel from the nonedible jojoba oil by using zeolite catalysts;
2. The feedstock was entirely converted to hydrocracked products in all conditions under investigation, except at a reaction temperature of 350 °C and 1 MPa hydrogen pressure. Nevertheless, the obtained yield is quantitatively and qualitatively dependant on the applied operating variables;
3. Sufficient high reaction temperatures and hydrogen pressures with a low LHSV are the most appropriate conditions for producing low-boiling petroleum-like fractions and vice versa;
4. It could be deduced from these results that the optimum operating variables, which lead to attaining 93.36 wt.% biofuel are an LHSV of 5 hr^{-1}, an H_2 pressure of 2.5 MPa, and a reaction temperature of 400 °C. The biofuel achieved in these circumstances is mainly composed of a fraction boiled in the gas oil range mixed with a considerable amount of kerosene. Upon raising the reactor temperature to 425 °C, the realized amount of biofuel is slightly diminished to 93 wt.% and is principally comprised of gasoline fraction, i.e. 86 wt%.
5. Petroleum-like fractions resulted from hydrocracking jojoba oil are mainly composed from alkanes as evidenced from the IR

spectra, which indicate a considerable elimination of O_2 from jojoba molecules;

6. As revealed from the GC chromatogram, a significant amount of iso-alkanes are detected, indicating that the isomerization of straight chain paraffins has also taken place along with hydrocracking reaction. Certainly this leads to a consequential improvement in the properties of the produce biofuel.

REFERENCES

[1] Kumar P., Barrett D. M., Delwiche M. J., and Stroeve P., "Methods for Pretreatment of Lignocellulosic Biomass for Efficient Hydrolysis and Biofuel Production," *Ind. Eng. Chem. Res.*, **2009**, *48*, 3713-3729.

[2] Patil P. D., Gude V. G., and Deny S., "Biodiesel Production from Jatropha Curcas, Waste Cooking and Camelina Stiva Oils," *Ind. Eng. Chem. Res.*, **2009**, *48* (24), 10850-10856.

[3] Huzayyin A. S., Badawy A. H., Rady M. A., and Dawood A., "Experimental Evaluation of Diesel Engine Performance and Emission Using Blends of Jojoba Oil and Diesel Fuel," *Energy Conversion and Management*, **2004**, *45*, 2093-2112.

[4] Hancsok J., Krar M., kasaza T., Kovacs S., et al., "Investigation of Hydrotreating of Vegetable Oil-gas Oil Mixtures," *Journal of Environmental Science and Engineering*, **2011**, *5*, 500-507.

[5] Priecel P., Capek L., Kubicka D., Homola F., et al., "The Role of Alumina Support in the Deoxygenation of Rapeseed Oil Over NiMo-alumina Catalysts," *Catalysis Today*, **2011**, *176*, 409-412.

[6] Pinzi S., Garcia I. L., Lopez-Gimenez F. J., Luque de Castro M. D., et al., "The Ideal Vegetable Oil-based Biodiesel Composition: A Review of Social, Economical and Technical Implications," *Energy & Fuels*, **2009**, *23*(5), 2325-2341.

[7] Lapuerta M., Fernandez J. R., Oliva F., and Canoiva L., "Biodiesel from Low- grade Animal Fats: Diesel Engine Performance and Emissions," *Energy & Fuels*, **2009**, *23*(1), 121-129.

[8] Radwan M. S., Ismail M. A., Elfeky S. M. S. and Abu-Elyazeed O. S. M., "Jojoba Methyl Ester as a Diesel Fuel Substitute Preparation and Characterization," *Applied Thermal Engineering*, **2007**, *27*, 314-322.

[9] Einloft S., Magalhaes T. O., Donato A., Dullius J., et al., "Biodiesel from Rice Bran Oil: Transesterification by Tin Compounds," *Energy & Fuels*, **2008**, *22*, 671-674.

[10] Knothe G., Cermak S. C., and Evangelista R. L., "Cuphea Oil as Source of Biodiesel with Improved Fuel Properties Caused by High Content of Methyl Decanoate," *Energy & Fuels*, **2009**, *23*, 1743-1747.

[11] Mahamuni N. N. and Adewuyi Y. G., "Optimization of the Synthesis of Biodiesel via Ultrasound-enhanced Base-catalyzed Transesterification of Soybean Oil Using a Multifrequency Ultrasonic Reactor," *Energy & Fuels*, **2009**, *23*, 2757-2766.

[12] Ooi Y., Zakaria R., Mohamed A., and Bhatia S., "Catalytic Conversion of Palm Oil-based Fatty Acid Mixture to Liquid Fuel," *Biomass and Bioenergy*, **2004**, *27*, 477-484.

[13] Belkacemi K. and Hamoudi S., "Law Trans and Saturated Vegetable Oil Hydrogenation over Nanostructured Pd/Silica Catalysts: Process Parameters and Mass-transfer Features Effects," *Ind. Eng. Chem. Res.*, **2009**, *48*, 1081-1089.

[14] Canoira L., Alcantara R., Garcia-Martinez M. J., and Carrasco J., "Biodiesel from Jojoba Oil-wax: Transesterification with Methanol and Properties as a Fuel," *Biomass and Bioenergy*, **2006**, *30*, 76-81.

[15] Yang Y., Wang Q., Zhang X., Wang L., et al., "Hydrotreating of C18 Fatty Acids to Hydrocarbons Sulphided NiW/SiO_2-AL_2O_3," *Fuel Processing Technology,* **2013**, *116*, 165-174.

[16] Jhan M., Sinha A., and Agnihotri P., "Hydroprocessing of Jatropha Oil to Produce Green Fuels," *International Journal of Chem. Tech. Research*, **2013**, *5*(2), 765-770.

[17] Tamunaidu P. and Bhatia S., "Catalytic Cracking of Palm Oil for the Production of Biofuels: Optimization Studies," *Bioresource Technology*, **2007**, *98*, 3593-3601.

[18] Corma A., Huber G., Sauvanaudm L., and Connor P.O., "Processing Biomass- derived Oxygenates in the Oil Refinery: Catalytic Cracking (FCC) Reaction Pathways and Role of Catalyst," *Journal of Catalysis*, **2007**, *247*, 307-327.

[19] Ooi Y., Zakaria R., Mohamed, A., and Bhatia S. "Catalytic Cracking of used Palm Oil and Palm Oil Fatty Acids Mixture for the Production of Liquid Fuel: Kinetic Modeling," *Energy & Fuels*, **2004**, *18*, 1555-1561.

[20] Sankaranarayanan T., Banu M., Pandurangan A., and Sivasanker S., "Hydroprocessing of Sun Flower Oil Gas Oil Blends Over Sulfide Ni-Mo-AL-zeolite Beta Composites," *Bioresource Technology,* **2011**, *102*, 10717-10723.

[21] Hancsok J., Krar M., Magyar S., Boda L., et al., "Investigation of the Production of High Cetane Number Bio Gas Oil from Pre-hydrogenated Vegetable Oils Over Pt/HZSM-22/Al_2O_3," *Microporous and Mesoporous Materials*, **2007**, *101*, 148-152.

[22] Egeberg R., Michaelsen N., Skyum L., and Zeuthen P., "Hydrotreating in the Production of Green Diesel," *Digital Refining*, **2010**, 1-11.

[23] Huber G. W., Oconnor P., and Corma A., "Processing Biomass in Conventional Oil Refineries: Production of High Quality Diesel by Hydrotreating Vegetable Oil in Heavy Vacuum Oil Mixtures," *Applied Catalysis A: General*, **2007**, *329*, 120-129.

22

Effect of Operating Conditions on the Dehydration of Toluene by Pervaporation using a Polyvinyl Alcohol Membrane

Ghasem Khatinzadeh[1]*, Mohammad Mahdyarfar[2], and Ali Mehdizadeh[1]

[1] Chemical Science and Technology Research Division, Research Institute of Petroleum Industry (RIPI), P.O. Box 14665-1998, Tehran, Iran

[2] Gas Research Division, Research Institute of Petroleum Industry (RIPI), P.O. Box 14665-1998, Tehran, Iran

ABSTRACT

The effects of operating parameters such as feed temperature and permeate-side pressure on the performance of a commercial PVA membrane in the dehydration of toluene by a pervaporation (PV) membrane system is studied in the present work. The results obtained indicate that increasing temperature in the range of 40 to 70 °C and reducing permeate pressure in the range of 1 to 10 mmHg increase membrane flux and selectivity, which consequently leads to the considerable modification of membrane performance in the reduction of water content of the feed.

Keywords: Pervaporation, Membrane, Toluene, Dehydration, Polyvinyl Alcohol, Processing Parameters

INTRODUCTION

In chemical processing industries, it is often necessary to remove water from solvents in order to avoid the deactivation of catalysts and/or the corrosion of equipment. Toluene is a widely used industrial aromatic solvent due to its capability of dissolving paints, rubbers, adhesives, and many chemical reagents. Some industrial uses of toluene include dealkylation to benzene and disproportionation to a mixture of benzene and xylene. When oxidized, toluene yields benzaldehyde and benzoic acid, two important chemical intermediates.

Adsorption and distillation techniques are usually used to remove water from toluene down to ppm ranges [1]. However, there are disadvantages associated with these processes including high energy consumption due to the regeneration of adsorbent and operation difficulties. In the past few decades, pervaporation has been developed as an alternative, promising separation technology for the dehydration of organic solvents including alcohols [2-4], especially ethanol [5-7], and other common organic solvents [8-9]. This process has characteristics such as high separation efficiency, simplicity of setting up and operation, and little energy consumption, making it favorable compared with other conventional methods.

In PV processes, the feed liquid contacts one side of a membrane through which a mixture enriched with one of the feed components

*Corresponding author
Ghasem Khatinzadeh
khatinzadehgh@ripi.ir

permeates. The separation in PV occurs because of the different rates of sorption and diffusion of the feed components through the membrane [10]. The permeate enriched in this component is removed as a vapor from the other side of the membrane. The vapor pressure difference between the feed liquid and permeate vapor is the driving force for the process. In the dehydration of various solvents by PV, hydrophilic polymers are selected as the membrane materials because water molecules are easily sorbed by such membranes [11]. In fact, the dehydration of aromatic solvents such as benzene using hydrophilic polymeric membranes in a pervaporation system has already been reported [12]. The study was performed at a constant temperature and pressure. However, the effects of feed temperature and permeate pressure, as two of the most important operating parameters which significantly influence the membrane performance, were not studied. Thus, it seems necessary to study the effect of these parameters on the optimization of PV processes [13].

In the present work, the influence of temperature and pressure on the performance of a hydrophilic polyvinyl alcohol membrane in the dehydration of toluene using pervaporation has been investigated.

EXPERIMENTAL

The schematic diagram of the experimental set up is shown in Figure 1. In the pervaporation experiments carried out, the feed (toluene) was pumped from a feed tank to the shell side of a 177 cm^2 Pervap 2201 membrane (Sulzer Chemtech). The membrane was a composite with a top layer of PVA (thickness=2 μm) on a polyacrylonitrile (PAN) porous support (thickness=80 μm) housed in a plate and frame module. Reduced pressure gradient was then applied using a vacuum pump maintained by a regulator and monitored using a vacuum gauge.

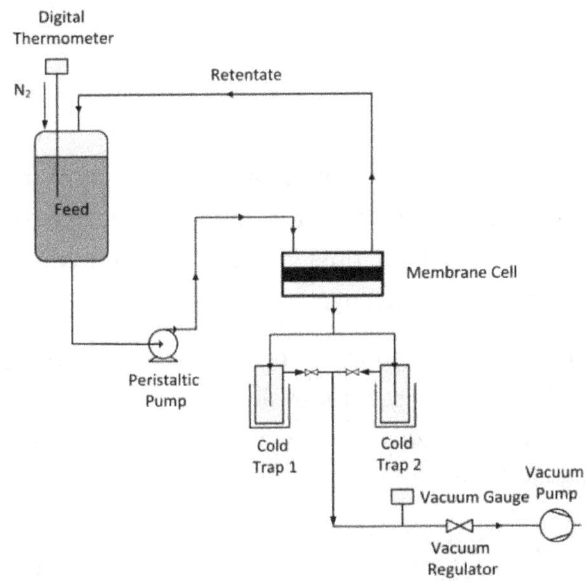

Figure 1: Schematic diagram of experimental set up

Before starting each PV experiment, the test membrane was equilibrated for about 3 hrs with feed solutions. Upon reaching the steady state, the permeate samples were condensed and collected in cold traps immersed in a liquid nitrogen bath. Toluene used in these experiments was provided by Isfahan Petrochemical Complex and contained 150 ppm of water. A heater and a temperature controller were used to keep the feed at a constant temperature. A temperature probe was attached to the feed tank and used to monitor the temperature during each run. The feed was circulated through the membrane at a feed flow rate of 1.1 l/min using a peristaltic pump. The water content in the retentate (product) was determined by the Karl Fisher method. Permeate fluxes were determined at different temperatures and permeate pressures by weighing the permeate obtained during a certain period of time by using Equation 1:

$$J = \frac{Q}{A.t} \quad (1)$$

where, J is the permeate flux (g/hm^2); Q, A, and t stand for permeate weight (gr), membrane area (m^2), and operation time (hr) respectively. Separation factor (α) was evaluated using Equation 2:

$$\alpha = \frac{Y_W / Y_T}{X_W / X_T} \qquad (2)$$

where, α is the separation factor and X and Y are the mass fractions of the feed and permeate respectively. Subscripts W and T stand for water and toluene respectively.

The pervaporation separation index (PSI) (g/hm^2) was used to evaluate the overall PV efficiency [14]. This was calculated using Equation 3:

$$PSI = J(\alpha - 1) \qquad (3)$$

All experiments were carried out under a blanket of nitrogen gas in order to prevent moisture from entering the system.

RESULTS AND DISCUSSION

Effect of Feed Temperature

PV experiments were performed at various operating temperatures in the range of 40 to 70 °C using feed mixtures containing 150 ppm water in order to investigate the effect of feed temperature on the membrane performance in the dehydration of toluene. The feed flow rates and pressures were kept constant at 1.1 l/min and 3 mmHg respectively for all the runs. Figure 2 shows water flux and the relative selectivity of the membrane to water as a function of temperature. As this figure clearly shows, water flux increases as the temperature rises.

Mass fluxes through the membrane are highly dependent on temperature. They increase exponentially according to the empirical Arrhenius law [15].

$$J = J_0 \exp\left(\frac{E_p}{RT}\right) \qquad (4)$$

where, J_0 is permeability constant; E_P represents activation energy, and T and R are the feed temperature and gas constant respectively. This may result from the increase in the diffusion coefficient of the solute as the temperature rises. In fact, the thermal motion of the polymer chains of the membrane increases by raising temperature, which may facilitate the permeation of the adsorbed molecules through the membrane. Furthermore, as the operating temperature rises, the vapor pressure at the feed side increases while the vapor pressure at the permeate side does not significantly change. Therefore, the vapor pressure at the feed and permeate sides of the membrane increases and consequently the permeation of the permeating components through the membrane becomes easier. Additionally, as shown in Figure 2, the selectivity increases at higher temperatures, indicating that an increased operating temperature favors the permeation of water rather than toluene.

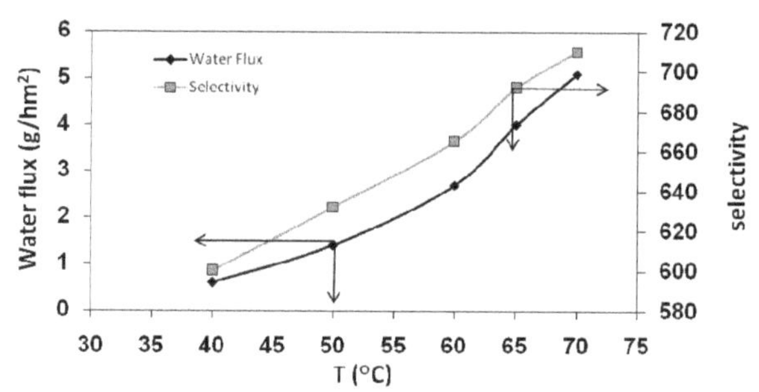

Figure 2: Water flux (◆) and selectivity (■) as a function of feed temperature (permeate pressure is set at 3 mmHg)

This is shown more clearly in Figure 3, showing the influence of temperature on the ratio of water to toluene flux. It can be observed that the ratio of water to toluene flux rises by increasing temperature. In other words, water concentration in the permeate is higher at high temperatures. Hence, increasing temperature leads to increased flux and selectivity, and thereby enhancing membrane performance in toluene dehydration by a PV system according to Equation 3.

Effect of Permeate Pressure

Water flux and selectivity are plotted versus

permeate pressure in Figure 4. According to the results, by varying permeate pressure from 1 to 10 mmHg, the membrane exhibits a considerable decrease in flux from 6.9 to 1.1 (g/hm^2) as well as a reduction in selectivity from 1122 to 153. Flux reduction depends on the thermodynamic properties of organic constituents such as saturated vapor pressure and activity coefficient.

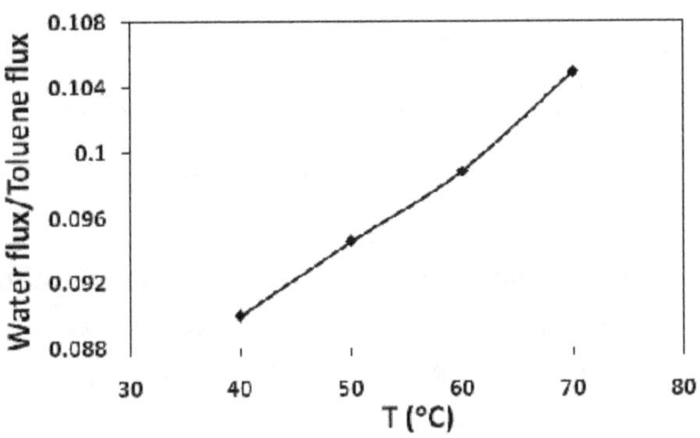

Figure 3: Variation in the ratio of water to toluene flux as a function of feed temperature (permeate pressure is set at 3 mmHg)

Drawing on this fact and according to Fick's law, increasing permeate pressure (lower vacuum) raises the activities of both permeants dissolved in the downstream layer of the membrane. Therefore, activity gradients across the thickness of the membrane decline and permeation flux drops [15]. Furthermore, increasing permeate pressure lowers the driving force, which slows down the desorption rates of the sorbed molecules. In such cases, the membrane relative selectivity to water is governed by the vapor pressures of the two components of the feed mixture. Toluene, which has a higher vapor pressure, permeates competitively with water, and thereby lowering the concentration of water in the permeate. On this basis, at pressures below 10 mmHg, reducing permeate pressure causes increased flux and selectivity and hence improves membrane performance.

The separation performance of the membrane is evaluated by calculating PSI, which shows the combined effect of selectivity and flux. As shown in Figures 5 and 6, dropping permeate side pressure and increasing feed temperature considerably raises PSI; in other words, the performance of a PV membrane system in the dehydration of toluene improves.

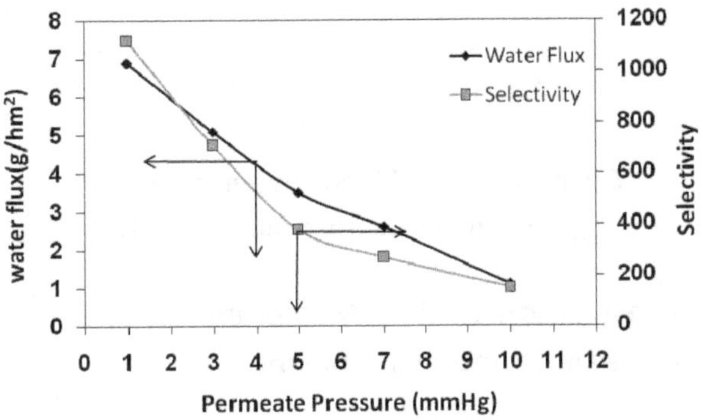

Figure 4: Effect of permeate pressure on membrane selectivity (■) and water flux (◆) (temperature is set at 70 °C).

Comparison of the Effects of Temperature and Pressure on PSI

It should be noted that PSI variations as a function of permeate side pressure 167-7735 (g/hm^2) are more considerable compared with the PSI variations resulting from temperature changes 335-3616 (g/hm^2). In other words, pressure variations have a greater influence on the membrane performance in toluene dehydration than temperature changes do. This is the result of the drastic influence of permeate side pressure on the membrane selectivity towards water molecules in the organic solvent.

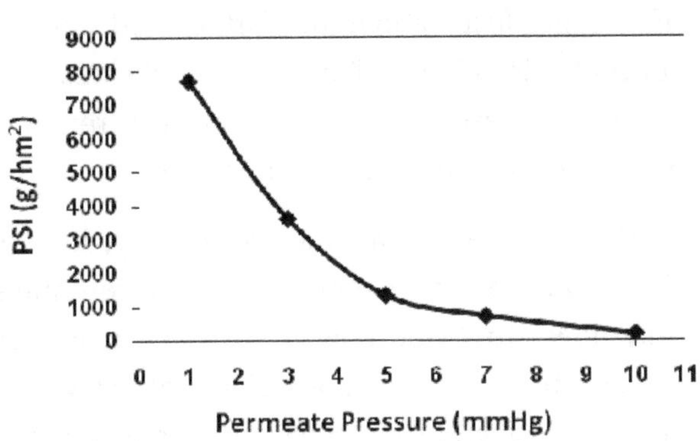

Figure 5: Effect of permeate pressure on PSI (temperature is set at 70 °C)

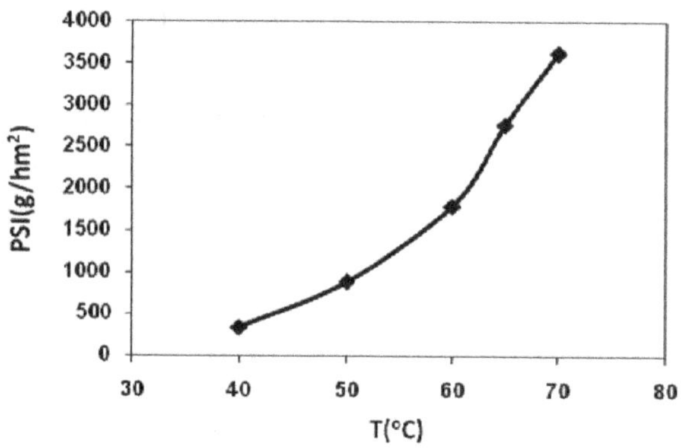

Figure 6: Effect of feed temperature on PSI (permeate pressure is set at 3 mmHg)

Investigation of PVA Membrane Performance in the Dehydration of Toluene

Figure 7 shows the variation of the water content as a function of time in the retentate (dried toluene) at various operating temperatures and a fixed permeate pressure of 3 mmHg. It can be observed that the final water content in toluene may experimentally drop to below 50 ppm. Therefore, the water content of toluene can be reduced to very low values by the application of polymeric membranes in a pervaporation system. It is well known that the separation characteristics of a membrane depend upon the interaction between the solvent to be separated and the membrane matrix [16]. Hydrophilic membranes such as PVA can develop hydrogen bonding interactions with water, leading to the preferential sorption and diffusion of water through the barrier membrane. The observation that the water content of the product cannot be further reduced to concentrations lower than 50 ppm indicates that water permeation into the membrane is remarkably decreased in such concentrations.

Furthermore, extending the process time lowers the amount of water remaining in the retentate only down to 40-50 ppm; nevertheless, longer process times do not appreciably affect water content reduction. This also depends on the degree of the interaction of water molecules with the membrane and its very negligible penetration at low concentrations. However, the reduction of water content to about 50 ppm shows the great potential of this process for the preparation of anhydrous toluene.

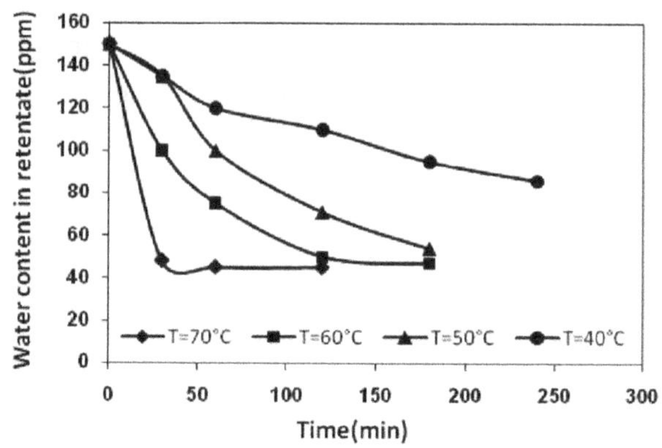

Figure 7: Variation of water content in the retentate against time at various operating temperatures (permeate pressure is set at 3 mmHg)

CONCLUSIONS

The water content of toluene can be reduced from 150 to below 50 ppm using a pervaporation system. Feed temperature and permeate pressure are two of the most important operating parameters, significantly influencing the membrane performance. Permeate flux and selectivity are enhanced by increasing feed temperature and dropping permeate pressure. This modifies the membrane performance and as a result favors the dehydration of toluene. Also, the effect of permeate pressure on PSI is more considerable than that of temperature. However, high temperatures can typically be reached more easily and/or economically than low permeate pressures do. Therefore, the optimization of permeate pressure based on balancing membrane performance and operational costs is necessary.

REFERENCES

[1] Joshi S., Fair J. R., "Adsorptive Drying of Toluene", *Ind. Eng. Chem. Res.* **1988**, 27, 2078-2085.

[2] Rachipudi P. S., Kariduraganavar M. Y., Kittur A. A., "Synthesis and Characterization of

Sulfonated-polyvinyl alcohol Membranes for Pervaporation Dehydration of Isopropanol", *J. Membr. Sci*, **2011**, 383, 224-234.

[3] Jiang L. Y., Chung T. S., "Homogeneous Polyimide/ Cyclodextrin Composite Membranes for Pervaporation Dehydration of Isopropanol", *J. Membr. Sci*. **2010**, 346, 45-58.

[4] Tang Y., Widjojo N., Shi G. M., Chung T. S., "Development of Flat-sheet Membranes for C1–C4 Alcohols Dehydration Via Pervaporation from Sulfonated Polyphenylsulfone (sPPSU)", *J. Membr. Sci*, **2012**, 415-416, 686-695.

[5] Lieu Le N., Wang Y., Chung T. S., "Synthesis, Cross-linking Modifications of 6FDA-NDA/DABA Polyimide Membranes for Ethanol Dehydration Via Pervaporation", *J. Membr. Sci,* **2012**, 415-416, 109-121.

[6] Bolto B., Hoang M., Xie Z., "A Review of Membrane Selection for the Dehydration of Aqueous Ethanol by Pervaporation", *J. Membr. Sci*, **2011**, 503, 227-235.

[7] Hyder M. N., Huang R. Y. M., Chen P., "Pervaporation Dehydration of Alcohol–water Mixtures: Optimization for Permeate Flux and Selectivity by Central Composite Rotatable Design" *J. Membr. Sci,* **2009**, 326, 343-353.

[8] Chapman P. D., Tan X., Livingston A. G., "Dehydration of Tetrahydrofuran by pervaporation Using a Composite Membrane" *J. Membr. Sci*, **2006**, 268, 13-19.

[9] Singha N. R., Parya T. K., Ray S. K., "Dehydration of 1,4-dioxane by Pervaporation Using Filled and Crosslinked Polyvinyl Alcohol Membrane" *J. Membr. Sci.* **2009**, 340, 35-44.

[10] Neel J., "Introduction to Pervaporation" in: Huang R. Y. M. (Ed.), "Pervaporation Membrane Separation Processes", **1991**, 36-37.

[11] Devi D. A., Smitha B., Sridhar S., Aminabhavi T. M., "Pervaporation Separation of Dimethylformamide/ Water Mixtures Through Polyvinyl alcohol/ Poly(acrylic acid) Blend Membranes" *Sep. Purif. Technol.* **2006**, 51, 104-111.

[12] Li J., Chen C., Han B., Peng Y., "Laboratory and Pilot-scale Study on Dehydration of Benzene by Pervaporation" *J. Membr. Sci.* **2002**, 203, 127-136.

[13] Lin L., Kong Y., Yang J., "Scale-up of Pervaporation for Gasoline Desulphurization, Part 1, Simulation and Design" *J. Membr. Sci.* **2007,** 298, 1-13.

[14] Nawawi M. G., Huang R. Y. M., "Pervaporation Dehydration of Isopropanol with Chitosan Membranes" *J. Membr. Sci*. **1997**, 124, 53-62.

[15] Cabasso I., Grodzinski J. J., Vofsi D., "Polymeric alloys of Polyphosphonates and Acetyl Cellulose: Sorption and Diffusion of Benzene and Cyclohexane", *J. Appl. Polym. Sci*. **1974**, 18, 2117-2136.

[16] Shao P., Huang R. Y. M., "Polymeric Membrane Pervaporation" *J. Membr. Sci.* **2007**, 287, 162-179.

PERMISSIONS

 Every chapter published in this book has been scrutinized by our experts. Their significance has been extensively debated. The topics covered herein carry significant findings which will fuel the growth of the discipline. They may even be implemented as practical applications or may be referred to as a beginning point for another development.

The contributors of this book come from diverse backgrounds, making this book a truly international effort. This book will bring forth new frontiers with its revolutionizing research information and detailed analysis of the nascent developments around the world.

We would like to thank all the contributing authors for lending their expertise to make the book truly unique. They have played a crucial role in the development of this book. Without their invaluable contributions this book wouldn't have been possible. They have made vital efforts to compile up to date information on the varied aspects of this subject to make this book a valuable addition to the collection of many professionals and students.

This book was conceptualized with the vision of imparting up-to-date information and advanced data in this field. To ensure the same, a matchless editorial board was set up. Every individual on the board went through rigorous rounds of assessment to prove their worth. After which they invested a large part of their time researching and compiling the most relevant data for our readers.

The editorial board has been involved in producing this book since its inception. They have spent rigorous hours researching and exploring the diverse topics which have resulted in the successful publishing of this book. They have passed on their knowledge of decades through this book. To expedite this challenging task, the publisher supported the team at every step. A small team of assistant editors was also appointed to further simplify the editing procedure and attain best results for the readers.

Apart from the editorial board, the designing team has also invested a significant amount of their time in understanding the subject and creating the most relevant covers. They scrutinized every image to scout for the most suitable representation of the subject and create an appropriate cover for the book.

The publishing team has been an ardent support to the editorial, designing and production team. Their endless efforts to recruit the best for this project, has resulted in the accomplishment of this book. They are a veteran in the field of academics and their pool of knowledge is as vast as their experience in printing. Their expertise and guidance has proved useful at every step. Their uncompromising quality standards have made this book an exceptional effort. Their encouragement from time to time has been an inspiration for everyone.

The publisher and the editorial board hope that this book will prove to be a valuable piece of knowledge for researchers, students, practitioners and scholars across the globe.

LIST OF CONTRIBUTORS

Ali Allahverdi and Shadi Soltani
Cement Research Center, School of Chemical Engineering, Iran University of Science and Technology Tehran, Iran

Ebrahim Najafi Kani
School of Chemical, Petroleum, and Gas Engineering, Semnan University, Semnan, Iran

Saba Karimi, Rohollah M. Keaei and Ahmad Tavasoli
School of Chemistry, College of Science, University of Tehran, Tehran, Iran

Armin Taheri Nakhost
Department of Petroleum Engineering, Sciences & Researches Branch of Islamic Azad University, Tehran, Iran

Seyed Reza Shadizadeh
Department of Petroleum Engineering, Abadan Faculty of Petroleum Engineering, Petroleum University of Technology, Abadan, Iran

Sidnei Cerqueira dos Santos, Leila Cristiane Silva das Virgens de Souza, Catia Larissa Santos Ramos, Luiz Lázaro Franco Batista, Josilene Borges Torres Lima Matos and Paulo Fernando de Almeida
Laboratory of Biotechnology and Ecology of Microorganisms, Department of Bio-Interaction Sciences, Federal University of Bahia, Brazil

Martin Gerald Liebensteiner
Laboratory of Microbiology, Wageningen University, Wageningen, The Netherlands

Cristina Maria Quintella
Laboratory of Kinetic and Molecular Dynamic, Institute of Chemistry, Federal University of Bahia, Salvador, Brazil

Yahya Zamani, Akbar Zamaniyan, Farzad Bahadoran and Mehrdad Shojaei
Research Institute of Petroleum Industry (RIPI), National Iranian Oil Company, West Blvd. of Azadi Sports Complex, Tehran, Iran

Hamid Soltanian and Ghorban Ali Sobhi
Exploration Department, Research Institute of Petroleum Industry (RIPI), West Blvd. of Azadi Sports Complex, Tehran, Iran

Abdolrahim Khojastefar
Department of Chemical and Petroleum Engineering, Sharif University of Technology, Tehran, Iran

Mohammad Parvazdavani, Seyed Mahdia Motahhari and Vahid Ansari
Reservoir Studies Research Division, Research Institute of Petroleum Industry (RIPI), Tehran, Iran

Zheng Liming, Liu Jing and Xu Jiaxiang
School of Petroleum Engineering, China University of Petroleum (Eastern China), Qingdao, China

Pu Chunsheng
School of Petroleum Engineering, China University of Petroleum (Eastern China), Qingdao, China
State Key Laboratory of Heavy Oil, Qingdao, China

Hamid Reza Bozorgzadeh
Research Institute of Petroleum Industry, West Blvd. of Azadi Sports Complex, Tehran, Iran

Mahmoud Shavandi and Majid Soheili
Microbiology and Biotechnology Group, Environment and Biotechnology Research Division, Research Institute of Petroleum Industry, Tehran, Iran

Shekufeh Zareian, Neda Akbari and Khosro Khajeh
Biochemistry Department, Faculty of Biological Science, Tarbiat Modares University, Tehran, Iran

Mansoor Bazmi and Mahmood Bayat
Research Institute of Petroleum Industry (RIPI)

Sayed Hasan Hashemabadi
Computational Fluid Dynamics (CFD) Research Laboratory, School of Chemical Engineering, Iran University of Science and Technology, Tehran, 16846, Iran

Farhad Khoshbakht
Department of Reservoir Study and Field Development, Research Institute of Petroleum Industry (RIPI), Tehran, Iran

Seyed Majid Hashemi and Gholamreza Bashiri
Institute of Enhanced Oil Recovery, NIOC, Tehran, Iran

Dariush Farkhani, Habibolah Rezaei, Nader Gholami, Mitra Sina, Ali Asghar Khalili and Ali Mehdizadeh
Chemical Science and technology Research Division, Research Institute of Petroleum Industry, West Blvd., Azadi Sports Complex, Tehran, 1485733111, Iran

Liu Xuan
SINOPEC Exploration and Production Research Institute, Beijing 100083, China
School of Geo-Sciences, China University of Petroleum, Qingdao 266580, Shandong Province, China

Li Yong and Ni Liangtian
School of Geo-Sciences, China University of Petroleum, Qingdao 266580, Shandong Province, China

Zhong Jianhua
School of Geo-Sciences, China University of Petroleum, Qingdao 266580, Shandong Province, China
Guangzhou Institute of Geochemistry, Chinese Academy of Sciences, Guangzhou 510640, Guangdong, China

Yang Yuping
School of Geo-Sciences, China University of Petroleum, Qingdao 266580, Shandong Province, China
SINOPEC Zhongyuan Oilfield Exploration and Development Institute, Zhengzhou 450003, Henan, China

Behruz Mirzayi and Mohammad Behruz -Chakan
Chemical Engineering Department, University of Mohaghegh Ardabili, Ardabil, Iran

Seyed Ali Mousavi-Dehghani
Research Institute of Petroleum Industry, Tehran, Iran

Li Yanjie
Development Research Department, China National Offshore Oil Corporation (CNOOC), Beijing, China

Zhang Yanyu and Chen Huijuan
School of Petroleum Engineering, China University of Petroleum (Eastern China), Qingdao, China

Wang Shichao
Research Institute of Western South China Sea Petroleum Administrative Bureau, CNOOC Zhanjiang Branch, Zhanjiang, China

Shi Xiaoying
Exploration and Development Research Institute of Huabei Oilfield Company, Renqiu, China

Bai Xuzheng
Gudong Oil Production Plant of Shengli Oil Field Co., Dongying, China

Yaser Pourmazaheri and Hamid Soltanian
Division of Petroleum Engineering, Faculty of Upstream Petroleum Industry, Research Institute of Petroleum Industry (RIPI), Tehran, Iran

Changhua Yang, Xinrong Wu Baolun Niu and Ruijian Deng
Zhongyuan Oilfield Branch Company, Sinopec, Puyang, Henan Province, 457001, P. R. of China

Yiqing Lin
School of Foreign Languages, Xi'an Shiyou University, Xi'an, Shaanxi Province, 710065, P. R. of China

Zhang Zhang
Oil and Gas Development Corporation, Zhongyuan Petroleum Engineering Ltd. Sinopec, Puyang, Henan Province, 457001, P. R. of China

Zhaomin Li and Shaoran Ren
School of Petroleum, China University of Petroleum, Qingdao, Shandong Province, 266580, P. R. of China

Mosayeb Shafiezadeh
Refuling Unit, NIOPDC, Kermanshah Region, Iran

Mansour Ziaee
Department of Petroleum Engineering, University of Shahroud, Shahroud, Iran

Behzad Tokhmchi
Department of Petroleum Engineering, University of Shahroud, Shahroud, Iran

Salwa. A. El Khatib, Samia Abas Hanafi and Eslam Fawzi Al-Amrousi
Egyptian Petroleum Research Institute, Nasr City, Cairo, Egypt

Mohamed M. H. Arief
Benha University, Faculty of Science, Department of Chemistry, Behna, Egypt

Ghasem Khatinzadeh and Ali Mehdizadeh
Chemical Science and Technology Research Division, Research Institute of Petroleum Industry (RIPI), Tehran, Iran

Mohammad Mahdyarfar
Gas Research Division, Research Institute of Petroleum Industry (RIPI), Tehran, Iran

Index

O

P

R

S

T

U

W

Y

www.ingramcontent.com/pod-product-compliance
Lightning Source LLC
LaVergne TN
LVHW060541210926
848485LV00005B/1236

* 9 7 8 1 6 3 9 8 7 2 9 4 7 *